Mathematical Mystery Tour

VOLUME I

Third Edition

Through the Looking Glass of the Human Mind: From Numbers to Reason and Beyond

By James P. Fulton, PhD
Suffolk County Community College

Dedication:

This book is dedicated to my father and mother, Raymond and Dorothy Fulton. Over the years, they have been very supportive, always believing in me, and for that I thank them. I love you both very much. I also dedicate this book to my wife Danise, who had to put up with me throughout this process. She is my best friend and the love of my life. Thank you for all your encouragement and support.

Motivation for a New Approach

Mathematics for the Liberal Arts can be one of the most difficult classes to teach. In fact, many mathematics professors go to great lengths to avoid teaching it. The goals and purposes of the course tend to be very loose and obscure. It is a course that supposedly "stresses critical thinking and reasoning" or provides a "survey of contemporary topics in mathematics designed to develop an appreciation of the power and significance of mathematics."

Typically there are two approaches adopted by the authors of the textbooks for this course. The first splices together a variety of ad hoc mathematical topics such as problem solving, logic, set theory, number systems, probability, statistics, algebra, geometry etc., to provide a series of hurdles and obstacles for the students to jump over and negotiate. There is no real connectivity between any of the topics and the professor can simply pick and choose topics at random.

The other approach tries to show the usefulness of mathematics in modeling the physical world around us. While this is a great and noble goal that we wholeheartedly support, it is we believe far beyond the grasp of most liberal arts students. Mathematical modeling requires a level of mathematical sophistication that has not yet been obtained by most of the students enrolled in this course. These books tend to be a sincere attempt at trying to show the vitality and utility of mathematics through applications, but aside from the difficulties of getting the students to understand and work with mathematical models, they do not provide any connectivity between the topics. It again becomes a hodgepodge of concepts and models loosely put together under the general heading of applications of mathematics.

There is a serious need for a new approach for this very important subject. For many students this will be their last exposure to mathematics. We don't want them to go away with the impression that mathematics is useless and that all that is taught are pointless truth tables or solving trivial problems with Venn diagrams. Especially since many of the students may graduate into jobs in government where the funding of mathematics is an issue that may come before them. With this in mind, we have pursued a totally different philosophy in this book. We treat mathematics as one of the liberal arts. It is more of a mathematics appreciation book with an underlying theme, purpose and perspective. It gives the mathematicians the opportunity to showcase the real beauty, elegance and relevance of mathematics. It focuses on the origins of mathematics and the fundamental questions related to its development. It is much more enjoyable to teach and the students tend to be more motivated and interested, especially when this is taken in the freshman or sophomore year, since they are often taking a philosophy or psychology course simultaneously and can readily see the interconnections.

One of the basic motivations for this book is that a person that is truly educated in the liberal arts needs to understand the place and relevance of mathematics in the world. The student should be able to read and understand the many books that find their way on to the NY Times bestsellers list that address these issues. They need to be both a critical reader and thinker. Our approach to critical thinking and reasoning is to have students write essays that discuss ideas and developments in mathematics related to philosophy, history, epistemology and psychology that are addressed in the book. They need to be able to put together a clear argument and be able to defend their position. However, more importantly, they should develop an appreciation for Mathematics!

TABLE OF CONTENTS

Preface:

Why Study Mathematics?

PART I: What is Mathematics?

Chapter 1: A Philosophical Perspective

1.1.	Introduction	1
1.2.	Defining Mathematics	2
1.3.	Eastern and Western, Philosophy and Metaphysics	4
1.4.	Philosophical Schools of Thought on the Origin of Mathematics	13
1.5.	Why Does Mathematics Describe the Physical World So Well?	22
	Exercises	25
	References	26

Chapter 2: Cognitive & Psychological Aspects Related to Mathematics

2.1	Introduction	31
2.2	The Brain	33
2.3	Consciousness	35
2.4	Reality and Perception	40
	Exercises	47
	References	48

Chapter 3: Knowledge & Mathematics

3.1	Introduction	50
3.2	Theory of Knowledge	50
3.3	Schools of Thought on How We Obtain Knowledge	53
3.4	Types of Knowledge	54
3.5	Mathematical Knowledge	55
	Exercises	56
	References	57

Chapter 4: Learning & Mathematics

4.1	Introduction	58
4.2	Theories of Learning	58

4.3	**Learning Mathematics**	62
4.4	**Dyscalculia**	65
	Exercises	68
	References	69

Part I Summary – Making the Connections 71

PART II: Historical Development of Numbers

Chapter 5: The First Numbers

5.1	**Introduction**	72
5.2	**Non–Positional Number Systems**	76
5.3	**Positional Number Systems**	92
5.4	**A Common Thread – Or Should We Say Bead?**	103
	Exercises	106
	References	112

Chapter 6: Fractions & Decimals

6.1	**Introduction**	113
6.2	**Egyptian Fractions**	113
6.3	**Greek Fractions**	119
6.4	**Roman Fractions**	119
6.5	**Babylonian Fractions**	120
6.6	**Hindu – Arabic Fractions**	124
6.7	**Decimals (Decimal Fractions)**	124
	Exercises	127
	References	129

Chapter 7: Zero

7.1	**Introduction**	131
7.2	**Why Was Zero Needed?**	132
7.3	**Why Was Zero a Difficult Concept to Understand?**	133
7.4	**Is Zero Like Other Numbers?**	138
7.5	**A Dangerous Idea**	140
	Exercises	142
	References	143

Chapter 8: Number Base Systems in General

8.1	**Introduction**	144

8.2   Base–2 or Binary Number System   144
8.3   Base–5 or Quinary Number System   148
8.4   Base–8 or Octal Number System   153
8.5   Base–12 or Duodecimal Number System   156
8.6   Base–16 or Hexadecimal Number System   160
       Exercises   164
       References   167

Chapter 9: Negative Numbers

9.1   Introduction   168
9.2   The History of Negative Numbers   169
9.3   Why Are Negative Numbers so Confusing? –
       A Web of Inconsistent Explanations, Symbols, and Analysis   173
9.4   Working With Negative Numbers   177
       Exercises   180
       References   181

Chapter 10: Exponential Numbers

10.1   Introduction   182
10.2   What Are Exponential Numbers?   182
10.3   Why Are Exponential Numbers Important?   190
10.4   Sustainability Versus Unsustainable Growth   193
10.5   Other Uses of Exponential Numbers   197
       Exercises   208
       References   211

Chapter 11: Irrational Numbers

11.1   Introduction   213
11.2   What Do You Mean, I Can't Measure This Length?   213
11.3   What is an Irrational Number Anyway, and Why Should I care?   215
11.4   Some Special Irrational Numbers   217
11.5   Do Irrational Numbers Really Exist?   219
11.6   Irrationals and Beyond   220
       Exercises   221
       References   223

Chapter 12: Infinity

12.1   Introduction   224
12.2   What is Infinity?   224
12.3   Set Theory – A Tool to Study the Infinite   230

12.4   Making the Incomprehensible, Comprehensible – Or Can We?   249

12.5   Paradoxes and the Infinite   255

Exercises   267

References   275

Chapter 13: Imaginary Numbers

13.1   Introduction   277

13.2   Brief History or Imaginary/Complex Numbers   278

13.3   Mathematics of Imaginary/Complex Numbers   279

13.4   The Physical Significance of Imaginary Numbers   283

Exercises   284

References   286

Part II Summary – Making the Connections

287

PART III: Historical Development of Reasoning (Logic)

Chapter 14: Introduction to Types of Reasoning, What is Proof?

14.1   Introduction   288

14.2   What is Proof?   289

Exercises   298

References   301

Chapter 15: Aristotle – The Father of Propositional Logic

15.1   Introduction   302

15.2   Historical Development of Propositional Logic   304

15.3   Fundamentals of Propositional Logic   305

15.4   Problems With Reasoning – Fallacies and Paradoxes   312

15.5   Having Fun With Meaning – Alice's Adventures in Wonderland   314

Exercises   315

References   321

Chapter 16: Beyond Aristotle – Symbolic Logic and the Insight of Boole

16.1   Introduction   322

16.2   Transforming Propositional Logic Into Symbolic Logic   322

16.3   Truth Values of Compound Statements   330

16.4   Truth Tables   333

16.5   More Complex Logical Expressions and Their Truth Tables   340

16.6   Valid Arguments – Tautologies   348
16.7   Logical Equivalence (Optional Section)   351
16.8   Related Conditional Statements (Optional Section)   352
        Exercises   354
        References   359

Chapter 17: The Search for the "Holy Grail" of Mathematics

17.1   Introduction   360
17.2   The Origin of the Quest   360
17.3   A Bump in the Road – Russell's Paradox   362
17.4   Kurt Gödel – The New Dawn of Mathematics   364
        Exercises   368
        References   369

Chapter 18: Fuzzy Logic

18.1   Introduction   370
18.2   Fuzzy Sets   371
18.3   Operations of Fuzzy Sets   380
18.4   Fuzzy Logic   386
        Exercises   390
        References   392

Part III Summary – Putting it All Together

                                                                         394

Answers to Selected Exercises

                                                                         396

Alphabetical Index

                                                                         414

Preface:

Mathematics is considered the queen of the sciences. It is one of the crowning accomplishments of humanity, and as such, it has and continues to touch every aspect of our lives. It comes from our necessity to understand and make sense out of the world around us. We create/discover mathematics so that we can begin to better understand ourselves, and the universe in which we live. It is this aspect of mathematics that we wish to capture and express in this book. It is something that every educated person should be aware of. Just as everyone should have exposure to great works of literature, art, and music, so too, should they get to know of and appreciate the real mathematics. Not simply the dry and mechanical mathematics from grade and high school, but the mathematics of mystery, inquisitiveness, uncertainty, wonder and beauty.

We will begin our journey by asking a fairly simple and straightforward question: What is Mathematics? As we shall see, this seemingly innocent question will take us down paths we never even considered. It will challenge how we think, what we know, and even who we are. It will lead us to many fascinating and important questions about truth, reality, knowledge, learning, consciousness and even life itself. We shall embark on a mathematical mystery tour that will take us through history, philosophy, psychology, science, pedagogy, epistemology (the study of knowledge), language and self–reflection. We will look at the intrinsic aesthetic value of mathematics that is independent of its practical results.

We often think of mathematics as being an isolated, complete and well–understood subject that is not open for debate, but as we shall see, this is far from the truth. Mathematics is the result of the efforts of many cultures and people throughout the world and throughout history. Yet, it is amazing that mathematics today is the same in Asia, Africa, Europe, and the Americas, unlike language and culture that developed differently in all these regions.

We will not focus a great deal of our attention on mathematical rigor and calculations, but rather look at mathematical concepts and gain an appreciation for them. The need to know how to use the quadratic formula, or factor a polynomial expression, or to use Descartes Rule of signs to find rational roots, while compelling to many mathematicians, will not be discussed, as it serves no purpose for the general liberal arts student, any more than learning to master paint mixing, canvas preparation, stroke and blending techniques, in an art history or appreciation course.

A large part of what we will concentrate on requires a great deal of careful thought and analysis that we can only communicate in words. Thus, the student will embark on a journey through critical thinking. Not just by defining it, but by doing it. A fundamental part of critical thinking is our ability to communicate our ideas and interpretations. Therefore, successful completion of this book will require numerous writing assignments by the reader.

This book is divided into II volumes with a total of four sections or questions. What is mathematics? – A philosophical perspective. How do we understand mathematics? – A question of psychology and cognition. How did mathematics develop? – A historical view. And finally – What is the impact of mathematics? However, as we shall see this or any other division is artificial, and is used simply as a matter of convenience, for we never truly leave a section. All the topics are related and interconnected, as we shall show, in this tapestry called mathematics.

Preface to the Instructor

The first part of the book is not typical for a conventional mathematics textbook. It focuses more on questions and concepts related to what mathematics is, and not how to apply it. If you wish, you can skip the first section of the text without loss of continuity, but you will miss out on a large part of the journey. You could also assign it as a reading assignment to the student, so at least they take a more thoughtful look at mathematics. The last option is to embrace the fact that mathematics can be much more than simply numbers, formulas, patterns and equations, and take the students through the journey yourself. It will be a rewarding exercise if you do, and even more rewarding if you look for additional resources and ideas to expand the material. The first part of the text does not contain any mathematics problems. There are only essays, and short answer types questions related to understanding what mathematics is, as an important idea, and not as a mathematical problem to be solved. The purpose is to give the student exposure to mathematics as a liberal art beyond the computational aspects they are all too familiar with. This part of the book is also meant to mirror other courses they are taking in the other disciplines to show that mathematics is really everywhere in the curriculum.

A word of warning. We do not approach the subject in a traditional mathematical way by introducing the finalized formality. Instead we approach the subject from a more humanistic approach, and go into the history and explore the "why" more than is traditionally done. Thus, as an instructor you should be more prepared to discuss the non–mathematical side as well as the mathematics. Discuss the ideas and the concepts instead of quickly jumping right into solving the problem or performing the required operation. This is the focus of the first part of the book.

After the first part of the book, the remainder is a bit more of the typical mathematics text, but from a more humanistic perspective, with a few exceptions.

Why Study Mathematics?

The question of why we should study mathematics and what mathematics everyone should know, has been looked at by many prominent thinkers throughout history. It has, however, never been answered to everyone's satisfaction. Even our teachers don't have the answers. They are just going through the motions and "teaching" any and every mathematical topic that appears before them in the text. Why?

Alfred North Whitehead a highly respected British Mathematician and educator had these views towards mathematics education in England in the 1940's:

…the teaching of mathematics has suffered, …because it was treated as a collection of mere uninteresting prolegomena (observations) to more advanced parts of the subject. But the mass of pupils never advanced to these further parts, and, in consequence, gained nothing but a set of purposeless dodges

We must conceive elementary mathematics as a subject complete in itself, to be studied for its own sake. It must be purged of every element, which can only be justified by reference to a more prolonged course of study. There can be nothing more destructive of true education than to spend long hours in the acquirement of ideas and methods, which lead nowhere. It is fatal to all intellectual vitality.

… we have to teach what logic is. I do not mean by this that we should indulge in the somewhat futile task of affixing names to elementary logical processes …

Mathematics [has been] divested of all discussion of ideas, and reduced to aimless acquirement of formal methods of procedure. (excerpts from Mathematics and Liberal Education)

In the 1960's the American Mathematician Morris Kline also addressed this question in his book, "Why Johnny Can't Add." Kline states that

Mathematics is the key to our understanding of the physical world; it has given us power over nature and it has given man the conviction that he can continue to fathom the secrets of nature. Mathematics has enabled painters to paint realistically, and has furnished not only an understanding of musical sounds, but also an analysis of such sounds that is indispensable in the design of the telephone, the phonograph, radio, and other sound recording and producing instruments. Mathematics is becoming an increasingly valuable tool in biological and medical research. The question – What is truth? Cannot be discussed without involving the role that mathematics has played in convincing man that he can and cannot obtain truths. Much of our literature is permeated with themes treating mathematical accomplishments. Indeed, it is often impossible to understand many writers and poets unless one knows what influences of mathematics they are reacting to. Lastly, mathematics is indispensable in our technology.

Knowledge is a whole and mathematics is a part of the whole. The subject did not develop apart from other activities and interests. To teach mathematics as a separate discipline is a perversion, a corruption and a distortion of true knowledge. If we are compelled for practical reasons to separate learning into mathematics, science, history and other subjects, let us at least recognize that this separation is artificial and false. Each subject is an approach to knowledge and any mixing or overlap where convenient and pedagogically useful, is desirable and to be welcomed.

Mathematics is not an isolated, self–sufficient body of knowledge. It exists primarily to help man understand and master the physical, the economic and the social worlds. It serves ends and purposes. We must constantly show that it accomplishes in domains outside of mathematics. We can hope and try to inculcate interest in mathematics proper and the enjoyment of mathematics, but these must be by–products of the larger goal of showing what mathematics accomplishes.

A feeling that mathematics is indeed a fundamental reality of the domain of thought, and not merely a matter of symbols and arbitrary rules and conventions. (E.H. Moore On the Foundations of Mathematics)

To teach students to pursue knowledge is a part of a liberal education. We should get students to want to pursue it and not proclaim it to be.

Let's not push facts down students' throats. We are not packing articles in a trunk. This type of teaching dulls the mind rather than sharpens them.

On one level, mathematics is a means to an end. One uses the concepts and reasoning to achieve results about real things. On another level, however, mathematics is also about who we are as a people and as distinct civilizations. It is a major achievement of humanity that should be studied, understood and cultivated for future use and societies.

Philosophy is written in this grand book, the universe, which stands continually open to our gaze, but cannot be understood unless one first learns to comprehend the language and interpret the characters in which it is written. It is written in the language of mathematics, and its characters are triangles, circles, and other figures, without which it is humanly impossible to understand a single word of it; without these, one is wandering about in a dark labyrinth. (Galileo Galilei, 1623)

Socrates once said that the unexamined life is not worth living. This is one of the guiding principles of this textbook. Higher education is a time for us to look more closely at things we previously took for granted, so that we can begin to try and understand the world and our place in it. Although it may not seem that way, mathematics as we'll see provides a wellspring of new and interesting revelations into understanding our world.

PART I: WHAT IS MATHEMATICS?

We begin with a look at mathematics from a non–traditional perspective. The first part of this books tries to show that mathematics is so much more than simply numbers, and shapes. It actually touches every area of humanity, from philosophy to psychology. We feel it is very important that the liberal arts student gets this greater exposure to mathematics. The topics we present in this chapter, however, are not presented in great depth. They are introduced more to highlight the connectedness of mathematics to other fields of study, rather than fully explain and define these themes. We encourage students to study beyond what we present here and use this part of the book as a road map for future inquiry, investigation, and exploration, as you take other liberal arts courses or as you study on your own. It is possible to skip this part of the book without loss of continuity, but we encourage you to include this section, as it enhances the mathematical experience.

CHAPTER 1

A Philosophical Perspective

Every good mathematician is at least half a philosopher, and every good philosopher is at least half a mathematician.

Gottlob Frégé

1.1 Introduction

What is mathematics? This is a question that, one would think, every high school graduate should be able to answer. After all, they have studied mathematics, in one form or another, for the last twelve years. However, if you ask any recent graduate this question, the diversity of responses might surprise you. Answer it yourself. Do some difficulties emerge? We have grown so accustomed to just trying to do, that we hardly ever stop to question what it is we are doing. First, you may have to deprogram yourself from the effects of secondary education. You are not trying to achieve some answer on a standardized test, or pick the right definition from a list of multiple–choice answers. Instead you are being asked to critically evaluate and analyze this entity we call mathematics.

We could start by giving some general descriptions of what it is. There are many compelling examples:

It is the language of the universe,
It is the subject that deals with measurement and form,
It is about proving things about numbers and shapes,

It deals with Arithmetic, Algebra and Geometry,
It is simply what mathematicians do.

These all give us some sense of what mathematics is, but do any really qualify as definitions? Go to any source – dictionary, encyclopedia or website, and try to find a universally accepted definition of mathematics, and you will see that it is not easy.

Mathematics is not something that can be defined so readily. It is, without doubt, one of the crowning achievements of humankind. It touches every aspect of our lives from the simple and practical to the abstract and intangible. For most people mathematics is just arithmetic. It's how we figure things so that we can buy our clothes and groceries, keep our bank accounts in balance, and perhaps even invest and save for retirement. Beyond this, it doesn't have a real need or purpose. Except maybe for those engineers and scientists. Even scientists and engineers oftentimes think of mathematics as just a bag of tools, and sometimes cute tricks, for solving problems, but not as a discipline itself. As we shall see, however, mathematics is so much more. In this chapter we take a deeper look into this question, and investigate the connection between mathematics and ourselves on many different and diverse levels. It is our hope that you will see mathematics in a way that you've not seen before.

1.2   Defining Mathematics

What does it mean to define mathematics? Any definition of mathematics should not only be able to tell us what it is, but also where it comes from, and why and how it works! If our definition cannot address these basic questions, then our definition is incomplete. We are missing something. To attempt to uncover a definition of mathematics, we will need to understand where it came from, and what role it plays in our culture and our lives. We need to be able to critically analyze, understand and express all aspects of this thing we call mathematics. We will also want to be able to present our ideas, and critiques of the opinions of others, in ways that are understandable and logical. In short, we want to apply solid critical thinking and analysis skills to this question.

Many groups and commissions have looked to define critical thinking. Frequently, the definitions they arrive at, are extremely long and overly complex. For us,

> **Critical Thinking** is defined as applying, analyzing, synthesizing, and/or evaluating ideas, events, and/or information.

Our ability to think critically is directly related to how deeply we question the information we receive. We should not accept everything we see on face value. This is a trait that has largely been missing in our educational system.

A century ago, we moved from teaching critical thinking skills to using the skills that students bring with them. We accepted that students, as human beings, are critical thinkers, and would display these skills if the classroom allowed such behavior. It seemed that we were not seeing critical thinking simply because we were preventing it from happening; through years of school, students were unwittingly "trained" not to think critically in order to succeed in school mathematics. Excerpt from the "International Encyclopedia of Critical Thinking," Danny Weil (ed). ABC–CLIO. 2003 (in press).

We now wish to reverse this trend as we take a more critical look at mathematics – what it is and what it means.

When we ask what mathematics is, we are also opening up the questions of – where did it come from? Does mathematics exist independent of people? Is mathematics the same everywhere in the universe? Why does it work? These are all questions of a philosophical nature. More specifically, they deal with an area of philosophy called "metaphysics." Philosophical questions probably arose when people first began looking up at the stars and contemplating what is out there, and wondering who they were, and how they got there. The earliest group for which we have documented writings and teachings in this area, is the Ancient Greeks. In fact, the word philosophy is from the two Greek words "philo" and "sofia" meaning "love of wisdom."

> *(Philosophy) is now widely used to designate the pursuit of knowledge or wisdom about fundamental matters concerning life, death, meaning, reality, being and truth.* Wikipedia

Looking at life in general and mathematics in particular, requires a philosophical perspective as we critically analyze them.

Putting on our critical thinking glasses we can provide a basic defintition of many of the other disciplines we are currently studying, e.g.,

- Language (English for example) is a form of verbal communication that is created by people.

- Physics is the study of how the physical world works.

- Biology is a study of how the living world works.

- Chemistry is a study of how atoms interact with each other.

- Philosophy is a study of what reality is, and what is our place in it.

- Psychology is a study of the emotions of the human brain and how it interacts with others.

- Sociology is a study of how societies and cultures grow and interact.

- History is an attempt to record and understand what has already happened.

- Art is a form of language for communicating different ideas, interpretations, and concepts.

Following this train of thought, we might say that:

Mathematics is a language created by people to describe the physical world around them.

4

This seems to be a reasonable definition, but let's examine it further to see if there are any questions or holes in this description.

First, if mathematics is created by people then it stands to reason that different people may and probably will come up with different mathematical descriptions to define a particular phenomenon, and that there is no reason for us to expect that any of these ideas would be related to each other. However, this is not the case when it comes to mathematics. Mathematics achieves its power and usefulness by being a consistent and connected system. If it continually led to distinct, or even worse, contradictory descriptions of the world, it wouldn't be as useful. The more interesting observation, though, is that mathematics can exist on its own, and at the same time be connected to the world around us. A simple way to illustrate what we mean by this is through the Pythagorean Theorem.

The Pythagorean Theorem, which states that the sum of the squares of a right triangle is equal to the square of the hypotenuse, owes its discovery to our trying to understand the physical world around us. The Pythagorean Theorem was initially used as a tool by the ancient Egyptians to build the impressive structures we still see standing today. It was a way to measure distances very precisely. For the Egyptians it was not a theorem, but a useful tool (formula) that allowed them to accurately build large structures with great accuracy. If humans had not taken it any further we might not have noticed a powerful new observation. It was the Ancient Greeks, Pythagoras in particular, that took this mathematics to a whole other level. Instead of being a formula that worked for a fixed number of triplets, it was always and universally true, and more importantly it could be proven true (we discuss the concept of proof in great detail in Chapter 14). The theorem is independent of the physical world where it came from. It seems to live in another world, different from our physical world.

This new Pythagorean world opened up a Pandora's box of new and frightening ideas and possibilities. Plato was deeply influenced by this unique aspect of mathematics. From his experience with geometry and the perfect geometric figures that could be described mathematically, but only approximated in the real world, Plato developed a unique view of reality. It was mathematics that began to push the boundaries of his philosophy.

Before we address philosophical and metaphysical questions regarding mathematics, we first need to understand and explain a little about what philosophy and metaphysics are. Only then can we consider questions regarding the origins of mathematics. Thus, we will need to take some time to briefly look at the historical development of philosophical thought.

1.3   Eastern and Western Philosophy and Metaphysics

The sixth century BCE is considered remarkable because it was in this century that humans seemed to have asked the definitive questions "Why?" and "How?" For the first time people looked for (had time for) natural explanations where previously there had been only myth and superstition. This didn't only happen in Greece with the Ionian philosophers and Pythagoras, but also in India (Buddha) and China (Confucius). Arthur Koestler says of this century, "A March breeze seemed to blow across this planet from China to Samos, stirring man[kind] into awareness, like the breath in Adam's nostrils." Koestler goes on to assert that "Every philosopher of the period seems to have had his own theory regarding the nature of the universe around him" and he likened the sixth century to an orchestra "expectantly tuning up, each player absorbed in his own instrument only, deaf to the caterwauling of the others." http://mathgym.com/history/pythagoras/pythintro.htm

As the above passage suggests, starting around the 6th century BCE and continuing for about 600 hundred years, a new way to look at the world began to emerge. A lively and colorful debate concerning the place and fate of humans in the universe was started. The start of the city–state provided some with the necessary free time to begin to think and speculate on our place in the world. Also, during the 6th century BCE the world had three major state societies, where most modern philosophical ideas started – Ancient Greece, India, and China.

Out of these three advanced cultures, two distinct philosophical perspectives arose. The Western Philosophy of Ancient Greece and the Eastern Philosophy of India and China. To gain a better understanding of what philosophy is and how it impacts our definition of mathematics, we need to look at both Western and Eastern Philosophies. We will not provide a full accounting of the development of these philosophical perspectives, as each would take up volumes of work. Instead we provide more of a summary of key people and events to help us better understand how it all "came together." We begin with the Ancient Greeks.

1.3.1   Western Philosophy (Ancient Greece)

Western philosophical thought has its origins in Ancient Greece. An overly simplistic definition of Western Philosophy, is a deductive cause–and–effect perspective, which sees the world as essentially physical in nature with only physically determined explanations on how the world behaves, being valid. Not all Greek philosophers agreed with this perspective, but one of the most respected philosophers of Ancient Greece, Aristotle, did. As it turns out, Aristotle, probably had the largest impact on Western thought and philosophy than almost any other Greek philosopher. Philosophy in Ancient Greece was quite diverse, but Aristotle had a view that stood out with future generations in the west. Aside from the fact that Aristotle was brilliant, his ideas seemed to have a common sense approach to the world, which resonated with people.

Aside from Aristotle, there were numerous philosophers in Ancient Greece that had an impact on western philosophy. We will focus on only a few, but keep in mind that these few were profoundly influenced by others at the time, and we will not be able to include them all. Some of the important philosophers contributing to western philosophy that we will discuss are Thales, Pythagoras, Socrates, Plato, and Aristotle. We have not, nor could we include all the important philosophers from ancient Greece in this short survey. There were many who had a significant impact on the development of philosophy and the ideas of the philosophers we have mentioned above. For a more complete picture you should look up people such as: Anaximander, Parmenides, Heraclitus, Democritus, and Zeno of Elea to name a few. Ancient Greece was a wellspring of ideas and opinions, and these ancient perspectives are also quite valid today.

We begin our discussion with Thales, who is considered to have begun the process we have come to know as western philosophy.

Thales

Thales (pronounced as THAY–leez, or less frequently as either thales or t–HAY–leez) is considered the father of modern western scientific thought by many. He is believed to have lived from about 624 BCE to 546 BCE in Miletus of Asia Minor. Much of what we know about Thales comes to us indirectly through the writings of others. Oftentimes these writings were done many years after his death, so we can't be sure of their accuracy, but we give greater credit to their authenticity when they come from different unconnected sources. From this information we feel the following is a reasonable summary of his life and accomplishments.

Thales is believed to have traveled extensively and is thought to have visited both Babylon and Egypt, and was greatly influenced by their mathematics, which was essentially geometry at that time. He was the first to abandon mythological explanations for natural phenomena, such as the weather and the motion of the stars and the planets. Instead, he searched for a rational, or what today we would call a scientific cause, for what he observed. To Thales everything must have a first cause, something that makes it what it is, or do what it does. Some of his explanations, or first causes, we might find humorous today, such as the belief that everything in the world was made out of water. However, since hydrogen is believed to be the most prevalent atom in the universe and there are two hydrogen atoms in every water molecule (H_2O), it may not have been as far fetched as one might think. Certainly water is the most critical component of all living things. The important point, however, is that we have the beginnings of rational causes for things, and not just myths.

Another very important contribution to western thought that Thales made, was to introduce us to the idea of **skepticism**, and the belief that we had to provide convincing evidence for our claims. This is the foundation of scientific reasoning today. Instead of simply using geometric formulas to get results, like the Egyptians did, Thales looked to what caused the formulas to be true. His search for a first cause introduced the idea of a proof to western reasoning (we discuss this concept in greater detail in Chapter14). His rational approach to the world led him to his great achievement/prediction of his day. Thales claim–to–fame, that is most probably true, is that he successfully predicted the occurrence of the eclipse of the sun in 585 BCE. An eclipse that ended a war at the time.

Thales is arguably the first true mathematician, since he pioneered the idea of proof. He is credited with many of the early geometric proofs used by Euclid in his book on geometry many centuries later. Thales was supposed to have been the teacher of Anaximander, another Greek philosopher, who took Thales' ideas and raised them to a higher level, and is considered by some to be the first scientist. He was the first to propose a mechanical model of the physical world, and is thought to have conducted the first scientific experiment: "by examining the moving shadow cast by a vertical stick he determined accurately the length of a year and the seasons."(Carl Sagan's Cosmos, 1980 Random House Publishing) Anaximander was also the teacher of Pythagoras.

Pythagoras

Pythagoras (pronounced 'puh–THA–guh–rus', or 'pi–THAG–er–uhs', or as 'puh–THAG–uh–ruhss') lived from around 570 BCE to about 490 BCE, and like Thales spent some time in Egypt, Babylon, and Phoenicia. Of this we are fairly certain, all else is open to debate and question, since there is no direct evidence that Pythagoras wrote down any of his beliefs or ideas. One school of thought presents Pythagoras, like Thales, as a man that was strongly influenced by mathematics and the role that it played in explaining and understanding the world. He is credited with everything from proving the Pythagorean theorem, to establishing a connection between mathematics and music, to helping (much to his dismay) to discover irrational numbers, and to believing the planets in the heavens made harmonious music. All of which had a deep connection to mathematics. For those from this perspective, he continued and further extended the mathematical tradition of Thales. He is said to have made the statement that "All is number." The belief being attributed to his view that the world was intimately connected to numbers, and that in fact numbers were more real than what our senses allow us to experience. They were not simply a creation of humans, but a fundamental entity that existed independently of us, with the power to affect and expose the world to greater understanding.

On the other hand, there are others that attribute this assessment of Pythagoras to unreliable or incorrect sources with little or no basis in fact. Four biographies of Pythagoras have survived from ancient times. They are from Diogenes Laertius (around 180 CE), Porphry (233–306 CE), Iamblichus (280–333 CE), and Phontius (820–891 CE), and they all provide a different picture of Pythagoras the man. The biography by Diogenes was written closest to the time Pythagoras was alive, but was still written nearly 700 years after his death! These biographies predominantly used the writings of Plato, Aristotle, Xenophanes, Heraclitus, Aristoxenus, Dicaerchus, and Timaeus of Tauromenium, to try and form a true picture of Pythagoras the man. However, neither Plato nor Aristotle was alive when Pythagoras was, so they only knew of him through the teachings of the society he had founded and that bore his name, called the Pythagoreans. Information about Pythagoras came to Plato and Aristotle through what were called *acusmata* (meaning things heard, or short expressions handed down orally). Those that did know him, did not provide enough details about him to form a complete picture as that written in the four biographies.

Recent projects were undertaken to re–examine the life of Pythagoras and to try and come up with a more complete picture of who he really was (see for example the Stanford Encyclopedia of Philosophy), since he was often presented as a "semi–divine–figure," and work that was obviously not his own is sometimes attributed to him. The consensus is that from the written record, Pythagoras was more a leader of a new way of living than he was a philosopher or mathematician. It appears that he probably never wrote anything, since nothing survives and there is no reference to his writings by others, around and shortly after his time.

Whatever Pythagoras' actual beliefs and teachings were, we will never know for sure, what is known is that he inspired a group of followers and disciples to create new mathematics and a philosophical vision that lasted for centuries after his death, and that through this group he had a tremendous impact on both mathematics and philosophy. One important disciple was Archytas. Archytas was a contemporary of Plato, and one of the first to look for more rigorous ways to prove results in mathematics. It is largely due to followers like Archytas that Pythagoras gets so much credit for "his" mathematics.

Aside from believing in reincarnation and adhering to a strict and regimented style of living, the Pythagoreans also believed that we would only come to know the world through our minds, and not exclusively by our senses. Furthermore, they believed there was a mathematical order to the universe, as they could see it everywhere – in music, the motion of the planets, the perfection in geometry, etc.

Socrates

After Pythagoras we come to Socrates (pronounced SAH–kruh–teez') who lived from 469 to 399 BCE. Socrates changed the path of Greek philosophy, and gave birth to the Socratic era in Greek philosophy. He was more concerned with ethics and living the "good" life than anything else. He is reported to have made the statement that "the unexamined life is not worth living." What we know of Socrates comes mostly from the writings of his student Plato. Socrates was very interested in knowledge and the "true" meaning of a word. He believed that if one knew what was right and just, then that person would not knowingly choose what was not right or unjust. He taught his ethics using a questioning process that came to be known as the Socratic Method. The basis of this methods is the assumption that we already possess all knowledge within us, and that through a questioning process we are able to recall this knowledge.

Socrates was not a mathematician, but believed that training in mathematics sharpened the mind for further explorations in philosophy, and as such was a necessary first step. Thus, while he saw its importance, he did not elevate it as his predecessors did. Mathematics was more of a means to an end. However, his belief in us, already having knowledge within us, goes along with the idea that mathematical knowledge is already complete within us. Socrates believed that true reality was not knowable by our senses alone, it could only be known through our minds.

Socrates was not wealthy, his father had been a sculpture (a tradesman to create the statues of the day), and when he died he left his financial estate to Socrates. Socrates was a soldier in the Athens military and was a devoted citizen of the great city–state. He lived an austere life and prided himself on how little he could get by on and how much he could endure. He constantly educated the youth of Athens, suggesting that one should question everything, for it is only through questioning that you arrive at the right answer. The process of questioning everything, even authority, gained Socrates many enemies in the influential people of Athens. Eventually Socrates was brought up on charges of corrupting the youth, when one of his students, Alcibiades, committed high treason against Athens. Ordinarily, Socrates crime would have only demanded the payment of a fine, and for Socrates to stop what he was doing and admit to his crime. However, the sentenced imposed was death. Some believe such a harsh sentence was handed down to try and convince Socrates to flee Athens, never to return. Socrates, however, did not leave and being a devoted citizen of Athens said he had no choice but to carry out the wishes of its citizens and to take hemlock as directed. He did so and died in 399 BCE. The story of his trial, imprisonment, and death was recounted by Plato in his works the Apology, Crito, and Phaedo, respectively.

Plato

Plato (pronounced 'PLAY–toe') was a student of Socrates who lived from 427 to 347 BCE. Plato wrote extensively on the subject of philosophy. He also founded the first university, called the Academy. Plato had the words "Let no one unversed (untrained, or ignorant) in geometry enter here" written over the entryway door.

Recently it has been proposed that the writings of Plato contain a hidden mathematical/musical structure (Dr. Jay Kennedy at the University of Manchester concerning "Plato's Code"). This, it has been suggested, tells us that Plato's beliefs may have been similar to the Pythagoreans, especially when it involved how mathematics is connected to the universe. Plato would not have wished to be associated directly with the Pythagoreans, as many were not held in high regard with the other Greek philosophers of the day. This is why Kennedy believes he wrote in "musical" code. This is a fascinating new view and we encourage interested readers to explore this in greater detail.

Fundamental to Plato's belief system is his theory of "ideas" or "forms." The ancient Greeks were confused by the idea of change. On one extreme we have Heraclitus with the belief that the only thing you can be certain about is that everything will change. On the other side of the debate we have Parmenides saying that motion and change are just illusions. In reality, says Parmenides, the universe does not change, it is all static. Plato reconciled these differences with his theory of forms. To Plato the world has two parts, the "visible" world of our senses, and the intelligible world of our minds. The mental world is unchanging and contains the forms (the things that are the ideal representation of the object in question). This is true reality. The visible world does change with varying copies of these ideal forms, but the ideal world does not change. Thus, ultimate reality is in ideas and forms, which we come to know through our minds and reason.

Aristotle

Aristotle (pronounced 'AIR–iss–tah–tuhl') was a student of Plato who lived from 384 to 322 BCE. Like Plato, Aristotle wrote extensively. Without question, Aristotle had the largest impact on western thought, and established the western scientific mindset. Aristotle believed that the ultimate reality resides in physical objects, which we come to know through our senses and experience. The world is made up of objects composed of matter and potential forms. For example, a block of marble (matter) has the potential to become whatever form the sculptor chooses. A seed (matter) has the potential to become a plant or a tree. An embryo has the potential to become a human or another form of animal. Also, the form of a living creature was characterized by its soul, and furthermore there was a hierarchy of souls. Inanimate objects did not possess souls. A plant had the lowest type of soul, animals had a higher level soul, souls that could feel, and humans had the highest level, souls that could reason.

Using these ideas Aristotle created an approach to systematically study animals, much like a biologist would today. He wrote many books describing his findings about animals, which he obtained in a methodical, what we would today describe as a, scientific, way. For this, Aristotle is sometimes called the father of the scientific method. It was an approach that made sense to people. If you wanted to understand what something was, you would take it apart and classify, and categorize its parts, so that you could understand what it truly is. You did not rely only on thought and reason to understand what it was, you used experience through your senses to understand what it was. Bertrand Russell once said that "Aristotle was Plato diluted with common sense."

Aristotle applied his same approach in his study of physical objects and phenomena, whether it be wind, weather, earthquakes, rainbows, comets, etc. His writings on his discoveries and beliefs sound quite modern when compared with other philosophers of his time. An excellent summary of Aristotle and his beliefs is provided by http://www.ucmp.berkeley.edu/history/aristotle.html. Many of the ideas above were borrowed from this source. An especially relevant passage describes how Aristotle's work became a central pillar of the western belief system, through the christian religion.

In the later Middle Ages, Aristotle's work was rediscovered and enthusiastically adopted by medieval scholars. His followers called him Ille Philosophus (The Philosopher), or "the master of them that know," and many accepted every word of his writings – or at least every word that did not contradict the Bible – as eternal truth. Fused and reconciled with

Christian doctrine into a philosophical system known as Scholasticism, Aristotelian philosophy became the official philosophy of the Roman Catholic Church. As a result, some scientific discoveries in the Middle Ages and Renaissance were criticized simply because they were not found in Aristotle. It is one of the ironies of the history of science that Aristotle's writings, which in many cases were based on first–hand observation, were used to impede observational science.

On the idea of a changing world, this is

Where Aristotle differed most sharply from medieval and modern thinkers was in his belief that the universe had never had a beginning and would never end; it was eternal. Change, to Aristotle, was cyclical: water, for instance, might evaporate from the sea and rain down again, and rivers might come into existence and then perish, but overall conditions would never change.

Aristotle, unlike Plato, and Pythagoras before him, did not see mathematics as intimately tied to the universe as they did. Mathematics was predominantly made up of "fictional entities grounded in physical objects" http://plato.stanford.edu/entries/aristotle–mathematics. It was a man–made tool to study the universe, nothing more, although it was a very useful tool.

In summary, Thales gave us the idea of rational reason for causes, and the beginnings of scientific thought. Pythagoras, through the Pythagoreans extended the idea of finding causes, unrelated to the gods for the world around them, in mathematics. They even created a belief system around number and mathematics. Socrates gave us ethics and an internal basis of knowledge, and the need to strive for it. Plato aside from documenting all the previous ideas, also provided a world view where mathematics too was at the very essence of the universe, through his ideas of forms. Aristotle took all the previous work and created what we know today as the western philosophy or a more scientific thought process. One theme that seems to emerge from this philosophy is that of a distinction between mind and body. Plato falls on the side where mind is the reality that gives rise to the physical world. Aristotle on the other hand seems to express the opinion that it is the physical world that gives rise to mind (or soul).

On the other side of the world, at about the same time, a similar, but at times quite different way of looking at the world was emerging.

1.3.2 Eastern Philosophy (India & China)

At the same time that philosophical ideas were being advanced in ancient Greece, great thinkers in India and China were also trying to make sense of the world. These beliefs and ideas have become known as eastern philosophy. Although in reality there isn't a true eastern philosophy, instead there are some common ideas that help us distinguish what has been characterized as western thought from eastern thought. We risk oversimplifying the discussion, however, and at times it may lead us to some controversial (not universally accepted, or overly–simplified) conclusions, but it will help us in trying to understand the differences, if we make these simplifications. We do recommend a more thorough investigation of these concepts and ideas through other available sources.

Chinese Philosophy

In China around the 6th century BCE, Lao–Tzu (also referred to as Lao–zi), was an individual or a group of individuals credited with creating the Chinese philosophy/religion of Taoism (pronounced dowism). We have no documented evidence of his life, which is why many believe he was not a person, but really a collection of people that were given the name Lao–Tzu. Taoism is one of the dominant belief systems of eastern thought.

In the east, mathematics developed more, as a tool than as an explanation for, or a model of reality. Reality to eastern thinkers could not be described or explained by language or mathematics. As soon as you tried to describe or explain it, you lost it. To many there was no distinction between mind and body, and you could not consider one without the other. You should not try and explain the world, but instead strive to experience it through meditation. For many this is a mystical approach to the world. The universe is not divided into different things, it is all just one thing. It is eternal and permanent. It may change its appearance to us and our senses, as it goes through cycles, but it is still just one connected entity. Consequently, you should strive not to explain or understand it, but to experience it in harmony. To experience it you needed guidance on how to live and proceed and this is what Taoism is. It is a path of behavior. A collection of ideas on how one should live in harmony with the world.

Around the same time that Taoism was developing, another Chinese thinker, Confucius was creating his vision of the world. Confucius lived from 551–479 BCE. Confucius, along with many others developed a philosophy which emphasized the proper way to live in society. Confucianism, as this system of beliefs came to be known, is a system of ethics. A way to live a good life. At times Taoism and Confucianism conflict.

In Chinese philosophy there is more a belief that the world is one, that it always was and always will be. It is everlasting. It goes through cycles which may appear to make it vanish and reappear, but that is just "the way." You cannot understand the world by studying it from the world of the senses, instead you must experience it from within, through the world of meditation. Notice both the similarity and difference with the Aristotelian point of view.

Indian Philosophy

India has had a long tradition of a consistent, but developing belief system in the form of Hinduism. Hinduism is a major religious belief system, that is not traceable to any one–person, but has a long history of development dating back to about 5,500 BCE. It is not, however, a unified belief system, but rather a plurality of different beliefs, and sects, rooted in traditions of the Vedic period in India (from the second and first millennia BCE up to the 6th century BCE). It is one of the oldest and most diverse religions of the world, deeply rooted in tradition, and for many it is more of a way of life than a religion.

Most Hindus believe in an eternal soul, and the goal is to end the continual cycle of birth and rebirth and to finally unite with "God," or the universe. To Hindus the concept of the void is a positive thing. The world was created out of the void, which is good. It also holds a pantheistic view of the world, considering everything, living and non–living as divine and sacred. God is in everything.

During the 6th century BCE another great thinker, Siddhartha Gautama (also called Buddha) founded a new religion/philosophy in India called Buddhism. Siddhartha is believed to have lived from 563–483 BCE, although the date of his death is very much in question, which also brings into question his birth date as well. In Buddha we see a parallel train of thought with the Chinese thinkers. There was a belief that we can only know the world through meditation. The main distinction is that Buddhism embraces the idea of emptiness.

The Buddhist belief system comprises the:

Three Universal Truths:
1. Nothing is lost in the universe,
2. Everything Changes,
3. Law of Cause and Effect (Karma),

and the

Four Noble Truths:
1. Life is suffering (Dukkha),
2. Suffering is caused by craving or desire (Tanha),
3. Suffering ends when craving or desire end (Nirvana),
4. Nirvana is reached by following the path to enlightenment.

East Versus West

Eastern philosophy is a way of life. Unlike the western view where we categorize things and separate them out from one another. The eastern viewpoint does not. Instead, you should simply strive to live in harmony with the world, and not look for its separateness, but its unity.

Eastern philosophy cannot be separated from religion. It is a belief system that encompasses both, whereas in the west, we separate religion from philosophy and can study both independently. Furthermore, in the west we can also study the material world independently from the mental world, while in the east they are not separate entities, but are one of the same. In the west we can have a separate discussion on religion and on philosophy, but in the east you cannot, because they again are one of the same. In the east a god or gods does or do not exist independent of the universe, and did not exist before the universe and create it, but are an integral part of the universe. All is one.

Western thought places the individual at the center, with the individual distinct from the universe, finding their way and discovering the separate world around them. Eastern thought puts the harmony of the universe first and the individual is simply a part of the universe, and the closer they get to not seeing a difference, distinction, or separateness, the closer they are to "seeing" the real universe. Also, if you can put it into words then you have not described it. It comes from simply experiencing it from within.

In the eastern viewpoint:
- There is no dualistic nature of the universe. Mind and body are one of the same.

- The subject and the object are one of the same.

- Knowledge comes from intuition, not from separate study.

- You cannot separate the observer from what is being observed. They are actually one of the same.

Eastern philosophy is a way of life where western philosophy is a system of beliefs that can be studied and understood separately.

In much of what follows we will take a decidedly western perspective on how to proceed. We do this because the western approach allows us to pose questions and then to try and answer them. It is also the approach we are most familiar with. When appropriate we will speak to the eastern perspective, and how the investigative process might be altered to examine the particular issue under discussion using this approach.

Metaphysics

A good portion of what the eastern and western thinkers were addressing above is what is called metaphysics.

Metaphysics: the branch of philosophy that deals with reality, existence and knowing.

We are particularly interested in how mathematics either influences this study, or how it is either related to or a part of reality.

In this next section we focus on mathematics, and in particular we look at different ideas that were put forth on what mathematics is and where it might have come from. These have been organized into different views or schools of thought.

1.4 Philosophical Schools of Thought on the Origin of Mathematics

What mathematics is and where it comes from has been an ongoing debate that has lasted over two thousand years and it will probably never be fully resolved. It is a metaphysical question that challenges human intellect. There is not one universally accepted definition of mathematics. In many ways it is like religion, a matter of personal belief and choice. Mathematics is different things to different people.

In one sense it is a language, a language that defines how the universe is put together. In another way it is a subject about quantity and measurement. It can also be viewed as a game that depends upon axioms (accepted or self–evident truths) and rules that lead to new results, which do not have any real meaning or connections beyond this. To some, mathematics is the framework of the universe. A framework that god created and humankind discovers. It exists independently of us. It may also be merely be a reflection of our culture, just like art, literature, and music. Thus, mathematics that is created elsewhere in the universe would have a different "look," "sound," or "feel" to it there, then it does here.

However an individual may view the answers to these questions, it is extremely important in our search for meaning and understanding, to see these differing views and perspectives. Although there is a wide assortment of beliefs on what mathematics is, they can loosely be assembled under five different philosophies or schools of thought, all of which present a very different perspective. They are Platonism, Formalism, Logicism, Intuitionism/Constructivism, and Humanism. There are others, but these are the most dominant ones. An excellent description of these and other philosophies of mathematics can be found in the book, "What Is Mathematics, Really," by Reuben Hersh, or online at http://en.wikipedia.org/wiki/Philosophy_of_mathematics.

1.4.1 Platonism (Realism): Mathematics is out there.

The mathematical philosophy of Platonism, also called Realism, is the belief that mathematics exists independently of the human mind. It is the "world of forms" of Plato. Mathematics resides in a world different from the physical world in which we live, and humans simply discover mathematics. We uncover preexisting truths that are already there rather than creating something from our own mental predispositions and bias'. Mathematics then becomes a search for absolute truth and ultimate reality.

> **Platonism**: The universe is mathematical and we discover mathematical truths.

As the name Platonism implies, the ideas behind this philosophy have their roots in ancient Greece. The ancient Greeks did not have a written algebra. Greek mathematics was predominantly geometric in nature. They would draw mathematical objects in the sand and construct proofs concerning properties and relations of these geometric objects. Thus, while nobody could actually draw a perfect circle or a line, many believed that there was a world in which these perfect entities did exist and that we can only approximate them in our imperfect physical world. Plato was a key figure in the development of this philosophical perspective, which is why his named is used. Plato believed in a "heaven of ideas," an unchanging ultimate reality that the everyday world can only approximate. It is in this other, perfect world that mathematical entities such as numbers, points, lines, and circles reside.

Plato's beliefs were almost certainly influenced by the earlier philosophies of Pythagoras (569–475 BCE) and his contemporaries. Pythagoras viewed numbers as perfect objects, and as we discussed earlier, he even went so far as to establish a religious cult called the Pythagoreans based largely upon the belief in the mystical properties of natural numbers.

ARGUMENTS FOR
Platonism explains why Mathematics works so well in describing the world. If mathematical entities exist in a perfect world for which our physical world is only a crude approximation, then it is only natural that mathematics should describe the world so well. Our physical world is just an inexact model of the perfect world of mathematical objects.

Another argument in favor of Platonism, was put forward by the philosophers W.V.O. Quine (1908–2000) and Hilary Putnam (1926–present). The argument is called the Indispensability Argument and states that:

> *Since mathematics is indispensable to all the empirical sciences, and if we believe in the reality of the phenomena described by the sciences, we ought also believe in the reality of those entities required for this description. The existence of mathematical entities is the best explanation for experience.*

A final argument is that mathematical entities seem to exist independently of human beings. This would explain why such vastly different people and cultures have "discovered" the same mathematical concepts and entities. Which is why any mathematicians view themselves as Platonists – discovers' of mathematical truths and ideas.

ARGUMENTS AGAINST
The major problem of Platonism is that it raises new unanswered questions – Precisely where and how do mathematical entities exist? Is there a world separate from the physical world that is occupied by these mathematical entities? If so, where is it?

SOME MATHEMATICAL PLATONISTS
Gotfried Leibniz, G. H. Hardy, George Berkeley, René Thom, Kurt Gödel, Paul Erdös, and Roger Penrose,

For nearly 2400 years the predominant philosophy of mathematics was Platonism. Then in the early 19th century two major events took place that gave rise to some new philosophies. In 1879 a German mathematician, Gottlob Frégé (1848–1925), building on the earlier work of Leibniz, published a paper titled *Begriffsschritf* (Concept–script), which established modern symbolic logic as a useful mathematical tool. With this came the birth of a new mathematical philosophy called Logicism.

The second event occurred in 1874 when Georg Cantor (1845–1918) published his paper *On a Property of the Collection of All Real Algebraic Numbers*, which created a fresh perspective on the age–old, mathematically troublesome, concept of infinity. Cantor created set theory, which in turn enabled mathematicians to better understand and "work with" the concept of infinity. These two events helped to give birth to a new perspective on the origins of mathematics.

1.4.2 Logicism: Mathematics is simply the by–product of rules of logic applied to numbers and shapes. (Circa 1884)

The philosophy of Logicism is based on the belief that mathematics is founded upon the rules of logic, and that all mathematical statements are just logical truths, also called "tautologies." In short, all of mathematics is reducible to formal logic. All mathematical truths and proofs can be translated into logical truths and proofs or, in other words, the vocabulary of mathematics is reducible to logic. Logicism claims that all of mathematics is derivable from logic without the need of any "mathematical" concepts.

> **Logicism**: The universe is based upon rules of logic and mathematics is derived from these rules as a byproduct.

Logicism was largely the creation of the German mathematician Gottlob Frégé, and the English mathematicians Bertrand Russell (1872–1970), and Alfred North Whitehead (1861–1947). Frégé was convinced that the truths of arithmetic are logical, analytic truths (an analytic truth is a truth that to deny its existence would lead to a contradiction, they are true by definiton), agreeing with Leibniz, and disagreeing with Kant, who believed that arithmetical knowledge was grounded in "pure intuition."

The purpose of Logicism was to show that classical mathematics (arithmetic, algebra, calculus, geometry) is really just a part of logic. Bertrand Russell had great hopes for Logicism. Russell wanted to use logic to clarify issues in the foundations of mathematics and philosophy as well. Russell

> wanted certainty in the kind of way in which people want religious faith. (He) thought that certainty is more likely to be found in mathematics than elsewhere.

Furthermore, he believed that Logicism would be

> a new field of mathematics, with more solid foundations than those that had hitherto been thought secure

and that

> mathematics is the chief source of the belief in eternal and exact truth. www-history.mcs.st-and.ac.uk/Quotations/Russell.html

Logicism, however, actually shifts questions regarding the origin of mathematics to logic and never really answers them. For we must now address the same questions, but under the logic umbrella. Do humans create logic, or is it already there for us to discover it? If the assumption is that the universe is based upon logic and reasoning, then logic is discovered, and mathematics is also discovered and not created.

ARGUMENTS FOR
Logicism would establish a solid foundation on which all mathematics could be based. It would not require another world for mathematical entities, but would still give meaning to mathematics in the sense that it flows naturally from logic and reason.

ARGUMENTS AGAINST
In Chapter 17 we discuss two important ideas that caused the failure of Logicism. They are Russell's Paradox, and Godel's Incompleteness Theorems. It was shown mathematically that this philosophical school of thought was flawed.

SOME MATHEMATICAL LOGICISTS
Gotlob Frégé, Bertrand Russell, Alfred North Whitehead

1.4.3 Intuitionism/Constructivism: Mathematics comes from within. (Circa 1908)

Intuitionism, also called Constructivism, is the belief that mathematics comes from the human mind. The mind intuitively possesses the notions of space and time. The natural numbers are already embedded in the brain and from this "intuitive" conception of number, all of mathematics is "constructed." Thus, mathematics is not discovered by studying the physical world, it is instead found within us – a free creation of the human mind, where an object exists only if it can be constructed by the mind.

> **Intuitionism**: Mathematics is the creation of the human mind. It does not exist without us. We construct it from our thoughts and ideas.

Some of the underlying ideas of intuitionism can also be traced back to the German mathematician Leopold Kronecker (1823–1891). Kronecker wanted to base all mathematical analysis on the natural numbers 1, 2, 3, 4, … and eliminate the need for irrational numbers such as π, or $\sqrt{2}$ (Again, as we'll see in Chapter 12, we have the concept of infinity rearing its ugly head, for remember that an irrational number has an infinite, non–repeating decimal representation.). Kronecker stated that:

The natural numbers come from God, everything else is man's work.

The Intuitionist philosophy of mathematics was created by the Dutch mathematician Luitzen Brouwer (1881–1966), by extending the ideas of Kronecker and partly in response to Cantor's analysis of infinite sets. Brouwer did not accept that an "actual" infinity existed. Other mathematicians, including Leopold Kronecker, and Jules Poincaré were also skeptical about the methods of proofs that used infinite sets. Intuitionist's argued that proofs to establish the existence of a mathematical object by showing that its nonexistence would lead to a contradiction (proof by contradiction) are valid only if all possible nonexistence candidates can be examined (the set of possibilities is finite). Since it is not possible to perform such an exhaustive analysis on infinite sets, they believed that proof by contradiction, was unacceptable. More to the point, the intuitionists rejected Aristotle's "Law of the Excluded Middle." This law states that a statement is either true or false, and there is not middle ground or truth–value in between. In classical mathematics, a proof by contradiction, says that if you show that nonexistence is false then existence must be true. For the intuitionist, this is invalid; the negation of the non–existence does not mean that it is possible to find a constructive proof of existence.

To avoid this difficulty, Brouwer created a new approach to mathematics. The essential idea is that mathematical proof can involve infinite sets only if it can be shown that there are finite methods to construct these entities. Any mathematical object is considered to be a construction of the mind. Thus, the existence of an object is equivalent to the possibility of it being able to be constructed. This contrasts with the classical approach, which states that the existence of an entity can be proved by negating its non–existence. Thus, instead of starting with the phrase "there exists" the intuitionist starts with "we can construct."

Brouwer's intuitionism is really an extension of the philosophy of Immanuel Kant (1724–1804). Kant believed that human knowledge is in part constructed by the mind. Brouwer used this idea and extended it to mathematics. He believed that mathematics is found in intuitive truths, and established two acts of intuitionism.

1. FIRST ACT OF INTUITIONISM: All mathematical truths are constructed by the human mind.

2. SECOND ACT OF INTUITIONISM: The way we construct a truth is through a proof, which is really just a mental process.

Putting (1) and (2) together, we get the intuitionist's view that: No mathematical claim is legitimate unless a successful proof is constructed for it. For Brouwer, mathematics is built into the human brain. It is more fundamental than language, logic and experience. Furthermore it rejects the idea of an actual infinity, i.e. it does not consider infinite objects such as the set of all natural numbers to exist as a completed set of objects (actual infinity).

In summary, the fundamental question is this: what does "existence" mean in a mathematical context? The classical mathematician believes that an object exists if he can prove the impossibility of its non–existence, however, the intuitive mathematician believes the object exists if there is an algorithm that constructs it. Thus, an intuitionist believes that mathematics is created by humans and not discovered.

ARGUMENTS FOR
Intuitionism avoids the problem of having to explain what other world mathematical objects exist in. It also avoids the complications inherent in working with infinite sets. It provides a "better" (cleaner/less–complex) notion of proof.

Intuitionism is free of contradictions. In fact, not only is it free from contradictions, but all other problems of a foundational nature as well receive perfectly satisfactory solutions in intuitionism.

ARGUMENTS AGAINST
Intuitionism does not provide us with any insight into the question – Why does mathematics work? Our understanding of the human brain is quite incomplete. Basic concepts such as love, hate, good, and evil as well as perceptions of color and appearances differ from person to person. Is it reasonable to assume then, that all people will have the same intuitive view of mathematics? Wouldn't it be more probable that mathematics is different from person to person?

With the law of the excluded middle rejected, one may wonder- "How can one even do mathematics?" Hilbert said that doing intuitionist mathematics is like doing astronomy without a telescope or boxing without fists.

SOME MATHEMATICAL INTUITIONISTS
Luitzen Brouwer (1881–1966), Hermann Weyl, Leopold Kronecker (1823–1891), Arend Heyting, Stephen Kleene (1909–1994), Michael Dummett (1925 – present)

1.4.4 Formalism: Mathematics is simply a game. (Circa 1910)

A formalist believes mathematics is nothing more than a "game." Mathematical statements or axioms are manipulated using certain specified rules to derive new statements. In one version of Formalism, mathematics is nothing more than the symbols themselves. It has no meaning outside the "game." It is nothing more than symbolic manipulation, without any meaning. This view of Formalism, however, is not widely held.

Another, more popular version of Formalism is called "deductivism." Here derived statements or theorems are not absolute truths, but instead relative truths. This means that they derive their meaning from the "game" and only have a meaning within the "game" and not outside of it. For example in Euclidean Geometry you have certain axioms and some rules of inference to generate new statements. Thus, it is possible to prove that the Pythagorean theorem, which states that the square of the length of the hypotenuse of a right triangle is equal to the sum of squares of the lengths of the other two sides, holds. This means that you can generate it from the given axioms and rules of inference. Pythagoras' theorem has a meaning within Euclidean Geometry, but not outside of this.

> **Formalism**: Mathematics is a "game," that is derived from a set of arbitrary rules, and it has no meaning outside the game.

Formalism was largely the creation of the German mathematician David Hilbert (1862–1943). Hilbert created the Formalist philosophy as an alternative to Brouwer's Intuitionism, which he believed was created by those who would *"seek to save mathematics by throwing overboard all that which is troublesome."* Furthermore, *"they would chop up and mangle the science. If we would follow such a reform as the one they suggest, we would run the risk of losing a great part of our most valuable treasure!"* (C Reid, p 155) Formalists set out to prove that any mathematical system with established axioms, definitions and theorems is "consistent." By consistent they meant that any given mathematical statement if true could be proven true, and if false, it could be shown to be false, i.e. no illogical or inconsistent conclusions could be derived. Formalism is similar to the ideas of the logicists, but the formalists do not believe that mathematics can be deduced from logic alone as the logicists do. Mathematics is a human creation, since we make up the rules, but it is played in a logical fashion, based upon the rules of the "game."

ARGUMENTS FOR
In the Formalist view you do not have to worry about where the mathematical entities reside. There is no other unexplained world of perfect mathematical objects. There are also no metaphysical questions to worry about. Mathematics is just a game and mathematical objects do not exist separately from us.

ARGUMENTS AGAINST
First, Formalism does not address the question – Why does mathematics work so well? Also, why do different people or groups of individuals derive or create the same "games?"

A greater flaw with Formalism was exposed by the Austrian mathematician Kurt Gödel. In 1931 Gödel proved a theorem, called the Incompleteness Theorem, that rocked the foundations of mathematics. We will consider this theorem in greater detail in Chapter 17, but for now we will just give some of the highlights. In short, Gödel's Incompleteness Theorem showed that the consistency of mathematical systems could not be guaranteed. In any "sufficiently" complex system there will be true statements that cannot be proven true and false statements that cannot be shown to be false.

SOME MATHEMATICAL FORMALISTS
David Hilbert (1862-1943), Rudolf Camap (1891-1970), Alfred Tarski (1901-1983), and Haskell Curry (1900-1982).

1.4.5 Humanism: Mathematics is a product of culture. (Circa 1997)

Humanism is a relatively new mathematical philosophy that claims to take its roots from the writings of many past philosophers and mathematicians. It uses the works of Aristotle, Euclid, John Locke, David Hume, Jean D'Alembert, John Stewart Mill, Charles Sanders Pierce, Henri Poincaré, John Dewey, Ludwig Wittengestein, and Imre Lakatos to create a new mathematical philosophy that sees mathematics as nothing more than a cultural creation. According to its founder the American mathematician/author, Reuben Hersh (1927– present):

> *Mathematics is like money, war, or religion – not physical, not mental, but social. Dealing with mathematics (or money or religion) is impossible in purely physical terms – inches and pounds – or in purely mental terms – thoughts and emotions, habits and reflexes. It can only be done in social–cultural–historic terms.*

> *Mathematics is another particular, social–historical phenomenon. It's neither physical nor mental, it's social. It's part of culture, it's part of history. It's like law, like religion, like money, like all those other things which are very real, but only as part of collective human consciousness. That's what math is.*

Humanism: Mathematics is a creation of human culture.

For Hersh, mathematics has existence or reality only as part of human culture. Despite its seeming timelessness and infallibility, it is a social–cultural– historic phenomenon. He strongly denies that mathematical objects and theorems have any reality apart from human minds. To him Mathematics is a:

> *human activity, a social phenomenon, part of human culture, historically evolved, and intelligible only in a social context. I call this viewpoint "humanist."*

ARGUMENTS FOR
Humanism, like intuitionism, avoids the problem of having to explain what other world mathematical objects exist in. It is free of contradictions, and also addresses the question of why mathematics seems to describe the physical world so well. Cultures, over time, made it work by modifying and adapting it.

ARGUMENTS AGAINST
The following excerpt from Martin Gardner's review of Hersh's book, provides an excellent summary on the issues with humanism as a mathematical philosophy.

Hersh grants that there may be aliens on other planets that do mathematics, but their math could be entirely different from ours. The "universality" of mathematics is a "myth." "If little green critters from Quasar X9 showed us their textbooks," Hersh thinks it doubtful that those books would contain the theorem that a circle's area is π times the square of its radius. Mathematicians from Sirius might have no concept of infinity because this concept is entirely inside our skulls. It is as absurd, Hersh writes, to talk of extraterrestrial mathematics, as it is to talk about extraterrestrial art or literature.

With few exceptions, mathematicians find these remarks incredible. If there are sentient beings in Andromeda who have eyes, how can they look up at the stars without thinking of infinity? How could they count stars, or pebbles, or themselves without realizing that two plus two equals four? How could they study a circle without discovering, if they had brains for it, that its area is π times the radius squared.

Why does mathematics, obviously the work of human minds, have such astonishing applications to the physical world, even in theories as remote from human experience as relativity and quantum mechanics? The simplest answer is that the world out there, the world not made by us, is not an undifferentiated fog. It contains supremely intricate and beautiful mathematical patterns from the structure of fields and their particles to the spiral shapes of galaxies. It takes enormous hubris to insist that these patterns have no mathematical properties until humans invent mathematics and apply it to the outside world.

Consider $2^{1398269}$ minus one. Not until 1996 was this giant integer of 420,921 digits proved to be prime (an integer with no factors other than itself and one). A realist does not hesitate to say that this number was prime before humans were around to call it prime, and it will continue to be prime if human culture vanishes. It would be found prime by any extraterrestrial culture with sufficiently powerful computers.

Social constructivists prefer a different language. Primality has no meaning apart from minds. Not until humans "invented" counting numbers, based on how units in the external world behave, was it possible for them to assert that all integers are either prime or composite (not prime). In a sense, therefore, a computer did discover that $2^{1398269}$ minus one is prime, even though it is a number that wasn't "real" until it was socially constructed. All this is true, of course, but how much simpler to say it in the language of realism!

No realist thinks that abstract mathematical objects and theorems are floating around somewhere in space. Theists such as physicist Paul Dirac and astronomer James Jeans liked to anchor mathematics in the mind of a transcendent Great Mathematician, but one doesn't have to believe in God to assume, as almost all mathematicians do, that perfect circles and cubes have a strange kind of objective reality. They are more that just what Hersh calls part of the "shared consensus" of mathematicians. Excerpt from Martin Gardner's Book Review of "What Is Mathematics, Really."

22

If mathematics is a product of culture, similar to money and religion, then why is the mathematics of the east and west identical? Perhaps this is due to the interactions of the cultures through trade and commerce. However, even cultures that were totally separated from each other did develop similar mathematical ideas and concepts. Different religions and forms of money developed in both eastern and western cultures, but only one form of mathematics has survived. Obviously, there is something more going on here than humanism can provide answers to.

SOME MATHEMATICAL HUMANISTS
Imre Lakatos (1922-1974), Reuben Hersh (1927-present)

1.5 Why Does Mathematics Describe The Physical World So Well?

Whatever philosophical view you may choose to accept, you are still faced with the question of – Why is mathematics so successful at predicting the behavior of the physical world? This is a question that has occupied many great minds throughout history.

Einstein made the following remarks:

> *The most incomprehensible thing about the universe is that it is comprehensible.*

> *How can it be that mathematics, being after all a product of human thought independent of experience, is so admirably adapted to the objects of reality? Is human reason, then, without experience, merely by taking thought, able to fathom the properties of real things?*
> Sidelights on Relativity

The renowned mathematician Eugene Wigner wrote a paper titled "The Unreasonable Effectiveness of Mathematics in the Natural Science," in which he said –

> *The miracle of the appropriateness of the language of mathematics for the formulation of the laws of physics is a wonderful gift, which we neither understand nor deserve. We should be grateful for it and hope that it will remain valid in future research and that it will extend, for better or for worse, to our pleasure, even though perhaps also to our bafflement, to wide branches of learning.*

The theme was further developed in "The Unreasonable Effectiveness of Mathematics" by R. W. Hamming in The American Mathematical Monthly (Volume 87, Number 2, 1980), which considered the predictive, as well as descriptive powers, of mathematics in relation to engineering.

> *From all of this I am forced to conclude both that mathematics is unreasonably effective and that all of the explanations I have given when added together simply are not enough to explain what I set out to account for. I think that we–meaning you, mainly–must continue to try to explain why the logical side of science–meaning mathematics, mainly–is the proper tool for exploring the universe as we perceive it at present. I suspect that my explanations are hardly as good as those of the early Greeks, who said for the material side of the question that the nature of the universe*

is earth, fire, water, and air. The logical side of the nature of the universe requires further exploration.

Two conclusions can be drawn from the papers of Wigner and Hamming:

(1) Although mathematics is a product of the human mind, it is also involved in some strange metaphysical way at the deepest levels of physical existence.

> *The miracle of the appropriateness of the language of mathematics for the formulation of the laws of physics is a wonderful gift, which we neither understand nor deserve. We should be grateful for it and hope that it will remain valid in future research and that it will extend, for better or for worse, to our pleasure, even though perhaps also to our bafflement, to wide branches of learning.* – Wigner

(2) There is no evolutionary explanation for the presence of mathematical abilities within the mind. The ability to understand physics could not have arisen by evolution. Although our bodies may well be the product of random mutation and selection all the way from amoeba to man, our minds have some "un–evolved" dimension.

> *But it is hard for me to see how simple Darwinian survival of the fittest would select for the ability to do the long chains that mathematics and science seem to require.* – Hamming

> *If you pick 4,000 years for the age of science, generally, then you get an upper bound of 200 generations. Considering the effects of evolution we are looking for via selection of small chance variations, it does not seem to me that evolution can explain more than a small part of the unreasonable effectiveness of mathematics.* – Hamming

> *Certainly it is hard to believe that our reasoning power was brought, by Darwin's process of natural selection, to the perfection, which it seems to possess.* – Wigner

We are then left with something of a mystery. According to the physical view of the world, the mind is just another aspect of matter, which has evolved solely to ensure the survival of the species. Yet why should mathematics, a creation of the mind, tell us anything about how the world works? Why should it be able to model the world way down to quantum mechanical levels?

Paul Davis, in his book "The Mind of God" says that he:

> *once asked Richard Feynman (A Noble Prize winning physicist) whether he thought of mathematics and, by extension, the laws of physics as having an independent existence. He replied: The problem of existence is a very interesting and difficult one. If you do mathematics, which is simply working out the consequences of assumptions, you'll discover for instance a curious thing if you add the cubes of integers. One cubed is one, two cubed is two times two times two, that's eight, and three cubed is three times three times three, that's twenty–seven. If you add the cubes of these, one plus eight plus twenty–seven– let's stop there – that would be thirty–six. And that's the square of another number, six, and that number is the sum of those same integers, one plus two plus three...Now, that fact which I've just told*

24

you about might not have been known to you before. You might say "Where is it, what is it, where is it located, what kind of reality does it have?" And yet you came upon it. When you discover these things, you get the feeling that they were true before you found them. So you get the idea that somehow they existed somewhere, but there's nowhere for such things. It's just a feeling...Well, in the case of physics we have double trouble. We come upon these mathematical interrelationships but they apply to the universe, so the problem of where they are is doubly confusing...Those are philosophical questions that I don't know how to answer.

What is the relation between mathematics and the physical world then? For example consider Newton's laws of motion. If we deduce results about mechanics from these laws, are we discovering properties of the physical world, or are we simply proving results in an abstract mathematical system? Does a mathematical model, no matter how good, only predict behavior of the physical world, or does it give us insight into the nature of that world? Does the belief that the world functions through simple mathematical relationships tell us something about the world, or does it only tell us something about the way humans think?

The most natural starting place, historically, for examining the relationship between mathematics and the physical world is through the views of Pythagoras. The views of Pythagoras are only known through the views of the Pythagorean School for Pythagoras himself left no written record of his views. However the views, which one has to assume originated with Pythagoras, were extremely influential and still underlie the today's science. Here we see for the first time the belief that the physical world may be understood through mathematics. Music, perhaps strangely, was the motivating factor, for the Pythagorean world view. They realized that musical harmonies were related to simple ratios. Moreover the same simple ratios hold for vibrating strings and for vibrating columns of air. The discovery of this general mathematical principle applying to many apparently different situations was seen to be of great significance. Pythagoreans then looked for similar mathematical harmonies in the universe in general, in particular the motions of the heavenly bodies. Their belief that the Earth is a sphere is almost certainly based on the belief that the sphere was the most perfect solid, so the Earth must be a sphere. The shadow of the Earth cast on the Moon during an eclipse added experimental evidence to the belief.

Plato followed the general principles of Pythagoras and looked for an understanding of the universe based on mathematics. In particular, he identified the five elements, fire, earth, air, water and celestial matter with the five regular solids, the tetrahedron, cube, octahedron, icosahedron and the dodecahedron. On the one hand there is little merit in Plato's idea: of course Plato's elements are not the building blocks of matter, and anyway his identification of these with the regular solids had little scientific justification. On the other hand, at least he was seeking an explanation of the physical world using mathematical properties, which history has shown is not far off the mark.

EXERCISES 1.1–1.5

ESSAYS AND DISCUSSION QUESTIONS:

Write an essay on each of the following, or be prepared to discuss these questions:

1. Write your own definition of mathematics.

2. What is metaphysics?

3. What is the difference between the eastern and western philosophy?

4. Compare and contrast the philosophical schools of thought on the origins of mathematics.

5. Search the Internet for definitions of mathematics. Give the definition and then state what is lacking in this definition.

6. Why do you think mathematics describes the physical world so well, or do you even believe that it does? If you don't think it does, give a reason or reasons why.

7. What is critical thinking/analysis?

8. To better describe how many people view mathematics they usually combine the above philosophies on the origin of mathematics into hybrid philosophies. Construct your own hybrid philosophy and defend it.

9. Which Philosophical School of Thought, do you most identify/agree with? Describe why.

10. We have mentioned the mind–body problem. What is it and provide some insight into how the west and east might view it.

11. Who was Plato and how did he influence the debate over the origin of mathematics.

12. What position might Socrates take on the origin of mathematics (support your conclusion)?

13. What position might Aristotle take on the origins of mathematics (support your conclusion)?

14. Do you believe that math is discovered or created? State why and defend your position.

15. Read Mario Livio's Book, *Is God a Mathematician* and summarize and discuss the main theme of the book.

26
References

GENERAL

1. Audi, Robert (ed., 1999). The Cambridge Dictionary of Philosophy, Cambridge University Press, Cambridge, UK, 1995. 2nd edition, 1999.

2. Balagner, M., (1990). Platonism and Anti-Platonism in Mathematics, Oxford.

3. Barker, S., (1964). Philosophy of Mathematics, Englewood Cliffs.

4. Benacerraf, Paul, and Putnam, Hilary (eds., 1983). Philosophy of Mathematics, Selected Readings, 1st edition, Prentice-Hall, Englewood Cliffs, NJ, 1964. 2nd edition, Cambridge University Press, Cambridge, UK, 1983.

5. Beeson, M., (1985). Foundations of Constructive Mathematics, Berlin.

6. Beeth, E. W., (1965). Mathematical Thought. An Introduction to the Philosophy of Mathematics, Dordrecht.

7. Benacerraf, P. and Putnam, H. (eds), (1983). The Philosophy of Mathematics. Selected Essays, 2. ed. Cambridge, Mass.

8. van Bendegem, J.P., (2000). Alternative Mathematics. The Vague Way. In: Synthese 125, p. 19-31.

9. Brouwer, L.E.J., (1975). Philosophy and Foundations of Mathematics, in: Collected Works Band 1, ed. by A. Heyting, Amsterdam-Oxford-New York.

10. Courant, R. and Robbins, H., (1948). What is Mathematics?, London-New York-Toronto.

11. Ernest, Paul (1998). Social Constructivism as a Philosophy of Mathematics, State University of New York Press, Albany, NY.

12. George, Alexandre (ed., 1994), Mathematics and Mind, Oxford University Press, Oxford, UK.

13. Hamming, R.W., (1980). The Unreasonable Effectiveness of Mathematics, The American Mathematical Monthly, Volume 87, Number 2.

14. Hardy, G.H. (1940). A Mathematician's Apology, 1st published, 1940. Reprinted, C.P. Snow

    (foreword), 1967. Reprinted, Cambridge University Press, Cambridge, UK, 1992.

15. Hart, W.D. (ed., 1996). The Philosophy of Mathematics, Oxford University Press, Oxford, UK.

16. Hatcher, W. S., (1982). The Logical Foundations of Mathematics, Oxford.

17. van Heijenoort, Jean (ed. 1967). From Frege To Gödel: A Source Book in Mathematical Logic, 1879-1931, Harvard University Press, Cambridge, MA.

18. van Heijenoort, J. (ed.), (1967). From Frege to Gödel. A Source Book in Mathematical Logic, 1879 - 1931, Cambrige, Mass.

19. Hellman, G., (1989). Mathematics Without Numbers, Oxford.

20. Kitcher, P., (1983). The Nature of Mathematical Knowledge, Oxford.

21. Kline, Morris (1959). Mathematics and the Physical World, Thomas Y. Crowell Company, New York, NY, 1959. Reprinted, Dover Publications, Mineola, NY, 1981.

22. Kline, Morris (1972). Mathematical Thought from Ancient to Modern Times, Oxford University Press, New York, NY.

23. Körner, Stephan, (1960). The Philosophy of Mathematics, An Introduction. Harper Books.

24. Lakoff, George, and Núñez, Rafael E. (2000). Where Mathematics Comes From: How the Embodied Mind Brings Mathematics into Being, Basic Books, New York, NY.

25. Lakatos, Imre (1976). Proofs and Refutations: The Logic of Mathematical Discovery (Eds) J. Worrall & E. Zahar Cambridge University Press

26. Lakatos, Imre (1968). Problems in the Philosophy of Mathematics North Holland

27. Livio, Mario (2009), Is God a Mathematician, Simon & Schuster, New York, NY.

28. Maddy, Penelope (1990). Realism in Mathematics, Oxford University Press, Oxford, UK.

29. Maddy, Penelope (1997). Naturalism in Mathematics, Oxford University Press, Oxford, UK.

30. Maziarz, Edward A., and Greenwood, Thomas (1995). Greek Mathematical Philosophy, Barnes and Noble Books.

31. Quine, W.V.O., (1951). Mathematical Logic, New York 1940, 2nd Ed. Cambridge, Mass.

32. Ramsey, F.P., (1931). The Foundations of Mathematics and Other Logical Essays, ed. by R. B. Braithwaite, London.

33. Resnik, Michael D. (1980). Frege and the Philosophy of Mathematics, Cornell University.

34. Russell, Bertrand (1919). Introduction to Mathematical Philosophy, George Allen and Unwin, London, UK. Reprinted, John G. Slater (intro.), Routledge, London, UK, 1993.

35. Shapiro, Stewart (2000). Thinking About Mathematics: The Philosophy of Mathematics, Oxford University Press, Oxford, UK

36. Strohmeier, John, and Westbrook, Peter (1999). Divine Harmony, The Life and Teachings of Pythagoras, Berkeley Hills Books, Berkeley, CA.

37. Tait, William W. (1986). Truth and Proof: The Platonism of Mathematics, Synthese 69 (1986), 341-370. Reprinted, pp. 142–167 in W.D. Hart (ed., 1996).

38. Triplett, T., (1994). Is there Anthropological Evidence that Logic is Culturally relative? Remarks on Bloor,b Jennings, and Evand-Pritchard. In: The British Journal for the Philosophy of Science 45, p. 749-760.

39. Weyl, H., (1990). Philosophy of Mathematics and Natural Science, Princeton, N. J. 1949. German original: Philosophieder Mathematik und Naturwissenschaft (Handbuch der Philosophie IV/V), 2 vols,, München 1926, 6. ed.

40. Wigner, Eugene (1960). The Unreasonable Effectiveness of Mathematics in the Natural Sciences, Communications on Pure and Applied Mathematics.

41. Wilder, R. L.: Introduction to the Foundations of Mathematics, New York-London 1967.

42. Wilder, R. L. (1980). Mathematics as a Cultural System, Pergamon.

43. Wittgenstein, L., (1980). Lectures on the Foundations of Mathematics, Cambridge, Mass.

44. Zalta, E.N., (2000). Neo-Logicism? An Ontological Reduction of Mathematics to Metaphysics. In: Erkenntnis 53, p.219-265.

INTRODUCTION TO PHILOSOPHY

45. The Cambridge Dictionary of Philosophy (1995).

46. Matter and consciousness (electronic resource) : a contemporary introduction to the philosophy of mind

47. The Consolation of Philosophy (1986).

48. The Encyclopedia of Philosophy (1967).

49. Handbook of World Philosophy: Contemporary Developments Since (1945).

50. Key Ideas in Human Thought (1993).

51. The Oxford Dictionary of Philosophy 3 (1994).

WESTERN PHILOSOPHY

52. Barnes, J. Editor, (1971). *The Complete Works of Aristotle, Vol. 1* Bollingen.

53. McGreal, I., Editor, (1992). *Great Thinkers of the Western World*, Collins Reference.

54. Kenny, A., (2001). *The Oxford Illustrated History of Western Philosophy*, Oxford University Press,.

55. Plato, Edwin, I., Editor, (1965). *The Works of Plato*, McGraw-Hill.

EASTERN PHILOSOPHY

56. Carr, B., Mahalingam, I. Editors, (1997). *Companion Encyclopedia of Asian Philosophy*, Routledge Press, London.

57. Collinson, D., Plant, P., Wilkinson, R., (2000) *Fifty Eastern Thinkers*, Routledge, London.

58. Feibleman, James Kern, (1976). Understanding Oriental Philosophy: A Popular Account for the Western World. New York: Horizon Press.

59. Hackett, Stuart Cornelius, (1979). Oriental Philosophy: A Westerner's Guide to Eastern Thought. Madison: University of Wisconsin Press.

60. McGreal, I., (1965). *Great Thinkers of the Eastern World*, Harper Collins, NY.

61. Reyna, R., (1984). *Dictionary of Oriental Philosophy*, New Delhi: Munshiram Manoharlal.

62. Riepe, Dale, ed., (1981). Asian Philosophy Today. New York: Gordon and Breach.

63. Schwartz, B., (1985). *The World of Thought in Ancient China*, Harvard University Press.

INDIAN AND SOUTHEAST ASIAN THOUGHT

64. Adikaram, E. W. Early History of Buddhism in Ceylon. Migoda: D. S. Puswella, 1946.

65. Bouquet, A. C. Hinduism. New York: Penguin, 1948.

66. Coomaraswamy, Ananda Kentish. Hinduism and Buddhism. New York: Philosophical Library, 1943.

67. Farquhar, J. N. A Primer of Hinduism. London: Faber & Faber, 1912.

68. Jayatilleke, K. N. Early Buddhist Theory of Knowledge. London: Allen and Unwin, 1963.

69. Klostermaier, Klaus K. A Survey of Hinduism. Albany, N.Y. : State University of New York Press, 1989.

70. Mohanty, Jitendra Nath. Reason and Tradition in Indian Thought. Oxford: Oxford University Press, 1993.

71. Nanananda. Concept and Reality in Early Buddhist Thought (2nd ed.). Kandy: Buddhist Publication Society, 1976.

72. Piatigorsky, Alexander. The Buddhist Philosophy of Thought. London: Rowman & Allenheld, 1984.

73. Zimmer, Heinrich. Philosophies of India. Princeton: Princeton University Press, 1968.

CHINESE THOUGHT

74. Allison, Robert, ed. Understanding the Chinese Mind. Oxford: Oxford University Press, 1990.

75. Chan, Wing-tsit, comp. and tr. A Source Book in Chinese Philosophy. Princeton: Princeton University Press, 1963.

76. Chang, Ch'eng-chi. The Buddhist Teaching of Totality: The Philosophy of Hwa Yen Buddhism. University Park: Pennsylvania State University Press, 1971.

77. Chang, Po-Tuan. The Inner Teaching of Taoism. Boston: Shambhala, 1986.

78. Chang, Po-Tuan. Understanding Reality: A Taoist Alchemical Classic. Honolulu: University of Hawaii Press, 1987.

79. Feng, Yu-lan. A History of Chinese Philosophy. Princeton: Princeton University Press, 1952- 53.

80. Li, Yen. Chinese Mathematics: A Concise History. Oxford: Clarendon Press, 1986.

81. Wilhelm, Richard, tr. The I Ching or Book of Changes. Tr. C. F. Baynes. Princeton: Princeton University Press, 1950.

CHAPTER 2

Cognitive & Psychological Aspects Related to Mathematics

Mathematics is the handwriting on the human consciousness of the very Spirit of Life itself.

Claude Bragdon

All science requires mathematics. The knowledge of mathematical things is almost innate in us. This is the easiest of sciences, a fact which is obvious in that no one's brain rejects it; for laymen and people who are utterly illiterate know how to count and reckon.

Roger Bacon

2.1   Introduction

In the previous chapter we took a look into what mathematics is from a philosophical vantage point. In our search for a definition, however, a number of other interesting and related questions emerged. Questions of whether or not mathematical concepts already exist in our brains when we are born. Are we hard–wired for mathematics? Or, do we learn it all? Other questions, such as, how do the mind and body interact? What is knowledge and how do we acquire it?, also remain. And, more to the point of this section, how do these questions help us in trying to comprehend another important aspect of this thing we call mathematics and that is – How is it related to our conscious perception of the world around us?

To begin to answer this question we will need to look into how the brain works and how we receive and process information. Essentially, we are trying to understand how our mind interacts with the physical world. We have decided to take a western view through this chapter. The eastern view does not allow for discussion or independent understanding, it instead leads to meditative experiential understanding, which by definition is not possible to discuss. Not that we are trying to argue that this perspective is incorrect, but we would have to end the journey here if we adopted this viewpoint, and we wish to continue.

The questions we posed above have aspects that are both philosophical and psychological in nature. In this section, however, we will focus more on the psychological aspects of understanding mathematics, but the deeper philosophical questions are never far from view, so we will from time to time discuss these notions both psychologically and philosophically.

We will begin with a basic mechanical description of how the brain functions. However, this description does not answer some of the more fundamental questions related to the interaction of the mind and the body and questions related to perception and consciousness. All of these aspects play a role in what mathematics is, and in comprehending how we come to learn and understand it.

The key to unraveling the mystery of what math is and whether it is either discovered or created, is related to what the universe is and how we are connected to it. Understanding the universe stimulates a deep philosophical discussion that was begun as soon as humanity began to think. It was one of the major themes of both the ancient Greek philosophers and the eastern sages, and while it will probably never be fully answered, it is the quest for an answer that is important.

On one side of the argument, there is only a physical world. The world of our senses. This is the modern scientific view, created by Aristotle, and perfected over many centuries by western cultures. The problem with this view is that we can't even define what physical is. The concepts of space, time, and matter we take for granted, but we can't even define them. They are simply things we measure. The view from modern physics, with quantum physics at its heart, has shown us that the so–called physical world behaves in odd and unexpected ways. Quantum physics tells us if we want to make accurate predictions about "physical things" on a small scale (the atomic level and smaller), we have to think of matter as something called a "probability wave" (a potential for existing). Furthermore, if we choose to try and follow a particular event, our probability wave collapses and becomes a "real" particle! Very strange indeed! This only goes to show us that things as fundamental as space, time, and matter are still not understood, even by those that adopt a purely physical model of the universe. This also serves to underscore the strange, but sometimes thought provoking explanations of eastern philosophy.

On the other side of the argument, we have those that believe it is all mental. This is the view of Plato and some forms of eastern mysticism. The mind creates the universe, and without it there would be no universe. Modern believers often cite quantum physics which says you need an observer before an event becomes physical. As I said above, if you observe a particular event, a physical particle emerges. One problem here, at least from the scientific perspective, is that we don't know what "mental" is, and that we don't know how the mental is connected to the physical world. From the eastern philosophical perspective, this is not a problem, since they believe the question itself is meaningless, since there is no distinction between mental and physical, which only underscores our complete confusion about the world. This seems to leave us at an impasse with each side unable to understand the other.

This is the age–old struggle between the mental and the physical, or the apparent mind/body duality of the world. We will NOT answer the question here in this book, but as we take our journey trying to understand mathematics, we should be aware of it. Even though we may not find the answers to all our questions, "It is the journey and not the destination that matters." For those who are interested, we have compiled a list of books at the end of this chapter, from which you can explore this subject further.

To understand how we interact with and experience the world, we need to look at the organ that provides the connection, the human brain.

2.2   The Brain

The physical brain becomes our next point of departure on this journey trying to understand what mathematics is. The brain is how we interact with the world (whatever that might be) and others around us. So little is known or understood about our brains. We know a bit about what it does, but how it does it is a mystery. We can say on a mechanistic level how we have our five senses, and how the brain coordinates this information to allow us to act and react. How it sends signals via electrochemical reactions that cause the synapses of neurons to fire, allowing our brains to "work."

There are over 100 billion neurons in our brain, roughly the same number of stars there are in the Milky Way Galaxy. Each neuron connects to 10,000 other neurons. That's a quadrillion (1,000,000,000,000,000) connections, roughly the same number of cells there are in the human body. These connections are organized into four major parts. The reptilian brain, also called the R–complex, is responsible for our basic instinctive behavior, and causes us to react. It controls our basic survival needs, such as movement, reproduction, digestion, breathing, circulation, and the "fight or flight" reaction.

The reptilian brain is surrounded by the mammalian brain. The mammalian brain, or limbic system, modifies our instinctive behavior through emotions. It comes into play when feelings, such as fear, anger, pity, and outrage occur, linking emotions with behavior. It serves to inhibit the reptilian brain in how we respond. It is associated with primal activities related to food, sex, sense of smell, and bonding/attachment and protection needs. It is also thought to control what long term memories are stored.

The mammalian brain is imbedded in the new brain. The new brain, or neocortex, is responsible for high–level cognition. It is divided into two halves, the left and right hemispheres. Many of the higher cognitive functions are shared between the two hemispheres. This interaction is called brain lateralization. However, it appears that one or the other side of the brain dominates certain types of higher cognitive functions.

The right hemisphere is primarily concerned with intuition, imagination, subjectivity, and empathy. It is involved with non–verbal forms of communication. It deciphers auditory and visual imagery. Its "thought" processes are related to rapidly identifying patterns. It is responsible for our spatial and musical abilities. It also appears to have special links with our emotions, expressing them and reading them in others, through our mammalian brain. The right hemisphere contributes to a "holistic," or "whole picture" way of thinking. It looks at the whole rather than the parts that created it. Information is processed in a random, or what is called a "non–linear" way.

The left hemisphere is primarily concerned with logic and reasoning. It provides us with a rational, analytic form of objective thinking. It is heavily involved with language, writing, and skilled movement. It looks at the parts needed to create the whole. It prefers lists and sequences of ordered steps. Information is processed in a sequential fashion. The left hemisphere is responsible for creating a "linear" way of thinking.

All parts of the brain communicate with each other through a two–way–network of nerves. At different times, different parts of our brain dominate our behavior. The relationships and functions of the parts of the brain are shown below:

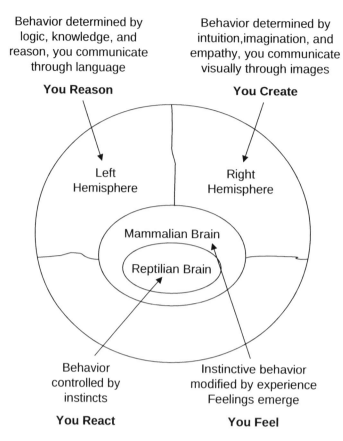

Behavior determined by logic, knowledge, and reason, you communicate through language
You Reason

Behavior determined by intuition, imagination, and empathy, you communicate visually through images
You Create

Left Hemisphere

Right Hemisphere

Mammalian Brain

Reptilian Brain

Behavior controlled by instincts
You React

Instinctive behavior modified by experience Feelings emerge
You Feel

The short description above provides basic ideas on how the brain functions and communicates with itself. The brain obtains data to communicate about through our five basic senses, allowing us to receive information about the outside world.

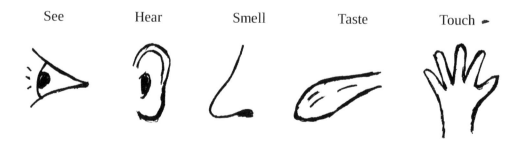

See   Hear   Smell   Taste   Touch

Our senses provide data to the brain that helps us to navigate through and interpret the physical world around us. They enable us to group and categorize things in our world.

Mathematics comes from a complex interaction between the two sides of the brain through lateralization. As we have said, the left side of the brain is associated with abstract, logical reasoning and language, while the right side of the brain has more to do with spatial and creative reasoning.

Mathematics requires the full capabilities of both sides of the brain to be fully developed and understood.

Although many believe that mathematics is a left hemisphere function, it is actually a function of both the left and right hemispheres of the brain. To be successful in the early mathematics courses (Arithmetic and Algebra) it is helpful to be able to memorize facts, and simple sequences. This is something that is taught more easily to the left hemisphere, which is why some say that people who are dominated by their left hemispheres are better at math. Most of the mathematics instruction we receive is tailored towards our left brain. This is probably why some suggest that you need to be a "left brain" person to be successful in mathematics. However, upper level mathematics courses such as, trigonometry, geometry, and calculus, require significant right hemisphere thinking skills related to seeing and understanding patterns in order to fully understand the material.

Our ability to think logically, in an organized way, needs the left–hemisphere of our brain, while our ability to see patterns and think creatively, calls for the right hemisphere of our brain. Thus, to be successful at mathematics you need to use all of your brain, as higher level mathematical concepts involve the entire brain.

Current research is involved with understanding how the brain develops, processes and works with mathematics (see Dehane and his Triple Code Model for example). This is a field that is continually being advanced, and we expect significant discoveries in the years to follow, which will enable better teaching and learning methodologies for mathematics education.

To summarize, we have provided a very mechanistic explanation of the brain. We have stated how different parts of the brain are responsible for different behaviors, and how it is through firing neurons and synapses that the brain "works." However, we actually don't know how this happens. What is a thought and how is it created? What is the mechanism that takes us from a bunch of electrical signals to a "thinking" brain. The thing that sets us apart from "machines" is something called consciousness. It is what establishes us in the world as sentient (thinking) beings.

2.3   Consciousness

What is consciousness, and how does it arise from our brains? Why is this important in our study of mathematics? Consciousness is the act of being aware of ourselves and the world around us (Note: Consciousness is different from conscience which is an awareness of a moral or ethical aspect to your conduct). A mind must be self–aware, before it can be aware of things outside itself such as its environment. This is why understanding consciousness is also critical for understanding mathematics. Mathematics is a way for us to describe the patterns we see in our environment, so it stands to reason that it must be a byproduct, or discovery of a conscious being.

Just examining the idea of consciousness leads us to many interesting questions and discoveries. Consciousness is something that we cannot explain. What is it, where did it come from? Is everything in the world conscious? Are there different levels or types of consciousness?

To shed some light onto this, David Chalmers (a mathematician/cognitive–scientist) has separated consciousness into two distinct types of problems. The "easy problems" of consciousness, and the "hard problem" of consciousness.

The easy problems/questions focus on:

- How can we discriminate sensory stimuli and react to them appropriately?
- How does the brain integrate information from many different sources and use this information to control behavior?
- How is it that subjects can verbalize their internal states?

The hard problem/question looks into:

- How physical processes in the brain give rise to subjective experience?

The easy problems are objective problems that current cognitive science is looking at, and while they are not trivial problems, it is expected that someday conventional science will be able to answer these questions to our satisfaction. These are the more mechanistic problems that western thinkers feel comfortable analyzing.

On the other hand, the hard problem of consciousness according to Chalmers is

(A) puzzle (that) involves the inner aspect of thought and perception: the way things feel for the subject. When we see, for example, we experience visual sensations, such as that of vivid blue. Or think of the ineffable sound of a distant oboe, the agony of an intense pain, the sparkle of happiness or the meditative quality of a moment lost in thought. All are part of what I am calling consciousness. It is these phenomena that pose the real mystery of the mind.

The hard problem of consciousness is about the inner subjective experience of things. The feeling you get when you see the color red, or the emotional images you get when you see flowers, or smell food cooking, are all examples of subjective experiences. It is what makes us human. Without this aspect of consciousness the world would be quite different, but why does it exist? Or as Chalmers questions, why are we not zombies? Why are there no zombies in the world? By zombies we mean creatures that can walk, talk, eat, and act exactly like humans, with one big exception. Zombies do not have an inner self. They don't dream or have an inner voice talking to them, they don't feel emotions. In essence they are biological robots, computers made of flesh and bones, and nothing more.

For a fascinating series of discussions and debates on consciousness and related questions go to www.closertotruth.com, and watch the videos involving Roger Penrose, John Searle, David Chalmers, and Fred Alan Wolf, among many others. It provides an interesting look at consciousness.

Schools of Thought on the Origin of Consciousness

There are several viewpoints on how consciousness arises in our brains. They include:

Dualism, which says that there are two separate aspects to the brain. A physical (body) part and a mental (mind) part. They are somehow connected to each other, but are distinct and different. Consciousness is part of the mental world, and that while consciousness is connected to the brain, it is not reducible to the brain. Meaning there is something, non–physical, that causes consciousness.

Consciousness will eventually be explained by understanding the connection between the physical and the mental aspects of the brain.

Philosophers/scientists like René Descartes, and more recently David Chalmers claim that consciousness must be a separate and distinct "thing." In Chalmer's opinion it does not come from other aspects of reality, but is a fundamental characteristic of reality in its own right. A true dualistic phenomenon.

Materialism/Physicalism, which believes that the world is just a physical world and that consciousness will ultimately be explained by a physical property of the brain, such as interacting neurons causing …, etc. To a materialist consciousness is a property/function of the brain, much like absorbing oxygen is a property/function of the lungs, or gastric secretion is a property/function of the stomach. Eventually consciousness will be explainable by physical causes alone. We will not need any mental "stuff" as Chalmers suggests to explain it.

Rationalism/Spiritualism/Mysticism, which believes, like dualism, that consciousness is different from the physical world, but it also asserts that consciousness can exist independently of the physical brain. It is a separate state that can exist on its own. Some of the more radical views in this group, even assert that consciousness actually gives rise to (creates) the physical world.

It is interesting that we again see the Mind–Body problem we saw in our discussion of philosophy back in play again. Is the world (reality) physical, mental, or both. Is the brain (consciousness in particular) physical, mental, or both. As you will see, this is a common theme throughout this book. You have two opposite ends of the spectrum, physical and mental, and then all views in between.

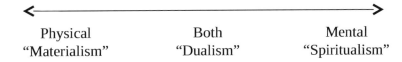

| Physical | Both | Mental |
| "Materialism" | "Dualism" | "Spiritualism" |

The study of consciousness creates a complex and ongoing debate. It is interesting that we are not alone, many living creatures have conscious brains, just like we do. There are others that may, but we cannot be sure of just how far down you can go in assigning consciousness. As the chart below illustrates there are many aspects to consciousness that can be explored and we invite the reader to do this on their own.

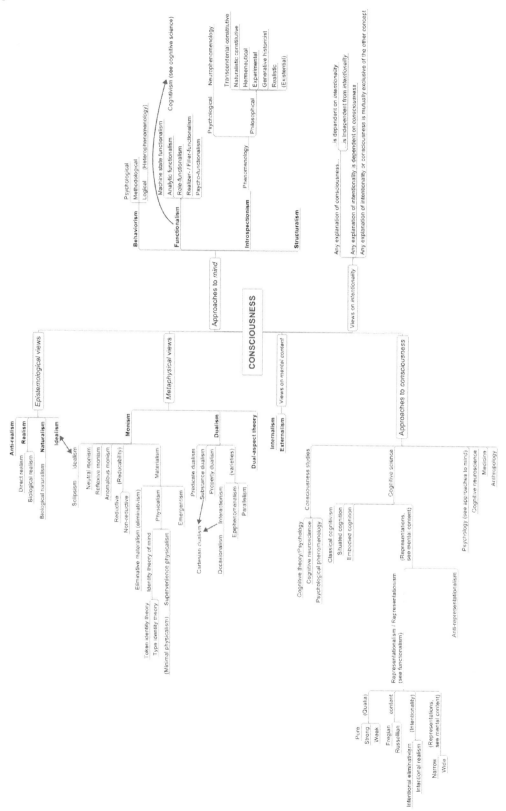

Author: Ostracon, Creative Commons Attribution–Share Alike 3.0 Unported license

Many living species have similar brains and senses, but we humans seem to be unique in our ability to understand and interpret the world at such a complex level. What seems to have separated us from the rest, is our ability to create a highly developed language for communication. Language has enabled a single brain with a quadrillion connections to expand even further as it interacts with thousands, millions, or even billions of other brains. It is through this collective brain that we observe and try to understand the universe. It is through this collective brain that mathematics has emerged, as either an elaborate human creation, a reflection of the world through our conscious human brains using a very complex language, or as a part of the very fabric of the universe, a blueprint for defining the world.

Consciousness and Mathematics

Now what does consciousness have to do with mathematics? On one level very little, but if mathematics is the creation of human beings, then quite a bit. Especially as it relates to what mathematics is from a philosophical perspective.

To a Humanist consciousness is everything. Without consciousness there can be no Humanist philosophy, since the social creation of mathematics depends upon it. You must have conscious beings interacting to culturally create mathematics. To a Platonist it could go either way. If mathematics is discovered, then it could only be found by a conscious being that is aware of itself and its environment. If there was no conscious being to discover it, it could still exists independently, but never be discovered. If mathematics is created, then a conscious being is required to create it. So it doesn't impact the philosophy either way.

The Formalist also requires a conscious being to create the game with its rules. A Logicist, you can argue, doesn't require a conscious being, since the universe arises through principles of logic and reason independent of humans.

Furthermore, our understanding of consciousness can lead to a deeper understanding of the philosophical questions about the origin of mathematics. On one level, mathematics and consciousness share a special bond. Mathematics, just like consciousness, seems to transcend humanity. It takes us to another level. It shows a richness and beauty to life, beyond the pale of simply existing. For those that do not see it, it means very little, but for those that have seen the power, beauty, and the connection, it is wonderful. Just as consciousness takes a stiff functioning zombie and transforms them into a vibrant living human being, so too does mathematics take a confusing and ad hoc universe and turn it into an understandable and coherent reality. It allows us to make sense out of it, and to feel that there is some underlying master plan. Life is not some random reality, but is governed by rules and laws, and these rules and laws are written in the language of mathematics.

In our brief study of the brain and consciousness we have tried to provide some insight into our uniquely human experience. We are conscious beings with a capability for a highly developed form of communication. We exist in this universe, and begin to formulate many fundamental questions about it. What is this thing we call "reality," is it physical or mental or both? How can we be sure we know what is real?

2.4   Reality and Perception

Questions regarding reality can again be traced back to ancient Greece. In the third century BCE Plato looked around him and saw different versions of the same "idea." For example, virtue was an idea, but some people were more virtuous than others. Justice was an idea, but some actions were more just than others. Thus, Plato thought there must be an ideal form of things such as virtue and justice. The perfect representation of each. Plato then extended his view of ideals to the physical world as well. For example, the physical world has many different trees. So there must be some ideal form of a tree that can be used to identify a tree. From this, Plato reasoned that there must be two "worlds," one of which is more real than the other (Plato outlined his views in his work, The Republic. In particular, look up his Cave Analogy as a metaphor for how he looks at what is real).

On the one hand you have the visible world. The visible world consists of what we experience through our senses. It consists of the physical objects and experiences derived from them. The visible world is constantly changing, and objects in it are merely temporary. It is a world of illusion.

Then you have the intelligible world. This is the world of the "mind's eye," and not the senses. Objects in this world are permanent and unchanging. It is the world that contains the ideal form of objects and their properties. It is in Plato's view, the real world.

Plato believed that the world of ideas, or ideals, is what is real. The world of our senses is where physical "things" reside. The world of the forms is where perfect ideal concepts reside. In Plato's view the world of forms is true reality. It is beyond space and time, eternal and unchanging, perfect and absolute, existing independently of the mind, but experienced only through our mind, not through our senses. The world of our senses, appears real, but it is forever changing. It exists within space and time, made of imperfect copies of the real world of ideals, and depends heavily upon our perception, which can be a problem.

Problems With Perception

Sometimes our perception can either lead us to wrong or inconclusive conclusions! What you see is not always what you get. Here are two simple examples to illustrate. Look at the picture below. Your mind will automatically try and make sense out of the picture by "seeing" a three–dimensional object. The problem is that there are two different ways to perceive this image. Therefore our perception leads to an inconclusive result.

Can you see the two different interpretations? In one perspective the curved lines could be curving out of the page, and the object appears to be standing up on the lines at the bottom of the image. In another perspective the curved lines could be curving into the page and the two straight vertical lines on either side of the figure in the image are resting on the page with the two other vertical lines rising above them.

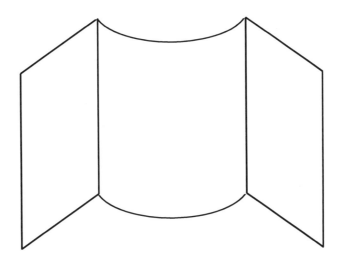

The purpose of this image is to highlight a problem we face when we must rely solely on perception and our senses to understand anything. Are we "seeing" correctly?

Another example is shown below with the checkerboard pattern. In the image to the left, we see that the boxes are all square, and all the lines are either perfectly horizontal or vertical. The image to the right, however, appears to have a "weave" pattern, and the lines in the pattern no longer look to be perfectly horizontal, but instead have alternating slight up and down slopes to them. Look carefully, though. The image to the right was actually formed from the left image by simply moving every other row of squares to the right by a half square. Thus, all the lines are still perfectly horizontal! If you still can't see this, turn the book sideways and look down the rows facing this other direction, and it should help. In this case our perception caused us to make a false conclusion.

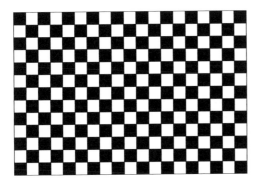

How we perceive the world is through our senses, but our senses can lead to false or inconclusive results. So how can we be sure what reality is? The western scientific view would say that we need to make different objective measurements or observations, and provided that the results are consistent and unchanging from test to test and experimenter to experimenter, then they must be true. Most of us would agree with this logical process (most of us being westerners by birth). However, it is not as easy as that, and this comes from one of the major achievements of western science called Quantum Physics (or Quantum Mechanics).

Quantum Reality and Mathematics

One of the fundamental pillars of quantum mechanics is that you can't know both the position and the momentum (mass times velocity, or roughly speaking the motion) of an object simultaneously. There is always an uncertainty with one or the other. If you measure the position, then its exact momentum is not known. In fact, we know the minimum level of uncertainty that can be achieved mathematically. This is related to a special number in physics called Planck's constant.

Werner Heisenberg in 1927 quantified this uncertainty with the following inequality.

$$\Delta x \Delta p \geqslant \frac{h}{4\pi}$$

The symbol Δ is the capital Greek letter delta. Δ usually means change in mathematics, but in this formula it stands for uncertainty (or more specifically something called the standard deviation of the position or momentum value). Thus, Δx is uncertainty in the position, and Δp is uncertainty in the value of the momentum. The letter h is called Planck's constant and has a numerical value of $h \approx 6.626 \times 10^{-34}$ *Joule seconds*. *Joule seconds* means energy expended over time. Now we don't have to understand exactly what this means to understand the significance of the inequality. What we should know, however, is that this number is very, very small. We can see this if we write it out in fixed decimal form:

$$h = 0.0000000000000000000000000000000006626 \quad \text{(ignoring the units for now)}$$

Now, this result from quantum physics doesn't have an impact on our day–to–day lives. The number h is far too small. It does, however, have meaning down at the atomic level, and also on a philosophical level as well.

First, let's consider the practical implications to see how just how insignificant they really are, and then we'll tackle the philosophical questions.

Example:
Consider driving in a car moving at about 55 mph, but let's say we can only measure this speed with a radar gun that has an accuracy of 0.25 mph. We also know that the car weighs approximately 3700 lbs. What is the limit on how accurately we can measure the location of the car based upon the uncertainty principle?

First we have to make sure we have the correct units. To keep our American (actually called the British or Imperial) system of units, we need to recalculate h in terms of foot–pounds and seconds. For simplicity, we did that for you and the value is given as:

$$h \approx 4.887 \times 10^{-34} \, ft \cdot lb_{force} \cdot s$$

Next we need to convert the weight of the car into a mass value. To do this we need to divide the weight 3700 lbs, by 32.2 feet per second squared. This is because the mass of an object is defined as the weight divided by the acceleration due to gravity.

Thus the mass of the car is:

$$\frac{3700 \ lbs_{weight}}{32.2 \ ft/s^2} = 114.9 \, lbs_{mass}$$

Next, we need to convert the uncertainty in speed, from miles per hour (mph) to feet per second (fps). We then multiply the uncertainty in our speed in mph by 1.4666... and replace our units with feet per second (fps). Thus, the uncertainty in our speed measurement is

$$0.25 \ mph \times 1.466 \, \frac{fps}{mph} = 0.3667 \ fps$$

The uncertainty in the momentum is simply the mass times the uncertainty in the velocity, or

$$114.9 \, lb_{mass} \times 0.3667 \ fps = 42.168 \ lb_{mass} \, fps$$

We can now solve the Heisenberg inequality for Δx the uncertainty in our position measurement.

$$\Delta x \geqslant \frac{h}{4\pi \ \Delta p}$$

Substituting for h, and Δp we have:

$$\Delta x \geqslant \frac{4.887 \times 10^{-34} \, ft \cdot lb_{force} \cdot s}{4 \times 3.1415 \ 42.168 \, lb_{mass} \, ft/s} = 9.222 \times 10^{-37} \, ft = 1.107 \times 10^{-35} \, inches$$

Thus, the best we can do is to be able to pinpoint the position of the car to within 1.107×10^{-35} inches. A distance so small that we cannot even come close to being able to measure it with any device known to humanity. Roughly one trillionth, of a trillionth, of a hundred billionth of an inch! A unit so small that it does not even have a designated name.

Obviously the Heisenberg uncertainty limitation does not have an impact on our day–to–day life. It is only at the atomic level or smaller that it has meaning. On the other hand, from a philosophical perspective it plays a huge role. Philosophically the uncertainty principle says that, regardless of what our senses may try to tell us, reality is more complicated and confounding then we had imagined. There are limitations on what we can know, not just from our senses, but these limitations are embedded in the very structure of the real world.

Ordinarily that would be challenging enough, but quantum mechanics goes even farther. It turns out that what we have come to know as the physical world, really isn't all that physical. The only way we can make any verifiable sense out of reality is if we abandon our notion of what our senses tell us is physical, and instead use mathematics alone to describe and understand our world. Mathematics tells us that what we thought was a physical particle is actually a wave at the same time. Arthur Eddington (1882–1944) a British physicist, coined the term "wavicle" to identify this strange new phenomenon. Matter can only be understood if we think of it as both a "solid" particle and a wave at the same time. Mathematics, it turns out, provides the only way for us to "understand" reality.

But wait, it gets even more bizarre. Reality further confuses and amazes us, for when we choose to look at the particle, it becomes a particle, but if we choose not to look, it is a wave. It seems to know we are looking at it! Objective reality depends upon the observer, thus, it is really subjective and not objective at all. This means that a tree falling in the forest does not make a sound if there is no one (conscious being) there to hear it fall. Physical particles only come into existence when there is a subject to observe them. Trees only make a sound when there is someone (or something) there to hear it. Otherwise, physical objects or events only have a possibility (probability to be more precise) of being real. You might ask, what are these waves made of. Again further confusion. It is a wave that represents all the possible physical states of a particle. It is also called a probability wave, since it gives the probability of a particle being in a particular state (location and momentum). So the wave isn't even something we can identify with a real substance. When we look for the particle, the wave collapses becoming the particle! If we don't look it remains a probability wave. So, what we call reality, only comes into existence when we look for it, otherwise, it is only a potential reality.

Now all this sounds rather fanciful and a bit like science fiction and not science fact. You could and should ask, what do we have to back up these bold unbelievable claims. The answer is, we have a series of repeatable western style experiments that confirm this. The first is called the "double–slit" experiment.

In modern physics, a fundamental constituent of matter is an electron. To us nothing could be more fundamental and real. It must be physical, since all of matter is made up of these electrons along with other particles. So let's test these particles in the following way. We will create a gun that can shoot out these electrons one at a time. We will shoot them at a barrier that has two vertical slits, or elongated narrow holes placed fairly close apart. Then we will fire our gun and let the electrons either "pass through" a slit or be blocked. If they pass through, we will have another barrier at the end that we'll place a film on, so that we can image the picture these electrons make as they hit the film. See the graphics below illustrating this.

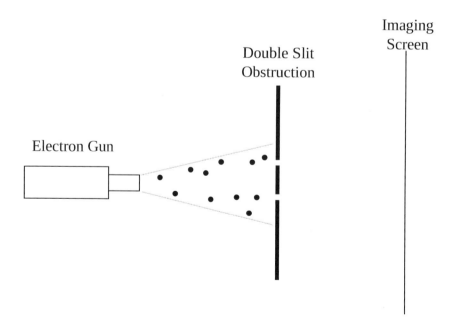

Running the experiment by following each electron or photon through the slits. We see this image

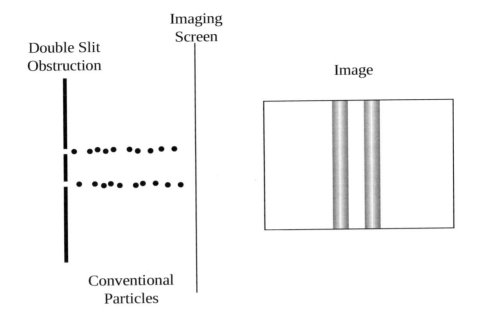

Double Slit
Obstruction

Imaging
Screen

Image

Conventional
Particles

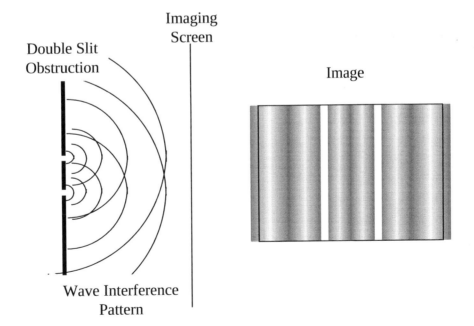

Double Slit
Obstruction

Imaging
Screen

Image

Wave Interference
Pattern

Running the experiment without following each electron or photon through the slits. We see this other different image.

If we look the electrons behave like particles. However, if we only look at the results, and not the trajectory of an individual electron, the image is an interference pattern that is similar to what we would see with waves of water in a pool with similar obstructions. Thus, the electrons are really waves. Or are they particles? I guess they are wavicles?!

Actually the two slit experiment has only been done with light and light particles called photons using a laser, since it is not possible to make these slits for electrons (they would have to be so close together and we can't physically construct them). However, a different type of experiment with electrons was first performed in 1927 by Davisson and Germer that proved the wave behavior of electrons, so the experiment above is not without substance. Davisson and Germer reflected an electron bean off a metal plate and used the lattice structure of the atoms in the metal to mimic the slits, and observed an interference pattern indicating the wave properties of electrons.

Other types of experiments have been performed and validated, and we cannot deny that matter is made up of particles that are also probability waves of reality. This sounds a bit like what the eastern mystics asked us to consider. We can't separate the objective world from the subjective viewer. Mind and body are one of the same. It is ironic that this was demonstrated using western scientific reasoning, though. The part that is also fascinating is that we can't describe this without mathematics. We could show the actual mathematics involved, but it is well beyond the scope of this book, so we shall just reference its existence here.

Why did we take a trip through this pathway through the brain, mind, and even quantum physics, which on the surface appear to be totally unrelated to mathematics? What if anything does this have to do with math? The reason we took this part of our journey is that it is through the mind that mathematics is either created or realized. Since the brain is the tool we use to either discover or create mathematics, we need to be aware of how it works, since this will directly affect the mathematics that comes out. We also needed to discuss what we consider reality, if we say that mathematics is used to describe it. Remarkably, even our detached and independent western science tells us that the mind has a stake in creating reality, and that mathematics is intimately involved in its description.

EXERCISES 2.1–2.4

ESSAYS AND DISCUSSION QUESTIONS:

Write an essay on each of the following, or be prepared to discuss these questions:

1. Briefly discuss the major parts of the brain, how they interact, and what they are responsible for?

2. What role do our senses play in understanding what is real?

3. What is consciousness? (Discuss the "easy" and "hard" problems as well as the "reductionist" versus the "mentalist" perspectives).

4. What is your view on the origin of consciousness? (provide supporting evidence)

5. How does the mind-body debate enter into the discussion on consciousness?

6. What was Plato's view on reality?

7. Is physical reality more real than mental reality?

8. What implications does quantum physics have on our interpretation of reality?

9. What is perception, and how does it influence our "perception" of reality?

10. Discuss how the different parts of the brain influence mathematics.

11. Is mathematics a left brain phenomenon? (Explain your answer)

12. Quantum physics suggests that objects are both particles and waves, what does that mean?

13. Quantum physics is a mathematical description of the world. If mathematics is simply a creation of humankind, why does a purely mathematical description work so well in predicting unknown and bizarre consequences of the theory?

References

1. Arntz, W., Chasse, B., and Vincente, M., (2005). What the Bleep Do We Know!?, Health Communications, Deerfield Florida.

2. Bailey, Ronald H., et al., (1975). The Role of the Brain, Time-Life Books.

3. Brians, P., (1999). Reading About the World, Vol. 1 & 2, 3rd edition, Harcourt Brace College Publishing.

4. Butterworth, B., (1999). What Counts: How Every Brain is Hardwired for Math. Free Press.

5. Blackmore, S., (2003). Consciousness: an Introduction. Oxford: Oxford University Press.

6. Calder, N., (1973). The Mind of Man, Penguin.

7. Chalmers, D., (1996). The Conscious Mind: In Search of a Fundamental Theory. New York: Oxford University Press.

8. Chalmers, D., (2002). The puzzle of conscious experience. Scientific American, January 2002. [2]

9. Dehaene, S., (2001). Précis of the number sense. Mind and Language, 16, 16-36.

10. Dehaene, S., & Cohen, L., (1997). Cerebral pathways for calculation: Double dissociation between rote verbal and quantitative knowledge of arithmetic. Cortex, 33(2), 219-250.

11. Dehaene, S., Piazza, M., Pinel, P., & Cohen, L., (2003). Three parietal circuits for number processing. Cognitive Neuropsychology(20), 487-506.

12. Dehaene, S., Spelke, E., Pinel, P., Stanescu, R., & Tsivkin, S., (1999). Sources of mathematical thinking: Behavioral and brain-imaging evidence. Science, 284(5416), 970-974.

13. Devlin, K., (2001). The Math Gene, Basic Books.

14. Fincher, J., (1981). The Brain - Mystery of Matter and Mind, U. S. News Books.

15. Harnad, S., (2005). What is Consciousness? New York Review of Books 52(11).

16. Klivington, K.A., (1989). *The Science of Mind*. MIT Press.

17. Neumann, Erich, (1954). The origins and history of consciousness, with a foreword by C.G. Jung. Translated from the German by R.F.C. Hull. New York : Pantheon Books.

18. Papineau, D., and Stein, H., (200). Introducing Consciousness, Totem Books.

19. Penrose, R., Hameroff, S. R. (1996). Conscious Events as Orchestrated Space-Time Selections, Journal of Consciousness Studies, 3 (1), pp. 36–53.

20. Searle, J., (2004). Mind: A Brief Introduction. New York: Oxford University Press.

21. Silverstein, A. & V., (1986). World of the Brain, William Morrow & Co. NY.

22. Sternberg, E., (2007). Are You a Machine? The Brain, the Mind and What it Means to be Human. Amherst, NY: Prometheus Books.

23. Velmans, M., (2000). Understanding Consciousness. London: Routledge/Psychology Press.

24. Velmans, M. and Schneider, S. (Eds.)(2006) The Blackwell Companion to Consciousness. New York: Blackwell.

25. Wooldridge, D., (1963). The Machinery of the Brain, McGraw-Hill.

CHAPTER 3

Knowledge & Mathematics

Mathematics is the only science where one never knows what one is talking about nor whether what is said is true.

Bertrand Russell

For the things of this world cannot be made known without a knowledge of mathematics.

Roger Bacon

3.1   Introduction

In the previous two chapters we have looked at mathematics from a philosophical perspective to see how it has influenced and been influenced by philosophy. We have also looked at how mathematics is related to our perception of reality. We have essentially addressed why we should want to know or question things, as well as how do we go about questioning, as it relates to how we perceive reality. In this chapter we look at what it is we are trying to learn. We try and answer the questions – What is knowledge and how do we know anything?

3.2   Theory of Knowledge

There are many possible propositions about the world. Distinguishing between fact and fiction can ofter be hard to do. What we often try to do is to acquire knowledge, so that we can understand and make informed decisions.

> **Knowledge** is justified true belief.

There are other definitions of what we call knowledge, but this one is reasonably clear and to the point. (This definition, however, has been challenged by Edmund Gettier (1963) with what are called Gettier problems or cases, but we will not go that deeply into the question here. For those that are interested, we suggest that you look further into this interesting subject.) The Venn Diagram below provides a nice graphical illustration of the relationship between, propositions, truths, beliefs, and knowledge (we'll discuss Venn Diagram in greater detail in Chapter 12).

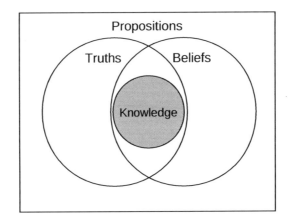

The study of knowledge is called epistemology.

> **Epistemology** is the study of what knowledge is, what we can know, and how we can know it.

Knowledge has three important aspects: the knower (who), ways of knowing (how), and areas or categories of knowledge (what). The who part is fairly obvious, that is us. How we acquire knowledge is through our senses (perception), our thoughts (reason), our emotions, and communicating through language. What knowledge we learn or acquire can essentially be categorized into six broad overlapping areas: the arts, ethics, history, the natural sciences, the human or social sciences, and mathematics. The image below summarizes the connection between the who, how and what of knowledge.

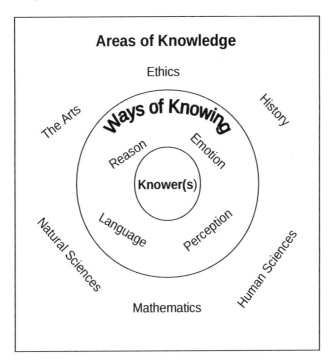

We have laid out a very organized description of knowledge. It appears reasonable and rational. It seems we understand it quite well. But do we? If we again take out our critical thinking cap and begin to analyze, question, and explore, we begin to see some holes in our supposed understanding.

If you want to find areas of uncertainty, the first place to look is connected to language. Language is NOT precise. We have defined knowledge as a justified, true, belief. The problem is with the words justified and belief. Truth on the other hand could be considered independent of the knower, and one could argue that something is either true or false (this may not always be the case, and would probably require further investigation, but for now we'll let it pass.) What is justified and what is believed, depends upon the knower.

First, what is considered justification? It is defined as:

> **Justify** means to demonstrate or prove something to be true

Do our senses count as justification? Because we see, hear, smell, feel, and/or taste something, is that considered justification? We have shown earlier that our sense perception can be wrong. What about the other ways of knowing? Are emotions and language any better or perhaps even worse than our perceptions? Lastly, what about reason, and our mind? Can reason alone give us justification?

What about belief? The typical dictionary definition is:

> **Belief** is a psychological state in which an individual holds a proposition to be true

Is this any better? What is a psychological state anyway? How certain is your belief? Obviously we have different levels of belief.

The problem is again in the definitions (much as it was in our trying to define mathematics). In the case of justify and belief, they are not absolute. There are obviously varying degrees of justification and belief. Perhaps we'd be better off if we talked about degrees of knowledge. We could then classify a range of justified–true–beliefs as knowledge, but not all of it would we call certain knowledge.

There are even those that believe that knowledge is not possible. They are skeptical that we can ever truly know anything. They are called skeptics. An illustration of skepticism was provided in the movie The Matrix (1999 Warner Bros. & Australian Village Roadshow Pictures.) The movie challenges our understanding of reality. Can we ever know for sure that the world we are living in is real? Is it possible that our brain is being tricked into believing we are experiencing what we think we are experiencing? Aren't dreams proof that we cannot be sure if we have certain knowledge or not?

Obviously, what knowledge is, is not easily definable. However, if we allow varying degrees of certainty in knowledge we can still proceed. The next question is – How do we obtain knowledge?

3.3   Schools of Thought on How We Obtain Knowledge

There are two main schools of thought on how knowledge is obtained. They are rationalism and empiricism. These systems lie at opposite ends of the spectrum with other hybrid philosophies in between.

Rationalism

Rationalism contends that knowledge is already innate (inborn, possessed at birth) within us. Innate knowledge is also called **a priori knowledge**, or knowledge prior to experience. A rationalist believes that it is reason, and not experience, that is most responsible for our acquisition of knowledge.

> **Rationalism** is the belief that knowledge comes from our minds and is innate within us

There are also those that believe in radical rationalism, where all true knowledge comes from within and not from our senses. While there are other non–radical rationalists who believe in a mixture of innate and experiential knowledge, but that rational, as a source of knowledge, dominates.

Some notable rationalists throughout history were Socrates (470–399 BCE), Descartes (1596–1650), Spinoza (1632–1677), Leibniz (1646–1716), and Kant (1724–1804).

Empiricism

Empiricism argues that all knowledge comes for our sense experiences. Experiential knowledge is also referred to as **a posteriori knowledge**, or knowledge after (post) experience. The mind is a "tabula–rasa," a blank slate, or blank paper, and it is sensory experience which leaves its mark on the brain, creating knowledge. We learn through perception, and knowledge without experience is not possible.

> **Empiricism** is the belief that knowledge comes from the experience of our senses

Again, you have radical empiricists that believe all knowledge is from non–rational experience, and others who believe in at least some basic level of innate knowledge. A radical empiricist would deny that religion or ethics has any place in the domain of knowledge. A more moderate empiricist would allow for certain facts to require rational thought to validate, but these are exceptions rather than the rule.

Some notable empiricists throughout history were Aristotle (384–322 BCE), Bacon (1561–1626), Hobbes (1588–1629), Locke (1632–1704), Berkeley (1685–1753), and Hume (1711–1776). Locke is the one credited with creating the designation empiricism, and in proposing the tabula–rasa explanation.

3.4 Types of Knowledge

In the introduction to the Critique of Pure Reason, Immanuel Kant (a rationalist) was one of the earliest people to study and write about the concept of knowledge in depth. Kant provided both insight and a common vocabulary for investigating it. He suggested that all propositions, from which we obtain knowledge, can be written in the subject–predicate form (Some argue this is too limiting, but we will not go into detail here. We suggest that you explore this on your own if you are interested.)

Examples of subject–predicate propositions:

Proposition	Subject	Predicate
All bodies take up space	bodies	take up space
Snow is white	snow	white
All bachelors are unmarried	bachelors	unmarried
John is tall	John	tall
All bodies are heavy	bodies	heavy
She is beautiful	she	beautiful
All rectangles have four sides	rectangles	four sides

The subject is what the proposition is talking about. The predicate is a characteristic of the subject.

Kant believed that there are two types of propositions. The first type are propositions that are true, due to their meaning. The predicate concept is contained in the subject concept. For example consider the propositions above, "All bachelors are unmarried," and "All bodies take up space." In each of these the predicate is already contained in the subject. Bachelors are defined as unmarried men, bodies are defined as objects that take up space. The concept "bachelor" contains the concept "unmarried," and the concept "body" contains the concept "take up space." The predicate is really part of the definition of the subject. Kant called these types of propositions analytic.

> **Analytic Proposition**: A proposition that is true by definition (or by virtue of its meaning)

The other type of proposition, according to Kant, are propositions in which the predicate concept is not contained in the subject. For example consider the propositions above, "John is tall," and "All bodies are heavy." In the first, the predicate tall is not automatically contained in the subject John, since John could be tall, short, or neither. In the second, the predicate heavy is not automatically contained in the subject bodies, since bodies can be light, heavy, or neither. Kant called these types of propositions synthetic.

> **Synthetic Proposition:** A proposition that is not analytic

Analytic propositions don't really lead to new knowledge, but synthetic propositions do. Synthetic propositions provide new knowledge about the subject that are not already part of the definition of the subject. They synthesize new information from other sources to expand your knowledge about the subject.

One important question that Kant was trying to address was – Were there any synthetic propositions that could be known without experiencing them, but we could know their certainty directly through thought and reason alone? Or stated in another way, are there any **synthetic a priori propositions** we are certain of. If so, then this would mean that knowledge could not be obtained exclusively through sense experience and perception. Kant concluded that there were, and that they can be found in mathematics.

3.5   Mathematical Knowledge

We have come to see that there are different types of knowledge and different views on how we come to acquire it. Considering the areas of knowledge carefully, you will notice that one category is rather unique. Mathematics is the one area where certain knowledge can be argued to be possible. Some of what we say here is controversial. It is not universally accepted, so we encourage the reader to explore this further on their own, by investigating some of the references at the end of this chapter.

Immanuel Kant was one of the first philosophers to suggest that in the world of knowledge mathematics is unique. It is the only area of knowledge where you can have certain knowledge as justified–true–beliefs. The justification in mathematics is through the process of proof, and proof leads to absolute certainty. With proof, you have full and total belief. Thus, the truth is believed and is not questioned, so it is certain knowledge. Even more importantly, Kant argues that mathematics is made up of synthetic a priori propositions that we can believe with certainty.

An example is Pythagoras' Theorem that says that the sum of the squares of the lengths of a right triangle is equal to the square of the hypotenuse. Once proven true, it remains forever true. We do not doubt it or have any level of belief that it will eventually shown to be false. It is certain knowledge. Can you think of any other type of knowledge that leads to such certainty?

> Aside: It has been argued that since we now know that we really don't live in a Euclidean world, that Pythagoras's Theorem is not true. And this, it is argued, shows that mathematics does not always yield certain knowledge when applied to the physical world. Instead it gives you possibilities. If the parallel line postulate of Euclid is changed we lose Pythagoras' Theorem on right triangles, but other propositions arise. We will not explore this in this textbook, but instead simply make note of it. We do not contend that all the answers are known on this subject, but we are focusing more on the fact that mathematics is at the core of the discussion. It is what adds to its mystery and our fascination with it.

Look at the sciences. In the sciences we have laws that over time become modified as our knowledge increases. For example in Physics, for centuries we believed in the Newtonian laws of motion. Then Einstein showed us that we had to modify these laws if two objects moved relative to each other (especially if the relative speeds were quite large). Then it was shown that at the atomic level Newton's laws were totally unworkable. Instead we needed a new set of laws explainable by quantum mechanics. Thus, in physics we do not have certain knowledge, it is instead limited knowledge. It is even worse when we consider the biological or the social sciences. Our knowledge in these areas is even less certain. Mathematics appears to be the gold standard of knowledge in the world. Why is that?

Mathematics is unique in this regard. It has caused some philosophers to put mathematics on a whole other level of reality: Plato for example. The fact that certain knowledge can exist in mathematics has led some philosophers to search for it in other areas. Perhaps not fully believing that they could attain certain knowledge in other areas, but rather believing that it must exist elsewhere, since the world is rational and logical. It seems that mathematics is mysteriously connected to the world in general, and knowledge in particular, in ways we cannot fully explain.

The focus of this book is on seeking and obtaining knowledge. In particular, knowledge about mathematics and its unique relationship to our world. Thus, it was reasonable for us to take a deeper look at what we call knowledge as we move forward in our journey. It is rather interesting that the common theme of mind (rationalism) versus body (empiricism) emerges again when we consider the subject of knowledge. Whatever we seem to look at, it appears we have to resolve the mind–body aspect of it.

EXERCISES 3.1–3.5

ESSAYS AND DISCUSSION QUESTIONS:

Write an essay on each of the following, or be prepared to discuss these questions:

1. What is knowledge? Discuss ways of knowing and areas of knowledge.

2. Describe the types of knowledge defined by Kant.

3. What is epistemology, and why is it important?

4. Explain the difference between rationalism and empiricism.

5. Explain how mathematical knowledge is unique.

6. Are you an empiricist or a rationalist? Explain and support your position.

7. How is the mind – body problem related to the study of knowledge?

References

1. Dancy, J., (1991). An Introduction to Contemporary Epistemology (Second Edition). John Wiley & Sons.

2. Descartes, R., (1641). Meditations on First Philosophy.

3. Kant, I.l, (1781). Critique of Pure Reason.

4. Keeton, M.T., (1962). Empiricism, in Dictionary of Philosophy, Dagobert D. Runes (ed.), Littlefield, Adams, and Company, Totowa, NJ, pp. 89–90.

5. Morton, A., (2002). A Guide Through the Theory of Knowledge (Third Edition) Oxford: Blackwell Publishing.

6. Rand, A., (1979). Introduction to Objectivist Epistemology, New York: Meridian.

7. Russell, B., (1912). The Problems of Philosophy, New York: Oxford University Press.

8. Russell, B., (1940). An Inquiry into Meaning and Truth, Nottingham: Spokesman Books.

CHAPTER 4

Learning & Mathematics

The only way to learn mathematics is to do mathematics.

Paul Halmos

I never teach my pupils; I only attempt to provide the conditions in which they can learn.

Albert Einstein

4.1 Introduction

We have had an interesting journey so far as we discovered new things about mathematics and its relationships with many diverse areas of study. We should now be aware that mathematics is not just about numbers and shapes, but it is intimately tied to philosophy, psychology, and epistemology. All of which have an impact on how we learn this fascinating subject. In this chapter we will draw on the previous ideas to present ways in which we learn this remarkable human creation/discovery.

Chapter 2 and 3 have provided a good starting point for our discussion on how we learn. Obviously, how we learn is related to the brain and cognitive processes, and what we learn is knowledge. The real question is–What is the most effective way we can learn?

This chapter will focus on the more practical applications of this subject, and as such will be useful for those looking to either teach mathematics at any grade level, or even those looking to discover ways they themselves can become more effective at learning mathematics better.

4.2 Theories of Learning

There are essentially two competing philosophies on how we learn. The first is Behaviorism, introduced by B.F. Skinner (1904–1990), and second is Constructivism which is related to the work of Jean Piaget (1896–1980). Again, we will see the mind–body dilemma at play even in how we learn.

4.2.1 Behaviorism

Behaviorism is a philosophical school of thought to teaching, that believes learning is essentially a change in behavior due to environmental causes. Learning is a new response or behavior to a specific set of stimuli. It aligns itself closely with the philosophy of empiricism on how we acquire knowledge. The behaviorist philosophy of teaching was primarily pioneered by the American psychologist Burrhus Frederic Skinner (1904–1990) or simply B. F. Skinner.

Skinner developed his approach to learning based upon a series of experiments he carried out using a process he called Operant Conditioning. This is where learning is believed to be a function of changing behavior using a reward (or punishment) to reinforce behavior. Changes in behavior are caused by rewards or punishments that occur. It is also called the stimulus–response approach to learning. Using this approach in the laboratory, Skinner could teach pigeons to do complicated multiple step activities. Skinner carried out most of his test in what is called a Skinner box, or operant conditioning chamber. (YouTube has a number of interesting and informative videos about Skinner that you should take the time to view).

The behaviorist essentially believes that all knowledge comes from our sensory experience, and that knowledge is fixed and finite. That means the world we gain our knowledge from is complete, and it is up to us to uncover and understand it. The most effective way for us to be taught, is for a skilled teacher to act as a conduit and teach the important points by lecturing and demonstrating to the student. It is the most common teaching approach used in our educational system today. This approach is illustrated graphically below.

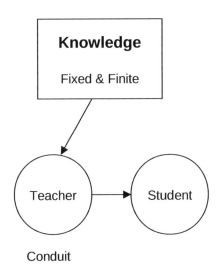

Conduit

Behaviorism is consistent with direct instruction, and curriculum based measurement (CBM) modes of teaching. Direct instruction uses lectures and demonstration as opposed to inquiry–based exploratory methods of teaching to educate the student. CBM is a method where testing, based upon a set curriculum, is performed on a regular basis to monitor the progress of the individual student.

60

This approach is used widely in public school systems, because it is one of the most cost effective ways of teaching large groups of students. It also lends itself well to assessing the progress of students using standardized tests.

4.2.2 Constructivism

Constructivism is a learning philosophy that believes all learning comes from within, and that the individual constructs their own knowledge of the world. Constructivism has its roots back to Socrates and his Socratic method of teaching. The Socratic method has the teacher asking the student questions until the student comes up with the answer themselves. It is teaching by asking and not telling. In this approach, learning appears as a process of remembering. The student discovers the concept from within, as if they knew the answer all along. (See www.garlikov.com/Soc_Meth.html for an example of using this approach in the classroom.)

The constructivist theory of teaching focuses more on this type of student–teacher interaction. The constructivist model that developed, considers the student (learner) as an important part of the learning process. The learner is a unique individual with specific needs and limitations. Learning was not knowledge–centric, but should be learner–centric. The teacher is not a conduit to simply pour knowledge into a students brain, but is instead a facilitator that helps the student learn for themselves by constructing their own knowledge through interacting with the world. Knowledge is no longer fixed and finite, but is constantly changing and growing. It is helpful to illustrate this relationship with the picture below.

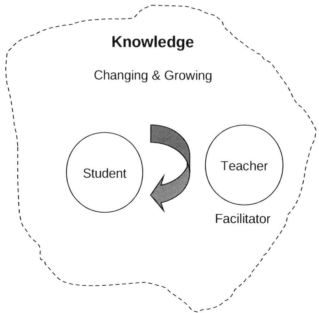

The theory of constructivism has a long list of people that have impacted its development including: Immanuel Kant, John Dewey, Maria Montessori, Jean Piaget, Lev Vygotsky, and Ernst von Glaserfield, to name a few.

Jean Piaget (1896–1980), a Swiss developmental psychologist, however, had a significant impact on formalizing and developing the constructivist theory of learning as we know it today. Piaget made extensive studies of the cognitive development of children, from birth to adulthood. In his studies of children, Piaget categorized learning into four different stages. We summarize their main features below, but encourage you to explore them further through other sources presented at the end of the chapter.

Piaget classified the learning process in humans into four stages of learning that we must all go through. They are:

1. **Sensorimotor stage** (birth to age 2): Children use their five senses and movement through their environment to explore and experience the world (there are five sub–stages in this learning stage), there is no sense for conservation of quantity

2. **Preoperational stage** (2–7 yrs): Magical (not logical) explanation for causation dominates, they are egocentric, they do not see conservation of quantity

3. **Concrete operational stage** (7–12 yrs): Logical explanation for causes and they are no longer egocentric, they can also see the conservation of quantity, but they need a practical aid to do so

4. **Formal operational stage** (12 onwards): Children develop abstract thought and can think logically and can see conservation of quantity in their minds without the need of physical aids

It is knowledge of which developmental stage a child is at, that dictates how and what the child should be taught. There are numerous publications discussing these stages along with many constructivist approaches to teaching at the different stages. We will not go in to greater detail here, but instead refer the interested reader to explore this fascinating subject on their own.

4.2.3 A Continuum of Approaches

Education today typically does not exclusively propose either behaviorism or constructivism, although you do have those that are both radical behaviorists and constructivists arguing for their approach to teaching over the other. There is instead a full continuum of approaches that educators use.

Behaviorism   Blend   Constructivism

The literature is filled with a variety of blended approaches to education. Some lean more to the behaviorist side, while some lean more towards the constructivist approach. We encourage the interested student to explore these ideas further, and have provided some references at the end of the chapter to help guide you.

It is interesting that the range of teaching theories also has the common theme of mind verses body, that we have seen throughout this journey, embedded in it. The behaviorist believing that all learning is tied directly to the physical senses, and through some form of physical–based (operant) conditioning (the body). The constructivist on the other hand, sees learning as more of an internal rational experience (the mind). Yes, we do use our physical senses to gather information, but it is mind that makes sense out of it, so it is the mind that is more important. The mind–body duality of our world seems to be present wherever we look.

4.3   Learning Mathematics

The previous section was an introduction to learning theories in general. In this section we concentrate on the learning of mathematics in particular. The first question that arises is – Is the brain already preconditioned to learn mathematics, or must we be taught everything? The empiricist/behaviorist would tell us that we must be taught everything, and we have no innate sense for number or quantity. We are born with a tabula rasa that must be written upon through education and learning. The rationalist/constructivist on the other hand believes that we do have an innate understanding of mathematics. Our brains are hard–wired for mathematics. However, there is disagreement on exactly how much mathematical understanding we are born with.

For a time people studied animals with the belief that animals too could learn all sorts of mathematics. At the turn of the 20th century (early 1900's), there was a horse in Germany called Hans the Wonder Horse or Clever Hans (der Kluge Hans). It was reported that Hans could do basic arithmetic with numbers and fractions, as well as some basic algebra. Hans would simply tap out the numerical answer with his hoof when asked a question. A panel of scientists (including the psychologist Carl Stumpf) were brought in and they confirmed the results. Many believed that this supported the contention that mathematics was an innate skill that even animals could do. However, it was quickly discovered by Stumpf's assistant, Oskar Pfungst, that Hans was not doing mathematics at all, but was instead reading the facial expressions and body language of his trainer in finding the answers. The trainer was not aware of this, and it wasn't meant to be a fraud or parlor trick. The horse, it was determined, simply reacted to the questioner/trainer who would subconsciously make facial or body movements when the correct answer was reached, causing Hans to stop pounding his hoof. This incident had a profound effect on the scientific investigation into this field. No self–respecting scientist would go near this and risk the ridicule endured by the others. Thus, this field of study languished for about twenty years.

In the mid to late 1920's, Jean Piaget conducted studies on infants and children that seemed to confirm the idea that when it comes to the fundamental ideas of mathematics, such as quantity, number or amount, humans did not have any core innate knowledge or understanding. It must all be learned. However, more recently many of Piaget's experiments have been called into question by other cognitive psychologist's (It seems the Hans effect has finally worn off and been forgotten). Stanislas Dehaene (1965–present), a mathematically trained neuro/cognitive scientist at the College of France, replicated many of Piaget's experiments, and created many of his own, only to come to a slightly different conclusion than Piaget regarding the learning of mathematics by humans. Dehaene believes that we ARE all born with a basic number sense. Although higher mathematics requires language to develop, the core of mathematics, the sense of different quantities and numbers, as well as a basic sense in conservation, are already a part of the human brain at birth.

Dehaene and others have carried out numerous experiments on infants and animals, and have found convincing evidence that humans, and animals, have a basic number sense. It could be related to a survival instinct, that is needed to know when it is better to fight or take flight, but we do not know for sure. What we do know is that there appears to be a basic understanding of quantity that does not have to be taught.

According to Keith Devlin, a mathematician at Stanford University, humans are uniquely suited to do mathematics. In his book, *The Math Gene*, he makes the argument that all humans have a "math gene" that uniquely enables us to think mathematically. Brian Butterworth, Professor of Cognitive Neuropsychology at University College, London, has suggested that this math gene might indicate why some of us are more predisposed to learning mathematics than others, and that a damaged gene could have important consequences, which we discuss further in the section below.

Butterworth has also proposed some radical changes on how we educate children in mathematics. Butterworth suggests that it may be better to not teach any formal mathematics (such as arithmetic) to children in the early years (up to the beginning of 7th grade) at all. Instead they should learn mathematics through other subject areas, and only when they are older should the more formal learning of mathematics take place. He bases this belief on an experiment that was done by a school superintendent in the public schools system in Manchester New Hampshire, USA back in the 1930's. The school superintendent, Louis P. Benezet, also believed that it would be better not to teach mathematics formally at a young age, so he decided to test this hypothesis. (This is a similar philosophy as expressed by Paul Lockhart in his book titled *A Mathematician's Lament.*)

Benezet created two groups of students, those educated using the current curriculum where formal arithmetic was taught in elementary school until the 6th grade, and another group that was exposed to an informal mathematics curriculum where the emphasis was on problem–solving that involved mathematical concepts. He published his findings in The Journal of the National Education Association in three articles in 1935 and 1936. It turned out that the students trained in the non–traditional approach did much better at problem–solving requiring mental math every time. In fact, they even did much better than the older 8th grade students in the Manchester School System. On the other hand, the students that received traditional lessons did better on the arithmetic drill tests, as expected, but after only a year the experimental group was performing better than the group that had been studying formal arithmetic manipulation exercises for three and one–half years the traditional way! Not only did this not hold back the group of students trained with the new approach, but it actually made them better at all mathematical skills and concepts in the long run.

In experiments in the late 1960's an American psychobiologist, Roger W Sperry, discovered that the human brain has two distinct ways of processing information. Prior to his work it was believed that the brain functioned as a single organ. Sperry's work showed that the brain was actually two distinct brains in one, with different functions for each side of the brain.

Using another approach to improve mathematical understanding, many scientists have looked to how the brain functions. Typically one side of our brain tends to dominate our behavior. As we have said earlier, the left side of the brain is focused more on logic, reason, and language, while the right side of the brain governs intuition, imagination, empathy and emotions, and visual images. In the early mathematics classes the left brain seems to have a stronger influence on your success, however, as you advance in mathematics the right side of the brain becomes equally, and probably more, important. Thus, if you did not find mathematics interesting early on in school, you may actually find it more enjoyable if you can advance to later courses. It is believed that Albert Einstein

64

was a right brained thinker. He often commented on the fact that he would "visualize" or "see" the solution.

When presented with a lesson in mathematics left brained, and right brained students prefer two different approaches to learning. Left brained learners prefer verbal instructions, and a very logical and structured step–by–step approach to solving the problem. They tend to be sequential thinkers who prefer sequence–based learning. The right brained learner needs to see the whole picture and to visualize the pattern or connectedness involved in the solution. They also learn better if there is a vibrant story, they can visualize, involved with solving the problem. They tend to be intuitive thinkers who prefer whole–picture–based learning.

Aside: After his death Albert Einstein's brain was preserved and the rest of his body was cremated. Possibly without the permission of Einstein or his family, his brain was studied to see if there were any notable differences between his brain and the brain of others of normal intelligence. It was found that the inferior parietal lobe was about 15% larger than other brains. This is the part of the brain that has been linked to higher level mathematical thought. The structure of his brain was also different in a few areas that might have made for a more visual rather than linguistic thinking brain. The point being that the makeup and structure of the brain may be an important part of how well we do in mathematics.

The typical mathematics course is taught for the left brain learner. Many right brain learners can learn through these methods, but this takes more energy, with which they could be using to learn additional concepts and ideas. For some right brain learners, however, they cannot adapt and find success in their early mathematics to be elusive at best. Even Einstein had trouble in his early mathematics courses.

Dianne Craft a teacher and proponent of right and left based learning techniques, provides an excellent illustration, which we provide below, of how a right–brained learner should be taught mathematics versus their left–brained classmates.

Multiplication fact memorization can be a real source of frustration for a right brainer and can keep him or her from going on to more difficult math because of this block. These facts can actually be very easy to learn when using a right–brain–friendly method. Right brainers learn anything easier when emotion, color, or stories are added to the learning method. For example, when learning the math fact "8x3=24," a picture story could be made, creating the number 8 as an eighth grader who has to babysit the neighbor's 3 year–old child while they go out for just an hour. He thinks he's too old to babysit, and besides, this 3 year old is a naughty little boy who doesn't listen to anybody. Put "hands" on the "hips" of the number 8, representing his indignation at the whole idea. When he goes to babysit the 3 year old, he jumps on the couch the whole time. Sketch a couch on which the number three is jumping, represented by lines going up. The eighth grader looks through the window on the door and sees the "mom and dad" 24 walking up. The number two is dad, with a hat on, and number four is mom, with a purse hanging from her "arm." He knows he's going to be in trouble because the three year old was jumping on the couch the whole time. If you draw the picture while telling the story, it's like "chalk talk" and makes a lasting impression on the child. You will find that with the combination of an emotion–filled story and the pictures he or she will remember it easily. Then, put the 24 in a division box with the 8 on the outside. They immediately know which one is missing. Do this same process putting the 24 in the division box with the 3 on the outside, and they will know that the 8 is missing because of the story. Thus, you have taught the multiplication and division facts at the same time. (Extended excerpt from: www.diannecraft.org)

This teaching illustration provides some valuable insight into how to approach mathematics education for right brain learners.

Do you know if you have a dominant side of the brain? For a possible answer go to the web site, *en.wikipedia.org/wiki/The_Spinning_Dancer*, and view the spinning dancer. Depending upon which way you see her spinning (clockwise is said to indicate a right brain dominance and counterclockwise a dominant left brain), MAY give you an indication as to which side of the brain dominates how you approach the world. Although the site argues this is pure speculation and that you cannot infer any brain dominance from this, it is interesting to see how people see it so differently. Also, the reference they cite to debunk this belief, is not a definitive or scientific study, so the jury may still out on the truthfulness of the claim or not. At the very least it is still entertaining and fun to explore, and it does make you think.

More recently, modern scientific methods such as Magnetic Resonance Imaging (MRI) have been used to study what parts of the brain respond to mathematical versus verbal or other stimuli. We are beginning to get a better functional picture of how the "mechanical" brain operates, and how it affects our learning of mathematics. We are currently at the early stages of studying and understanding how the brain works and scientists continue to examine it from every possible perspective, so we expect new and interesting developments in the coming years.

We have presented a variety of perspectives and views on how we learn mathematics. This is not a definitive review of the subject. This is a survey course and our goal is to provide an overview and to help stimulate the interested student to seek out more information on their own. Without doubt, these innovative and groundbreaking studies deserve more notice, as they could lead to better ways in teaching mathematics that will help to curb the high mathematical illiteracy rate in this country. They can also be used to help identify learning bias' as opposed to actual learning disabilities in mathematics. In the next section we introduce a little known, but very important, mathematics learning disability.

4.4 Dyscalculia

In 1974 Ladislav Kosc published a paper in the Journal of Learning Disabilities, titled "Developmental Dyscalculia." This was one of the first articles to not only make a connection between math learning disabilities and the brain, but to also give it a name – Dyscalculia (the Latin word for "counting badly"). Today dyscalulia is recognized by the World Health Organization and the DSM (Diagnostic and Statistical Manual of Mental Disorders). However, many professionals, including teachers and psychologists, have not even heard of it, even though it is believed to affect up to 6% of the population.

Dyscalculia has also been called "number blindness," where those that are afflicted by it have problems recognizing numerals, doing basic arithmetic, or even memorizing the times–tables. You may have problems telling time, making change, following directions, as well as some other basic number related functions. Interestingly enough though, these people can have normal or even above average IQ's. It is a learning disability that only affects their numerical abilities. Just as people can be born color–blind, dyscalulia is a sort of number–blindness. It can, however, have a devastating impact on a person's life.

There is no known treatment or cure for dyscalculia, although there are some teaching and learning strategies that can help in coping with it, especially if it is diagnosed early.

Take the dyscalculia test below.

1. I have difficulty making correct change.
2. I find it really hard to copy a list of numbers from the board onto a piece of paper.
3. Sometimes I see signs like + or ÷ but I can't remember what they are called. If someone says "divide" I can't think of the symbol.
4. I will sometimes look at a number, but copy it in the wrong order.
5. I can't remember numbers, even when I use them often (for example telephone numbers, combinations, codes, addresses)
6. I always find adding and subtracting numbers difficult.
7. I have never been able to do the "times tables".
8. I can't understand fractions at all.
9. I can't quickly recognize odd and even numbers, and I have to think very carefully to work out which is which.
10. The 24 hour clock totally confuses me.
11. I have never been able to subtract large numbers.
12. Sometimes I know the answer to the math question, but I can't explain how I got it.
13. It seems that everyone else in my class understands what "square root" means, but I really have no idea.
14. Most of the people I work with can use a calculator, but I hardly ever get the right answer.
15. Sometimes when I am faced with a question that has to do with numbers I just cannot cope and become very anxious.
16. When I start a math problem I often forget where to begin and frequently cannot finish the problem.
17. Sometimes I forget the names of shapes like triangle, pentagon, semi–circle, etc.
18. When I work out my math homework the work is always very messy.
19. I get really confused with big numbers such as 100,000 and 99,999 and I can't determine which is larger
20. When I travel to foreign countries I can never get the hang of their money system, so I always let someone else figure it out for me.
21. I don't understand percentages at all.
22. I can never do problems like; "If it takes John an hour to drive 50 miles, how long does it take him to drive 60 miles?," but other people in the class can.
23. Math frightens me. I really don't understand it at all.

If you answered yes to 12 questions or more, you could have dyscalculia, and should probably get tested.

Note: This is a modified version of a test on the Dyscalculia Centre (UK) website at: www.dyscalculia.me.uk

The British mathematician Keith Devlin, wrote a book titled *The Math Gene: How Mathematical Thinking Evolved and Why Numbers Are Like Gossip.* In his book, Devlin contends that everyone is born with a "math gene." We all have an innate capacity for mathematical reasoning. Humans are born with a basic number sense, the ability to think logically and abstractly, as well as an awareness of cause and effect, and the capacity to construct and follow causal chains of events. He argues that these traits, along with our complex social structure of using language to talk about (gossip) events and relationships, almost assures us that humans would develop mathematics. Now some people may have a better math gene than others, and some will most certainly have a damaged math gene. This could account for the rather large variation we see in people's mathematical abilities, from the child prodigy, to the number–blind dyscalculiac.

Brian Butterworth is another pioneer in promoting awareness and understanding of dyscalculia. He also believes a person's basic numerical ability is connected to their "math gene," and that those with dyscalculia have a damaged gene. He provides support for his belief in his book, *What Counts: How Every Brain is Hardwired for Mathematics.*

It has been suggested that dyscalculia is believed to be caused by lesions (damaged or abnormal tissue) between the temporal and parietal lobes of the brain (the gray area indicated in the image below.) Perhaps this is the "math gene" Devlin and Butterworth speak of, or what Dehaene cites as our number sense. The recognition and understanding of this phenomena has come a long way in a short time, and I'm sure we'll see a greater awareness and understanding of how and why this happens in the very near future.

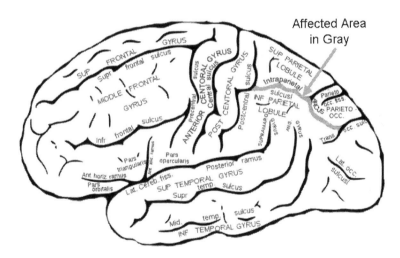

EXERCISES 4.1–4.4

ESSAYS AND DISCUSSION QUESTIONS:

Write an essay on each of the following, or be prepared to discuss these questions:

1. Compare and contrast the theories of learning mathematics of Piaget, Dehaene, and Butterworth.

2. What are Constructivism and Behaviorism (Objectivism)? Also discuss how each views the source of knowledge.

3. What is Dsycalculia, and how does it impact those that have it?

4. Discuss the difference between left and right brain learners?

5. Discuss the study performed by Benezet. What is your opinion?

6. How does the mind–body question enter into the real of learning mathematics?

7. Which approach to learning do you think is better? Explain and support your choice.

8. Do you believe that we all have a math gene? Support you position either way.

References

1. Benezet, L.P., (1935a). The Teaching of Arithmetic I: The Story of an Experiment, Journal of the National *Education Association* **24**: 241–244.

2. Benezet, L.P., (1935b). The Teaching of Arithmetic II: The Story of an Experiment, *Journal of the National Education Association* **24**: 301–303.

3. Benezet, L.P., (1936). The Teaching of Arithmetic III: The Story of an Experiment, *Journal of the National Education Association* **25**: 7–8.

4. Berman, E., (1935). *The Result of Deferring Systematic Teaching of Arithmetic to Grade Six as Disclosed by the Deferred Formal Arithmetic Plan at Manchester, New Hampshire*, Masters Thesis, Boston University, USA.

5. Butterworth, B., (1999). What Counts: How Every Brain is Hardwired for Mathematics, The Free Press.

6. Carpenter, T.P., Lindquist, M.M., Matthews, W., and Silver, E.A., (1983). Results of the Third NAEP Mathematics Assessment: Secondary School, *Mathematics Teacher* **76**: 652–659.

7. Dehaene, S., (1999). *The Number Sense: How the Human Mind Creates Mathematics*, Oxford University Press USA.

8. Devlin, K., (2000). *The Math Gene: How Mathematical Thinking Evolved and Why Numbers Are Like Gossip*, Basic Books.

9. Dougiamas, M. (1998). A journey into Constructivism, http://dougiamas.com/writing/constructivism.html.

10. Hardy and Taylor, (1997). Von Glasersfeld's Radical Constructivism: A Critical Review, Science and Education, 6, pp 135-150, Kluwer.

11. Lockhart, Paul, (2009). *A Mathematician's Lament*, Bellvue Literary Press, New York.

12. Manchester School Board, (1936). (September 11), Minutes of the School Board, Manchester, New Hampshire. Manchester School Board, (1938). (May 10), Minutes of the School Board, Manchester, New Hampshire.

13. Polya, G., (1977). *Mathematical Methods in Science*, Mathematical Association of America, Washington, DC.

14. Reusser, K., (1988). Problem Solving Beyond the Logic of Things: Contextual Effects on Understanding and Solving Word Problems, *Instructional Science* **17**: 309–338.

15. Schoenfeld, A.H., (1989). Teaching Mathematical Thinking and Problem Solving, in L.B. Resnick and L.B. Klopfer (eds.), *Toward the Thinking Curriculum: Current Cognitive Research*, Association for Supervision and Curriculum Development, Alexandria, Virginia.

16. Schoenfeld, A.H., (1992). Learning to Think Mathematically: Problem Solving, Metacognition, and Sense Making in Mathematics. In D.A. Grouws (ed.), *Handbook of Research on Mathematics Teaching and Learning*, Macmillan, New York, pp. 334–370.

17. Skinner, B.F., (2002). Beyond Freedom & Dignity, Hackett Publishing Co, Inc.

18. Skinner, B.F., (1938). The behavior of organisms. New York: Appleton-Century-Crofts.

19. Skinner, B.F., (1945). The operational analysis of psychological terms. Psychological Review. 52, 270-277, 290-294.

20. Skinner, B.F., (1953). Science and Human Behavior (ISBN 0-02-929040-6) Online version.

21. Skinner, B.F., (1957). Verbal behavior. Englewood Cliffs, NJ: Prentice-Hall.

22. Skinner, B.F., (1969). Contingencies of reinforcement: a theoretical analysis. New York: Appleton-Century-Crofts.

23. Skinner, B.F., (31 July, 1981). Selection by Consequences, Science 213 (4507): 501–504.

24. Staddon, J., (2001). The new behaviorism: Mind, mechanism and society. Philadelphia, PA: Psychology Press. Pp. Xiii, 1-211.

25. Sowder, J., (1992). Estimation and Number Sense, In D.A. Grouws (ed.), *Handbook of Research on Mathematics Teaching and Learning*, Macmillan, New York, pp. 371–389.

26. Whitney, H., (1986). Coming Alive in School Math and Beyond, *Journal of Mathematical Behavior* **5**: 129–40.

27. Von Glasersfeld, E., (1990). An exposition of constructivism: Why some like it radical. In R.B. Davis, C.A. Maher and N. Noddings (Eds), Constructivist views on the teaching and learning of mathematics (pp 19-29). Reston, Virginia: National Council of Teachers of Mathematics.

Part I Summary – Making the Connections

The first part of this book focused on trying to define what mathematics is. To do so we looked at it from several different perspectives. We examined mathematics through the lens of philosophy, neurology and cognition, and epistemology, as well as how it is learned. We showed that mathematics truly is a Liberal Art. It touches just about all the other liberal arts in one way or another, and enriches our experience with them. You must first let some of your bias' against mathematics subside, and approach the subject with a fresh, open, and inquisitive mind.

Hopefully along the way you developed a greater appreciation for mathematics beyond the mechanical aspects of working out solutions to problems in arithmetic and algebra. We are still not able to provide a definitive definition of what mathematics is, but we have at least provided some interesting viewpoints and perspectives on the subject. It is truly a wonderful mystery tour.

ESSAYS AND DISCUSSION QUESTIONS:

Write an essay on each of the following, or be prepared to discuss these questions:

1. What is mathematics? Incorporate some of the ideas discussed in the first part of the book.

2. Is mathematics discovered or created? Incorporate some of the ideas discussed in the first part of the book.

3. Has this opened up your view on what mathematics is? If yes, then how, and if no then why not.

PART II: HISTORICAL DEVELOPMENT OF NUMBERS

From the concrete to the abstract, as humankind developed so did mathematics. Although we cannot say with certainty that mathematics is either discovered or created, we can say that mathematics has developed into what it is today due to the influences and contributions of many different cultures throughout history. Mathematics grew out of our need to explain and control our environment. These two needs went hand–in–hand and as history evolved so did mathematics. Mathematics and history are intimately connected, with each impacting and influencing the development of the other. As we examine how mathematics developed we will at the same time be taking a deeper look into how humanity developed.

We will also look at challenging how we think. Thus, exposure to different number systems is not intended to teach a new and useful way to do computations, rather it is instead needed to train us to look at things differently. It helps us to see more clearly and get beyond our natural bias' as we look at the world. We learn to look at another perspective, or more importantly how to look at something we have grown accustomed to with fresh eyes.

The focus here will be placed equally on the abstraction of the number concept, together with the rules for computation that arise when the extended number concept is developed.

CHAPTER 5

The First Numbers

God made the natural numbers, all the rest is the work of man.

Leopold Kronecker

5.1   Introduction

What are numbers? Are they simply names for made–up objects, or are they as real as the sun, the earth, an ocean, a plant, or any other tangible physical thing we might experience in the world? If humans had never emerged out of this universe, would the concept of number still exist? This is a question that many have tried to answer, but as we have seen from Part I in this book, it is not easily accomplished. In this chapter we'll examine the concept of number from a historical perspective to see if that can provide us with any additional insights.

The concept of number probably came from our need to quantify and compare. The earliest evidence of any form of counting is from about 20,000 BCE, during the Paleolithic period, when humans used tally marks on bones to count items (see image below of the Ishango Bone courtesy of the Science Museum of Brussels.) In an attempt to understand the origin of this attribute, psychologists and cognitive scientists have looked into the concept of number and here is what they have found. It appears that many species of living things are able to distinguish between different quantities of objects. For most this distinction stops at three or four objects, but for others, humans uniquely in particular, it seems to have no bounds. Why is this?

Ishango Bone showing tally marks circa 20,000 BCE

We know that as humans developed the number concept, they did so in a very basic way. People began to "count like the animals"; observing quantities of objects like one, two, three, four, and then simply many. Either we never initially saw the difference between larger numbers, or we didn't see a need to try and distinguish between them. It seems that the concept of many, for any quantity greater than four, was used universally.

Unlike most species of living things, however, humans developed a highly evolved form of communication in their language. Some early human civilizations, however, never developed numerical concepts, or even words for quantities of five, or higher. In fact, even today there are tribes in the Amazon rain forest, that have been isolated from the outside world, and they too, count in this limited way today. So while it seems language was a necessary condition for developing larger numbers, it was not sufficient. Something else caused the giant leap beyond the many. Whatever that catalyst was, it did happen in some tribe or culture, and as people interacted it spread, largely due to language.

As we've discussed earlier in the book, basic counting seems to be an innate skill – something many living creatures are born with, and not an acquired or learned skill. This may be due to a natural survival mechanism that exists in living things, but nobody knows for sure. The ability to recognize quantities beyond four, however, seems to be related to an ability to develop language, and it is a learned skill.

In the early stages of number development it appears that humans did not distinguish the quantity from the object they were counting. Both concepts were connected in the minds of early humans. Thus two ducks, or a brace of ducks, two oxen, or a yoke of oxen, and two hawks, or a cast of hawks, may have required a separate word for the same quantity of two. The concept of "twoness" had not been understood as being independent of what was being counted. A number did not exist independently of objects to count. At some point, though, the idea of a number, independent of what was being counted, came to stand alone. No one knows when this occurred, but at that time the concept of number came into being. Words and symbols were now needed to represent this new abstract concept. A hunter might use tally marks on a bone to keep track of his kill, or a herdsman might use pebbles in a pouch to keep track of his sheep. From these basic symbols and representations, new more complex written symbols were eventually created. The image below shows some more elaborate symbols used to represent numbers.

Eventually two important ways to represent numbers developed. Today we refer to these as **positional** and **non–positional** number systems. A non–positional number system uses symbols to represent numbers without regards to any place value associated with the location (position) of the symbol. This is probably how numbers and their symbols first developed. This system has its drawbacks, though, as we'll show later.

A positional number system, on the other hand, has a place–value associated with the location of the number symbol. Both forms of numbers systems emerged at about the same time in ancient cultures. Different systems emerged for different civilizations. See the table below for some common early civilizations, the approximate time during which they began to flourish, and their early type of number system. Some of the civilizations that used the non–positional number system, would sometimes use a positional number system, whenever complicated arithmetical calculations had to be made, especially regarding calculations involving the planets, sun, and moon.

Non–Positional	Positional
Egyptian (3,000 BCE)	Babylonian (2,000 BCE)
Sumerian (3,000 BCE)	Mayan (400 CE)
Chinese (1,300 BCE)	Indian/Hindu (500 CE)
Ancient Greek (500 BCE)	
Roman (0 CE)	

The figure below is a world map showing the geographical relationship between these civilizations. As the map shows, some were more isolated from others.

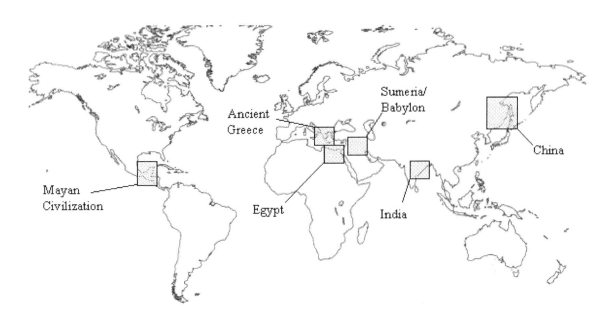

World Map of Ancient Civilizations

The Chinese and the Mayan number systems developed in isolation from the number system of other civilizations. It is interesting that all but the Mayans used groups of ten, in some form, for their counting system. The Mayans used groups of twenty, with groupings of five as a subgroup. The Babylonians used groups of sixty in counting, but they also used subgroups of ten within their group–sixty system. Nobody knows for sure, but it is widely speculated that since we have ten fingers, humans began counting in groups of ten. As for twenty, some have suggested that the ten toes were added to the ten fingers to provide a group of twenty. As for the Babylonian group–sixty system, that is still a mystery, but we'll discuss how that might have come about in more detail below. We will address many of the questions above as we take a historical tour through the development of numbers and processes associated with them. We begin with the early non–positional numbers.

We should also note that there are two different uses for numbers. The first is to be able to count objects, and these we call the Cardinal Numbers. The second is to identify the order in which things appear (first, second, third, etc.), and these are called Ordinal Numbers. In the discussion that follows, we will be concentrating on Cardinal Numbers. What we develop for cardinal numbers can be used for ordinal numbers as well, but with a different conceptual understanding. The quantities we are investigating now are truly human creations, since they are names for the numbers we count with, which is different from the number concept itself.

5.2   Non–Positional Number Systems

A non–positional number system relies on a symbol alone to represent a quantity, and not on the location of that symbol. We'll illustrate these types of number systems by presenting several different examples: the Egyptian, the Ancient Greek, the Chinese, and the Roman number systems.

5.2.1   Egyptian Number System (Hieroglyphics)

One of the most widely known systems for representing non–positional numbers is the Egyptian number system. This number system was used in Egypt for about 3,000 years, until about 1 CE. The Egyptians drew pictures to represent the numerical value they were trying to keep track of. The table below shows the some of the Egyptian symbols, also called hieroglyphs, for various numerical quantities. Notice how the Egyptians developed symbols for groups of ten, ten–ten's or a hundred's, ten–hundred's or a thousand's, etc.

Hieroglyph	Description	Value
|	Tally Mark	1
∩	Heel Bone	10
℮	Snare (coil of rope)	100
(lotus flower)	Lotus Flower	1,000
(pointing finger)	Pointing Finger	10,000
(frog)	Frog or Tadpole	100,000
(god with arms raised)	God with arms raised above head	1,000,000

In the Egyptian non–positional number system, the value of a number was only dependent upon the observed symbols, i.e., the number 2,314, is made up of two thousand–symbols, three hundred–symbols, one ten–symbol, and four one–symbols, as shown below. Furthermore, the location or the order in which you put the symbols does not matter. Examples are given below to demonstrate how to write numbers using this system.

Examples:

1.   2,314

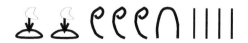

or

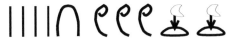

or

or even

(but not ever really done this way)

2. 123,021

3. 41,238

4. 99

5. 1,111,000

Typically the Egyptians would write their numbers from left to right, but they also wrote them right to left and top to bottom. They could have written them in any random order, as this is possible in a non–positional number system, but they did not.

78

Rules to Write a Number using Egyptian Hieroglyphics
1. Determine how many units, tens, hundreds, etc. your number has.
2. Choose to start from left–to–right, right–to–left, or top–to–bottom.
3. Depending upon the quantities from 1 above, and the order in 2, write the number of figures from largest to smallest for each of the groupings of numbers.

Rules to Read a Number in Egyptian Hieroglyphics
1. Determine how many units, tens, hundreds, etc. your number has, by counting the symbols for each grouping.
2. Multiply the number of symbols for that particular grouping by the group value.
3. Add all the multiplied values together

Arithmetic of Egyptian Hieroglyphs

To add, subtract, multiply or divide using hieroglyphs, you had to come up with a conceptual way to think about what it was you were trying to accomplish, and then devise a method for doing it.

Addition
The addition concept of hieroglyphs is fairly straight forward. You simply write the numbers together, as you would if you wanted to write a single number. Addition is already equivalent to how they wrote their numbers, since it was an "additive" system.

The method for computation involves grouping all the like symbols and replacing groups of smaller symbols with larger symbols, starting with the smallest and proceeding to the largest, as needed.

Examples:
1. Add

and

We add by combining the two numbers together

Group like symbols

Simplify symbols by combining ten smaller symbols into one larger symbol

ℓ ℓ∩∩ ∩I

2. Add

ℓℓℓ∩∩∩∩∩IIIIIIIII

and

ℓℓℓℓℓℓℓℓ∩∩∩∩∩∩IIIIII

Combine the two numbers together

ℓℓℓ∩∩∩∩∩IIIIIIIIIℓℓℓℓℓℓℓℓℓ∩∩∩ ∩∩IIIIIII

Group like symbols

ℓℓℓℓℓℓℓℓℓℓℓℓ∩∩∩∩∩∩∩∩∩∩∩ IIIIIIIIIIIIIIIII

Simplify symbols by combining ten smaller symbols into one larger symbol

 ℓ ℓ∩ ∩IIIII

Subtraction
The Egyptians thought of subtraction in terms of the addition concept. Thus, 21 – 7 would be thought of as – What number when added to seven equals twenty–one, 7 + ? = 21?

The method would be to write the number you were subtracting and then continue to add to this number until you reach the number you were subtracting from. The final answer is the numbers you added to reach the number.

Examples:
1. Subtract

∩III

from

Rewrite the problem as

ℓℓℓ∩∩∩|||||

∩||| + (?) = ℓℓℓ∩∩∩|||||

We see that we need to add, ℓℓℓ plus ∩∩∩ plus || to obtain the desired result, so our answer is:

ℓℓℓ∩∩||

2. Subtract

ℓℓ∩∩||||

from

ℓℓℓ∩∩||||||||

Rewrite the problem as

ℓℓ∩∩∩|||| + (?) = ℓℓℓ∩∩|||||||||

We see that we need to add, ℓ plus |||| to obtain the desired result, so our answer is:

ℓ||||

Multiplication

The Egyptians thought of multiplication as adding powers of two times the number, until the correct multiplying factor was reached.

The method required that the Egyptians know how to write doubles of numbers quickly and easily. They would construct a table and add together only the doubled numbers that were associated with powers of two that added up to the multiplier.

Example:

We illustrate the approach using our familiar number system, and then redo the problem using hieroglyphs.

1. Multiply 13 times 21
    a. Construct a table by doubling 21, and its result in each previous row.
    b. Only use the rows containing 1, 4, and 8, since they sum up to 13, the multiplier.

c. Add the numbers associated with the rows from b above, and this is the final answer.

Yes	1	1×21	**21**
No	2	2×21	42
Yes	4	$4\times21 = 2\times42$	**84**
Yes	8	$8\times21 = 2\times84$	**168**
	1+4+8=13		**21+84+168** $= 273$

We choose the numbers in the second column that sum to 13. Then just the multiples of 21 in column four that are associated with the factors we chose in column two, are added together to find the product.

Thus, 13 times 21 equals $21+84+168 = 273$.

Same example using hieroglyphs

Keep as a Factor?	Powers of two	Multiple	Simplified Multiple
Yes			
No			
Yes			
Yes			

82

Note: Column three is obtained by simply doubling the hieroglyphs in the previous row. The fourth column is used, as needed, to combine symbols in the previous column using addition, to simplify the final calculation.

Now, pick the powers of two that add up to the multiplier. Then, sum up the multiples associated with these multipliers.

The result equals

ᗉᗉᑎᑎᑎᑎᑎᑎᑎ|||

2. Multiply 27 times 43
    a. Construct a table by doubling 43, and its result in each previous row.
    b. Only use the rows containing 1, 2, 8, and 16, since they sum up to 27, the multiplier.
    c. Add the numbers associated with the rows from b above, and this is the final answer.

Yes	1	1×43	43
Yes	2	2×43	86
No	4	4×43 = 2×86	172
Yes	8	8×43 = 2×172	344
Yes	16	16×43 = 2×173	688
	1+2+8+16=27		43+86+344+688 = 1,161

We choose the numbers in the second column that sum to 27. Then just the multiples of 43 in column four that are associated with the factors we chose in column two, are added together to find the product.

Thus, 27 times 43 equals 43+86+344+688 = 1,161 .

Same example using hieroglyphs

Keep as a Factor?	Powers of two	Multiple	Simplified Multiple
Yes	|	ᑎᑎᑎᑎ|||	
Yes	||	ᑎᑎᑎᑎ||| ᑎᑎᑎᑎ|||	

No	||||	∩∩∩∩||| ∩∩∩∩||| ∩∩∩∩||| ∩∩∩∩|||	ℓ ∩∩∩∩∩ ∩∩ ||
Yes	|||| ||||	ℓ ∩∩∩∩∩∩∩ ∩ || ℓ ∩∩∩∩∩∩ ∩ ||	ℓℓℓ ∩∩∩∩ ||||
Yes	∩ ||||||	ℓℓℓ ∩∩∩∩ |||| ℓℓℓ ∩∩∩∩ ||||	ℓℓℓℓℓℓ ∩∩∩∩ ∩∩∩∩ |||||||

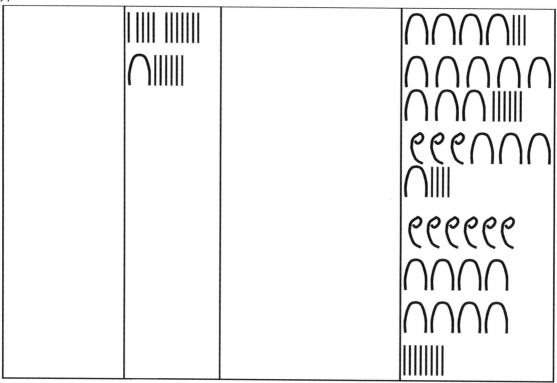

Note: Column three is obtained by simply doubling the hieroglyphs in the previous row. The fourth column is used, as needed, to combine symbols in the previous column using addition, to simplify the final calculation.

Now, pick the powers of two that add up to the multiplier. Then, sum up the multiples associated with these multipliers.

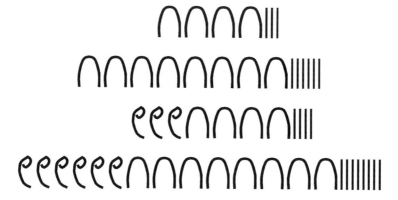

Add these together, and the result equals

Division

Division was thought of conceptually as the inverse of multiplication. We should point out that division only worked if the final answer was a natural number. Thus, the number we are dividing into our other number, must divide into it "evenly." That is, there is no remainder. Egyptian division simply did not give a fraction.

The method they used to do division was to construct the multiplication table, as above. Then starting with the largest multiple in the third column, add the previous row values in that column until you have the divisor. All rows may or may not be included. Then just add up the powers of two, used in completing the divisor, and this is the answer.

Examples:

We only show the process in our number system to illustrate the approach. It is a bit too confusing using hieroglyphics.

1. $\dfrac{275}{11}$

Yes	1	11
No	2	22
No	4	44
Yes	8	88
Yes	16	176
	1+8+16=25	11+88+176 = 275

Since the sum of 11, 88, and 176 (power of two multiples of 11) equals 275, the divisor, the sum of the powers of two that make up this number, 1, 8, and 16, which equals 25 is the quotient (answer).

$$\frac{275}{11} = 25, \text{ since 25 times 11 equals 275.}$$

2. $\dfrac{693}{33}$

Yes	1	33
No	2	66
Yes	4	132
No	8	264
Yes	16	528
	1+4+16=21	33+132+528 = 693

Since the sum of 33, 132, and 528 (power of two multiples of 33) equals 693, the divisor, the sum of the powers of two that make up this number, 1, 4, and 16, which equals 21 is the quotient (answer).

$$\frac{693}{33} = 21, \text{ since } 21 \text{ times } 33 \text{ equals } 693.$$

5.2.2 An Early Chinese Number System (Shang Numerals)

The Chinese number system developed independently from the Egyptian number system, starting in and around 1,300 BCE, during the Shang Dynasty. They did, however, share their affinity for breaking things into groups of ten, with the Egyptians. The table below shows some of the early representation of Chinese numerals. Typically they would write their numbers from the top downward. The Chinese system would always write the larger number on top.

There is also a similarity with the Chinese number system and positional numbers, which we'll discuss in further detail later on in this chapter, in that their symbol for two–hundred would use a symbol for two and a symbol for one–hundred in the same symbol. The same holds true for three–hundred, four–hundred, etc. This is different from the Egyptian Hieroglyphics where two, three or more symbols for a hundred, thousand, etc. were used. A similar, but slightly different pattern was used for numbers such as, fifty, sixty seventy, eighty, and ninety. Now instead the symbol for ten is on top, and the symbol for how many tens is beneath it, and this is very similar for the thousands symbols, as you can see in the table below.

Number	Symbol	Number	Symbol	Number	Symbol	Number	Symbol
1		10		100		1000	
2		20		200		2000	
3		30		300		3000	
4		40		400		4000	
5		50		500		5000	
6		60		600		6000	
7		70		700		7000	
8		80		800		8000	
9		90		900		9000	

Examples:

1. The number   54   would be written as

2. The number   987   would be written as

<u>Rules to Write a Number using Shang numerals</u>
1. Determine how many units, tens, hundreds, etc. your number has.
2. Depending upon the quantities from 1 above, write the number of figures from largest to smallest for each of the groupings of numbers from top to bottom. Then largest being on top and the smallest on the bottom.

<u>Rules to Read a Shang numeral</u>
1. From top to bottom write the numerical equivalent of the number.
2. Add all the multiplied values together

5.2.3 The Ancient Greek Numbering System

In 500 BCE the Greeks, like the Egyptians, had developed a non–positional number system. The Greeks used their alphabet system to double as their number system. One problem is that the Greek alphabet had twenty–four letters, but there were twenty–seven symbols required for numbers in their system. This was because they had unique symbols for the numbers one thru nine, ten thru ninety, and one–hundred thru nine–hundred (nine for each group) Thus, three new (actually old) letters/numbers needed to be added. These were vau or digamma (for six), koppa or qoppa (for ninety) and sampi (for nine hundred).

The table below shows the numbers, symbols, and their names.

Number	Symbol	Name		Number	Symbol	Name		Number	Symbol	Name
1	α	alpha		10	ι	iota		100	ρ	rho
2	β	beta		20	κ	kappa		200	σ	sigma
3	γ	gamma		30	λ	lambda		300	τ	tau
4	δ	delta		40	μ	mu		400	υ	upsilon
5	ε	epsilon		50	ν	nu		500	φ	psi
6	F	vau/ digamma		60	ξ	xi		600	χ	chi
7	ζ	zeta		70	ο	omnicrom		700	ψ	phi
8	η	eta		80	π	pi		800	ω	omega
9	θ	theta		90	ς	koppa/ qoppa		900	⋀	sampi

Obviously, this sort of system can cause confusion between words and numbers. To avoid this confusion, the Greeks would put an accent line above and to the right of the last letter to the right, when they wanted to indicate a number. As you can see this is a very complicated system, and you had to memorize twenty–seven different symbols for numbers! Also, this system only allows them to count up to 999. To go beyond 999 they added another accent mark in the lower left–hand corner, right before the number. This meant that you would multiply the numerical value of the next character by 1,000.

This is a fairly complicated systems and requires that we demonstrate it through some examples:

Examples:

1. The number   37   would be written as   λ ζ '

   Note: λ for thirty and ζ for seven, which you add together to obtain 37. The accent mark at the end is used to signify it is a number and not a word.

2. The number   720   would be written as   ψ κ '

   Note: ψ for seven–hundred and κ for twenty, which you add together to obtain 720. Again, the accent mark at the end is used to signify it is a number and not a word.

3. The number   879   would be written as   π ο θ '

   Note: π for eight–hundred and ο for seventy and θ for nine, which you add together to obtain 879. Again, the accent mark at the end is used to signify it is a number and not a word.

4. The number   3,704   would be written as   ͵γ ψ δ

   Note: γ with the sub–scripted accent mark in front, indicates three times a thousand, or three–thousand. ψ is for seven–hundred, and the δ for four, which you add them all together to obtain 3,704. Whenever an accent mark is used before the number you eliminate the accent mark at the end. You only need one accent mark to indicate it is a number and not a word.

Rules to Write a Greek Numeral
1. Use one of twenty–seven symbols to represent the one's, ten's, or hundred's value in the number.
2. Write the number from left to right.
3. If the number is greater than nine–hundred–and–ninety–nine, use an accent mark as a subscript in front of a number that tells you how many thousands you have.
4. If the number is less than one–thousand you need to put an accent mark at the end indicating this is a number and not a word.
5. When complete, the values of all the symbols should add up to the number you are representing.

Rules to Read a Greek Numeral
1. If there is an accent mark in the lower left hand corner of the number, determine how many thousands are in the number by multiplying the value of the character to the right of the accent mark by one–thousand.

2. With the remaining symbols, determine how many units, tens, hundreds, etc. your number has, by counting the symbols for each grouping.
3. Add all the values together.

The Greek number system is very complicated, with many symbols that needed to be learned, as well as many rules to know and understand. It is not surprising that this system did not catch on. The Greeks made numerous important contributions to mathematics, the number system was not one of them!

As you can see, both the Chinese and the Greek number systems had the beginnings of assigning a place–value to the location of the symbol for the numeral, but neither civilizations made the leap to change their systems until much later.

5.2.4   The Roman Number System (Roman Numerals)

The Romans were great at conquering, but not at mathematics. Even though their number system came a long time after the Chinese and the Greek systems, it wasn't any better, and in some ways it was worse. The Roman Numbering System is also a non–positional numbering system, but the location of a number does have a meaning, but does not take on a place–value due to its location. We'll explain what we mean by this further below.

The symbols used to represent numbers are as follows:

Number	Symbol
1	I
5	V
10	X
50	L
100	C
500	D
1,000	M

These symbols have a long history of development, and changed their form many times before settling in to those shown above.

The symbol for 1 (I), is believed to have come from the shape of a notch made on a tally stick. Then after making four notches on the stick, the fifth notch was a double cut (similar to the V shape) and every tenth cut was a cross–cut (similar to an X). It is believed that since these markings resembled the shapes of the letters I, V and X in the Roman alphabet, these symbols were eventually adopted for use as the Roman Numerals. The other symbols (L,C,D, and M) changed their form many times before being finalized into the form above. The C character took its final form from the Latin word centum, which means a hundred, and M was eventually adopted from the Latin word mille for a thousand. To make very large numbers, they would put a line over the number, signifying that the number was to be multiplied by a thousand. This process is similar to what the Greeks did for their numbers.

The Romans would write the numbers from left to right in descending order. In general, to obtain the value of the number you simply add the values of all the symbols together and that would be the value of the number. However, if the numerals are not all descending from left to right, you will need to use something called the "**Subtractive Principle**."

The Subtractive Principle is used whenever a smaller number symbol precedes a larger number symbol. If this occurs, you always subtract the smaller from the larger. While the rule does not change they did, however, restrict which smaller numbers could be placed before a larger number.

Restrictions for smaller numbers before larger numbers
- In front of a V you could write an I, to represent the number 4.
- In front of an X you could write an I, to represent the number 9.
- In front of an L you could write an X, to represent the number 40.
- In front of a C you could write an X, to represent the number 90.
- In front of a D you could write a C, to represent the number 400.
- In front of an M you could write a C to represent the number 900.

Can you see the pattern?

You can use the subtraction principle to represent the 4 and 9 digits of any number, but no others. This means that the smaller numbers are either I, X, or C, and the numbers you are subtracting from, are no greater than ten times the smaller number you are subtracting.

This essentially comes down to writing a single symbol for each digit in our number system, except if the digit is either a 4 or a 9. Then you must use the proper pair combination to generate that digit with the proper value: 4, 9, 40, 90, 400, 900.

For the most part the Romans did not violate their restrictions. Thus 999 would be thought of as 9 hundreds, 9 tens, and nine ones, and not as 1,000 minus one! This is why you would never see IM for 999 and you'll always see either CMXLIX or CMXLVIIII (the former is the correct way, but the later was frequently used on their buildings, it seems that they too even broke their rules from time to time).

Examples:

Writing Decimal Numbers as Roman Numerals

1. Write the number 974 as a Roman Numeral.

   You notice that we have the digits 9 and 4 in this number, so we will use the subtractive principle for these two numbers.

   We proceed from left to right, starting with the 900 which we write as CM, or 100 subtracted from 1,000.

   The 70 will be represented as a fifty (L) plus two tens (XX) or LXX

The 4 is given as IV.

We then write the number in order from left to right.

The Roman number we are looking for is: CM LXX IV
(the spaces in the numbers are not needed, we just wanted to highlight the procedure). Thus, we would simply write this as: CMLXXIV.

2. Write the number 3,749 as a Roman Numeral.

Again we see a 4 and a 9 in this number, so well will need the subtractive principle for these.

We proceed as described above from left to right. We write 3 one-thousand symbols or MMM.

We will need three symbols to represent 700, the symbol for 500 (D) and two 100 (C) symbols or DCC.

We next write the 40 using the subtractive principle, XL, and

Finally, we write the 9 using the subtractive principle as well, IX.

The Roman Numeral we are looking for is written left to right as: MMMDCCXLIX.

Writing Roman Numerals as Decimal Numbers

3. Write the Roman Numeral MCDXXXII, as a decimal number.

Reading from left to right we see an M which is 1,000,

then we have a C before a D (a smaller before a larger, so we must subtract) which is 400, then three X's which represents 30,

and finally II which is two one's.

Adding everything together we have: 1,000 + 400 + 30 + 2, or 1432.

4. Write the Roman Numeral MMCMXCIX, as a decimal number.

Reading from left to right we see two M's which is 2,000,

then we have a C before an M (a smaller before a larger, so we must subtract) which is 900, then a X before a C (a smaller before a larger, so we must again subtract) to obtain 90,

finally we see an I before an X (a smaller before a larger, so we must again subtract), 9.

Adding everything together we have: 2,000 + 900 + 90 + 9 or 2999.

What we have not shown is just how difficult many of the above numerical systems are when it comes to multiplying or dividing numbers, and even addition or subtraction that requires "borrowing" are not easy. All non–positional number systems make basic arithmetic very awkward. Lucky for us, other civilizations found a better way, and the positional numbers were created.

5.3   Positional Number Systems

Both the Chinese and the Greeks seemed to be struggling with how to write large numbers in a compact, and easy way to remember and work with. The Chinese would have a symbol for groups of ten, hundred and thousand, and combine it with their symbols for unit values to create multiples of these groupings, e.g. 20, 30, 40, … , 90, or 200, 300, 400, … , 900, or 2000, 3000, 4000, … ,9000 etc. The Greeks used an accent mark before a number to indicate you should multiply that number by a thousand. However, neither culture made the connection that was made by the Babylonians centuries earlier.

The Babylonians were the first civilization to create what we now call a positional (place–value) number system. In a positional number system we have symbols to represent numbers just like the non–positional system, but the location of these symbols in the number are now important. The position of the symbol within a number identifies the place–value of that symbol. By that we mean the amount we must multiply the value of our symbol by to compute the value of that place in the number. This important concept needs further review and clarification, so instead of starting with the Babylonian system, we will instead review our own number system, since we are already familiar with it, and we can use this to guide us in understanding other positional number systems.

5.3.1   Hindu–Arabic Number System

The system the world uses today is called the Hindu–Arabic Number System. It came from the Hindu's of India, and was modified and preserved by the Arabs of Mesopotamia before it was brought to Europe by Fibonacci in the 1200's. It is a base–10 positional number system. Even though this came after the Babylonian system, we'll talk about this first, since it is something we are already familiar with. It will help us later, when we look at other positional number systems.

The origin of the Hindu number system in India is widely disputed. Some feel it was developed in India alone, while others feel that it came to India from the Babylonian's by way of Greek mathematics. There was also quite a bit of trading going on between India and China. It is possible that the Indian merchants brought back a Chinese abacus to India sometime in the fifth or sixth century, since shortly after this period the Hindu's dropped their usage of number symbols greater than 9 and adopted a positional number system.

So, did the Chinese inspire the Hindus to create a new number system? Furthermore, the dating of many of the early documents containing the usage of this new number system are often in dispute. We may never know for sure how it came about, unless some historical finds are uncovered to help us unravel the mystery. However it came about, the one thing we can be sure, is that it eventually became the dominant number system throughout the world.

The Indians eventually adopted a decimal (base–10) number system, with nine symbols to represent the first nine counting numbers (one less than the base). Over the years the symbols themselves have changed many times until they became the one's we are all familiar with today. Some examples of the evolution of the western styled symbols are shown below.

The first is the set of early Indian numerals called Brahmi numerals. This form of the numerals dates from about 100 CE.

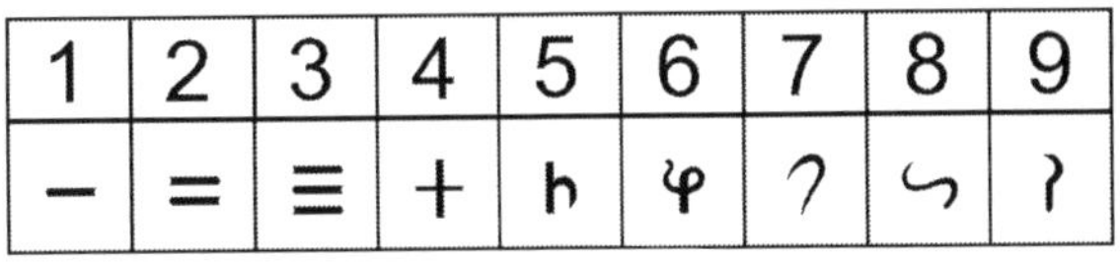

1	2	3	4	5	6	7	8	9
—	=	≡	+	Һ	५	?	ᕋ	?

Below is a set of numerals similar to those used by the Indian mathematician Brahmagupta in around 628 CE.

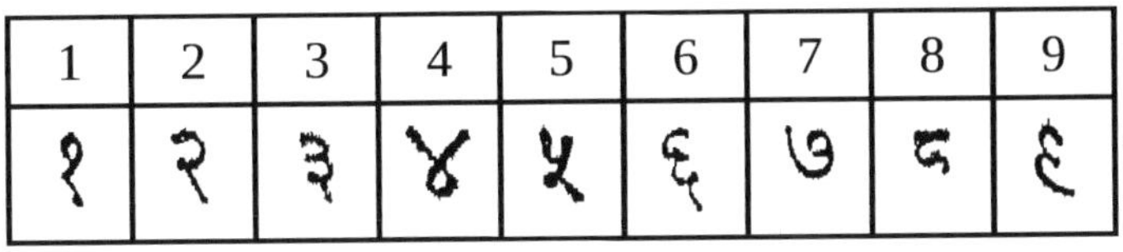

1	2	3	4	5	6	7	8	9
१	२	३	४	५	६	७	८	९

Some 350 years after Brahmagupta and beyond, we see the symbols being transformed as they moved through the Arab nations and on to Europe, and eventually their more familiar form.

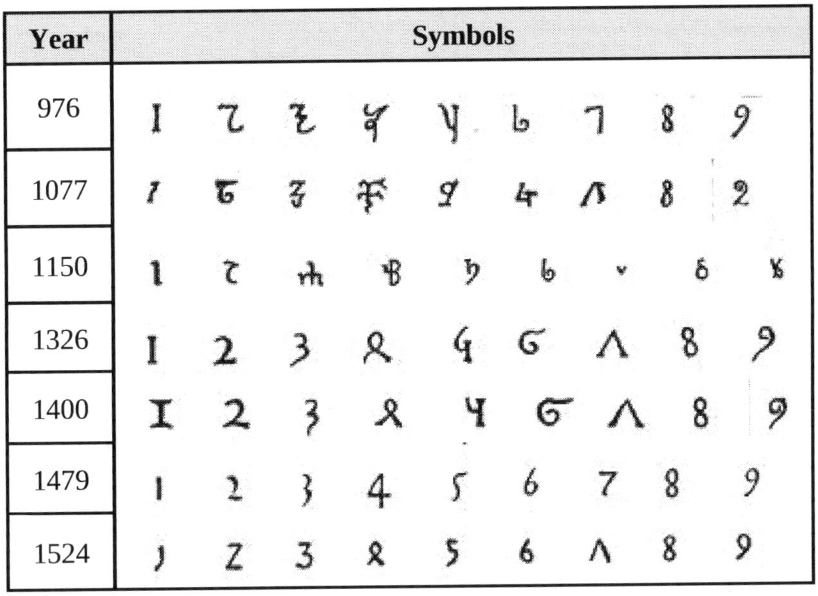

Year	Symbols								
976	I	ᒿ	ᘐ	४	Ꮙ	Ⴆ	ᒣ	8	᧙
1077	ᑺ	Ꮽ	ᘗ	ᙩ	᧙	4	Λ	8	2
1150	1	ᒿ	ᵣᐱ	8	ᑐ	Ⴊ	˅	ᱍ	ᙏ
1326	I	2	3	ᒿ	᧙	Ꮆ	Λ	8	᧙
1400	I	2	᧙	ᒿ	᧙	Ꮆ	Λ	8	᧙
1479	ı	2	᧙	4	ᒣ	6	7	8	9
1524	ı	Z	3	ᒿ	5	6	Λ	8	9

Source, "Development of Arabic Numerals in Europe," George Francis Hill, 1915.

The chart below shows some other forms of the numerals. As you can see they went through many changes and transformations over the years, until they finally settled into the ones we use today.

Source, Jean–Étienne Montucla, Histoire des Mathématiques, 1758
(1798 second edition), Tome 1, Planche XI,

Representing and Interpreting Numbers in the Hindu–Arabic Number System

When we see the number 4578, we see the symbols 4, 5, 7, and 8.

What is not seen, but must be understood, is that the order in which we place the numbers is significant, since each location has a group value associated with it.

What we really have is $4 \times 1000 + 5 \times 100 + 7 \times 10 + 8 \times 1$, or 4 groups of 1000, plus 5 groups of a hundred, plus 7 groups of ten, and 8 one's, but we write it in a compressed form as 4578.

Also, as we read the number, we specifically identify these place–values by saying 4 *thousand*, 5 *hundred*, seventy (7 *tens*), 8 (eight *ones*).

This is what we mean by a positional, or place–value number system. The symbols do not standalone like they do in the Egyptian hieroglyphics, or the Greek and Chinese numerals, they require us to know a place–value to associate with each symbol.

Although we have avoided saying it until now, all number systems are based upon a specific sized grouping to count by. We call this grouping the **base** of our number system. In all the non–positional number systems we discussed previously, the base was 10. Since their counting was "based" upon groups of 10 and multiples of it, such as 10 times 10, or a hundred, and 10 times 10 times 10, or a thousand, etc. Number systems that use this type of grouping scheme are called **base–10** or **decimal** systems. Again, the speculation for a base–10 system has to do with our having 10 fingers. It is more natural for us to count in groups of 10, since we have 10 fingers to count with.

One thing that the Hindu–Arabic number system eventually had, was a symbol for zero. This is important and didn't happen overnight, in fact it took thousands of years to develop. We'll explore this controversy further in Chapter 7. Right now we'll simply state that because of the zero concept we could write numbers such as 100 or 100,000 and see the difference, without a symbol for zero they would both look the same, 1 and 1.

We would give some examples here, mainly to show the procedure, but we assume that all students are familiar with the decimal system, so we will not provide them here. Instead we will use this system to illustrate computational problems in the other bases below.

Before there ever was a decimal place–value number system, there was a sexagesiaml system, i.e. a base–60 number system. This was the first positional numbers system, and we present it in the next section.

5.3.2 Babylonian Number System

As the above discussion would suggest, we don't necessarily have to count in groups of 10. We could use any grouping we want. We could count in groups of 2, 3, 4, 5, 6, etc. With the predominance of the base–10 system it is rather amazing that the first positional number system ever developed was not a base–10 system, but a base–60 system instead. A base–60 system is also called a sexagesimal number system. The civilization that provided the foundation of a base–60 system were the Sumerians who lived in Mesopotamia (now Iraq) (See ancient world map above). The Sumerians developed a non–positional base–60 system which was adopted by the Babylonians and then converted by the Babylonians into a positional number system in and around 2000 BCE.

Nobody knows for sure why the Sumerians counted in groups of 60, but several possible reasons have been given. Reasons such as, 60 is the number that has the most factors, which means it easy to break into many different sized pieces without having a remainder to worry about. Also, that there are approximately 360 days in a year and 6 times 60 is 360, or the use of other body parts in addition to the fingers to count in groups of 60, or to ideas related to weights and measures of the day, as well as geometry. Some have even speculated that it was the joining of two or more civilizations (the Sumerians, Akkadians, and the Babylonians) with different base systems (maybe base–10 and base–6, or base–12 and base–5), that a hybrid base–60 system was created. Perhaps some day we'll discover some historical artifacts that will provide evidence for one of these claims, but for now we can only speculate. History has a major role to play as we seek to find the reasons of why things are as they are.

All that being said, we cannot at this time state with any certainty why they chose base–60, we can only state that they did, and the rest is history. So what is a base–60 system and why and how does it work? First, the base of a positional number system tells us how many symbols we'll need to represent our numbers. In the earliest positional number systems, this was always one less than the base (we'll increase this number by 1 in Chapter 7). Thus, for the base–60 system we'll need 59 symbols to represent our counting numbers from 1 to 59. Just as in the base–10 we have the counting numbers from 1 to 9. Any guesses on what the missing symbol is? Hint: I don't see anything!

The 59 symbols used by the Babylonians for their numbers from 1 to 59 are shown in the table on the next page.

Number	Symbol		Number	Symbol		Number	Symbol
1			21			41	
2			22			42	
3			23			43	
4			24			44	
5			25			45	
6			26			46	
7			27			47	
8			28			48	
9			29			49	
10			30			50	
11			31			51	
12			32			52	
13			33			53	
14			34			54	
15			35			55	
16			36			56	
17			37			57	
18			38			58	

Number	Symbol		Number	Symbol		Number	Symbol
19	𒌋𒌋𒐖		39	𒌍𒐖		59	𒐐𒐖
20	𒌋𒌋		40	𒐏			

Can you see another pattern in these symbols? You might have noticed that contained within the base–60 system, there is also a base–10 system. Notice that there are symbols for 1 thru 9 and then distinct symbols for 10, 20, 30, etc. So even in this base–60 system, there is still a base–10 sub–system!

Now the next question, how do we use these symbols to create a base–60 positional system? The answer is that we must now create place–values. Again, let's use our familiar base–10 system as a guide. In the base–10 system the place–values start at the one's position and then increase as we move to the left by powers of 10, i.e..

BASE–10 PLACE–VALUES				
etc.	Ten–Thousand	Thousand	Hundred	Ten	One
...	10,000	1,000	100	10	1
...	$10,000 = 10 \times 10 \times 10 \times 10 = 10^4$	$1,000 = 10 \times 10 \times 10 = 10^3$	$100 = 10 \times 10 = 10^2$	10^1	1

You start with one and continue to multiply by the base value of 10 to get the next place value. We can follow this same pattern for our base–60 system. However, instead of multiplying by 10 we multiply by 60, i.e.,

BASE–60 PLACE–VALUES				
etc.	Two Hundred and Sixteen Thousand	Three Thousand Six Hundred	Sixty	One
...	216,000	3,600	60	1
...	$216,000 = 60 \times 60 \times 60 = 60^3$	$3,600 = 60 \times 60 = 60^2$	60^1	1

In a base–60 system the first column tells us how many one's we have, the next column to the left how many groups of 60's we have, and the next column how many groups of 3,600's we have, and the next column how many groups of 216,000's we have, and we could continue with this pattern indefinitely. As you can see this system was designed for very large numbers. With only a few symbols you could represent extremely large numbers. It is no wonder that the Babylonians were known for their calculations in Astronomy where very large numbers were often needed.

The Babylonian number system was a very effective number system, except for one major problem. The Babylonian system had no symbol for zero. Thus, a number like 3 would be written exactly the same as 30, or 300, or 3,000 or even 30,000 (actually they would use different numbers in their sexagesimal system than we've given here, but we're using base–10 numbers for convenience in our explanation). This caused the Babylonians difficulties, which they got around by either explicitly stating beforehand in words what the number represented, or by assuming that within the

98

context of the problem the reader would infer or know the correct value. This inevitably could cause a great deal of confusion. Eventually the Babylonians developed a symbol to serve as an empty place–holder, so that you would know if the place–value was empty. This symbol was similar to their symbol for one except that it was rotated to the left slightly, or sometimes two of these symbols together as shown below, were used.

$$\measuredangle \quad \text{or} \quad \cancel{A}$$

This is a very significant event in the development of number systems, and we'll look at it more closely in Chapter 7.

Examples: Reading Babylonian Numbers

Rewrite the following Babylonian numbers as decimal numbers.

1. YY <YYY

The space between the two symbols indicates two separate characters. The first symbol on the left, YY, indicates two one's or 2, but it is in the 60's place-value. The symbol, <YYY, has a ten and three one's, or 13, in the 1's column. Putting this in a table we can easily find the value.

Symbols	YY	<YYY
Symbol Value	2	13
Place Value	60	1
Net Place Value	$2 \times 60 = \mathbf{120}$	$13 \times 1 = \mathbf{13}$
Final Value	$120 + 13 = 133$	

Thus, YY <YYY = 133

2. ≪Y ᵠ ≪ᵡYY

The space between the three symbols indicates three separate characters. The first symbol, on the left, ≪Y, indicates two ten's and one 1 or 21, but it is in the 3,600's place-value. The next symbol, ᵠ, has four one's, or 4, in the 60's column. The last symbol, ≪ᵡYY, has five ten's and two one's, or 52 in the 1's column. Putting this in a table we can easily find the value.

Symbols	⟪Y	▽	⟪YY
Symbol Value	21	4	52
Place Value	3600	60	1
Net Place Value	21×3600 = **756,000**	4×60 = **240**	52×1 = **52**
Final Value	756,000+240+52 = 756,292		

Thus, ⟪Y ▽ ⟪YY = 756,292

3. ⟨ ⟪▦ ⟨▦

The space between the three symbols indicates three separate characters. The first symbol, on the left, ⟨, indicates one ten or 10, but it is in the 3,600's place-value. The next symbol, ⟪▦, three tens' and six one's, or 36, in the 60's column. The last symbol, ⟨▦, has one ten and seven one's, or 17 in the 1's column. Putting this in a table we can easily find the value.

Symbols	⟨	⟪▦	⟨▦
Symbol Value	10	36	17
Place Value	3600	60	1
Net Place Value	10×3600 = **36,000**	36×60 = **2160**	17×1 = **17**
Final Value	36,000+2,160+17 = 38,177		

Thus, ⟨ ⟪▦ ⟨▦ = 38,177

We could start with decimal numbers and find their Babylonian equivalent, but this is a bit complicated, so we won't demonstrate the process. It is a bit easier if the base value is not so large, so we'll demonstrate this process with other numbers systems below..

5.3.3 Mayan Number System

The Mayan civilization flourished around 400 CE. For the most part, the Mayan number system developed independently of the other number systems discussed above. This may explain that while the Mayan number system was also a positional system, it was unique in that it was not a base–10, but a base–20 or a vigesimal system instead. The Mayans counted in groups of 20! One possible reason given for this is that the Mayan's used their fingers and their toes to count on.

The Mayans had a keen interest in calendars and the passing of time. Their rulers were astronomical priests who were highly skilled in creating calendars and using mathematics to make advanced calculations on the position of the planets. Mayan mathematics was quite advanced. In fact, their calendar was even more accurate than the calendar we use today! Today we use the Gregorian Calendar. It approximates the Solar or Tropical year as 365.2425 days. The actual length is closer to 365.24219 days. The Mayan calendar estimates it to be 365.242036 days, which is slightly closer to the actual value than our calendar. This means that in 10,000 years our calendar will be ahead of where it should be by a little over three days, but the Mayan calendar will be behind by only about a day and a half! Not bad for an ancient civilization!

The one very distinct feature about the Mayan number system, other than the base–20, was the fact that the Mayan's were probably the first civilization to create a symbol for zero. This may not sound that important, but we have reserved a whole chapter for this in the book, because conceptually it was. Zero helped to remove a major stumbling block related to the positional number system.

Now the Mayan number system was not a "true" positional number system, because they modified their base–20 system to accommodate their calendar. In a normal base–20 system the first place after the one's position would be reserved for the number of 20's the number had. The next place, or the third symbol in the number, would be the 20×20 or the 400's position, but for the Mayan number system this third place–value was 18×20 or the 360's position. Since 360 was close to the number of days in a year, it is believed that this was why the number system was altered, essentially to make calendar calculations easier. This can cause problems with numbers between 7,600 and 8,000.

The Mayan's, like many cultures, had two different types of written numbers. Symbols they used every day in their basic calculations, and more elaborate symbols used by their priests in religious ceremonies, or on special buildings and other items. The more common and practical set of numerals is given in the first table below. The more elaborate and ornate symbols are shown below the common numerals.

The practical day–to–day system had three symbols: A "bead," a "stick," and an "empty shell."

Number	Symbol	Number	Symbol	Number	Symbol
0		11		22	
1		12		23	
2		13		24	
3		14		25	
4		15		26	
5		16		27	
6		17		28	
7		18		29	

Number	Symbol	Number	Symbol		Number	Symbol
8	●●● (─)	19	●●●● (≡ over lines)		30	● over ═
9	●●●●	20	● over ⬭		360	● over ⬭ over ⬭
10	═	21	● over ●		8000	● over ⬭ over ⬭ over ⬭

Below are examples of the more ornate Mayan numerals used for religious ceremonies. These numerals were typically different images of their gods. This shows the high importance the Mayan's placed upon their numbers, as they associated them with their gods.

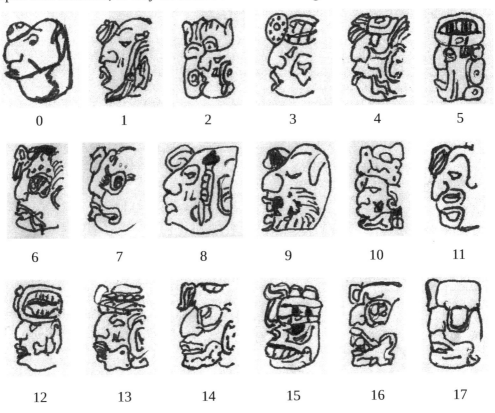

18	19	20

Examples:

Rewrite the following Mayan numbers as decimal numbers.

1.

This Mayan number has three symbols stacked above each other. The top symbol, •, represents a value of one in the 360's place. The next symbol, ••, represents a value of seven in he 20's place, and the last symbol, ⬭, is a zero in the one's location.

Putting this in table form we have:

Symbols	•	••	⬭
Symbol Value	1	7	0
Place Value	360	20	1
Net Place Value	$1 \times 360 = \mathbf{360}$	$7 \times 20 = \mathbf{140}$	$0 \times 1 = \mathbf{0}$
Final Value	$360 + 140 + 0 = 500$		

The answer is 500 in decimal form.

2.

This Mayan number has three symbols stacked above each other. The top symbol, ••••, represents a value of nine in the 360's place. The next symbol, ——, represents a value of five in he 20's place, and the last symbol, ≣, is a nineteen in the one's location.

Putting this in table form we have:

Symbols	●●●● ‾‾‾	‾‾‾‾‾‾	●●●● ≡≡≡
Symbol Value	9	5	19
Place Value	360	20	1
Net Place Value	$9 \times 360 = \mathbf{3{,}240}$	$5 \times 20 = \mathbf{100}$	$19 \times 1 = \mathbf{19}$
Final Value	$3{,}240 + 100 + 19 = 3{,}359$		

The answer is 3,359 in decimal form.

3.

This Mayan number has two symbols stacked above each other. The top symbol, ●●●●, represents a value of four in the 20's place. The last symbol, ≡≡, represents a value of eleven in the 1's location.

Putting this in table form we have:

Symbols	●●●●	≡≡
Symbol Value	4	11
Place Value	20	1
Net Place Value	$4 \times 20 = \mathbf{80}$	$11 \times 1 = \mathbf{11}$
Final Value	$80 + 11 = 91$	

The answer is 91 in decimal form.

5.4   A Common Thread – Or Should We Say Bead

One thing that all the above civilizations had in common, was a device for doing calculations quickly and efficiently. Oftentimes the numbers used for these calculations were different from the numbers they counted with, especially if their number system was non–positional. Their non–positional numbers made arithmetic calculations almost impossible, or highly awkward at best. So, even though their number system may not have handled arithmetic in an effective manner, their calculating device did. They simply translated their numbers into their device, did the calculations and then translated the result back into their number system so that they could write the answer.

This calculating device was called an abacus. The word abacus is derived from the Greek words for a table covered in dust or sand. There were essentially three types of these devices. The first device was a table that was covered in dust or sand along with a stylus (finger or writing stick) to be able to quickly write and calculate with numbers and rewrite. The two other devices used as an abacus, later in history, were the table with loose counters, and the more familiar table or rack with counters fastened to built in lines or wires. Be careful if you search the internet for the history of the abacus as the vast majority of what you will find is unsubstantiated, and probably wrong.

We cannot be sure who or even which culture developed the abacus, because the history is very spotty and unclear on this. Some people say that it was originally developed in China in 3000 BCE, but there is no written record of its existence in China before 1175 BCE. Others claim that the likely place is in Mesopotamia (Sumeria/Babylon, modern day Iraq) because of the sophisticated number system they developed. We will side with the fairly recent conclusions of Georges Ifrah, who cites writings in the Sumerian language that seems to establish that an abacus of some form was in use in Mesopotamia in and around 2700–2300 BCE. This is the earliest reference to an abacus that can be found. Thus, until other possible more definitive discoveries in the future, this seems to be the most likely beginning of the abacus.

Who or how the abacus developed is not as important as the impact its development had on mathematics. It was a very important invention that spread throughout the world. The users of the abacus were typically merchants and businessmen that needed a fast, accurate, and effective way to do money and trade calculations. They even created pocket versions of the abacus that could be carried with them to do quick on the spot calculations, much like the modern day calculator. It was an indispensable business tool, and as merchants traveled the world, so too traveled the abacus.

It seems probable that as a new culture was exposed to the abacus, by a visiting merchant, they might purchase a copy from the merchant, receive some instructions on its use, and then began to make their own version. They may have kept certain parts of the other merchants number system if they found it easier to do a calculation on the abacus with it. We cannot underestimate the impact of technology in joining and assimilating different cultures. The abacus was the equivalent of the modern electronic calculator or computer in its day. Not only did it help to carry out commerce and trade deals, but it also helped to connect the mathematics of one culture to another. It was probably more instrumental in spreading mathematics and standardizing our current number system than almost anything else.

The Egyptians, the Greeks, the Indians, the Chinese, and the Romans, all used a form of the abacus extensively in their businesses and trade. Many of these cultures added something to modify and improve upon the design of the abacus. The Chinese created the first abacus with beads on a string called the *saupan*, which eventually became the modern standard. From China the abacus moved on to Japan and Korea, and became ingrained in the Asian culture, even to this day.

We cannot overstate that one of the most important outcomes of the development of the abacus was not the assistance it gave to the business calculations of merchants, but rather its utility in spreading a common number system throughout the world, and providing a common link to the mathematics of the day that helped to foster further advances in mathematical thinking.

 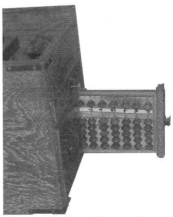

Examples of the Chinese abacus: A Standalone Abacus, and A Pull–Out Abacus Insert in Another Device

Summary

In this chapter we have explored the many names and representations of the things we call the natural or counting numbers. Obviously, the names and symbols are human creations, but what they represent is another story. Does a number, regardless of how it is written have some type of existence beyond humans? Quantities of objects obviously do exist without the presence of humans to see them, but do their numerical values have some sort of cosmic significance beyond us?

Regardless of which number system they are written, the numbers 3, 5, 7, 11, 13, …, are all prime numbers (numbers divisible by one and themselves only). Numbers such as 2, 4, 6, 8, … are even numbers, and numbers like 1, 3, 6, 7, 9, 11, … are odd numbers, regardless of how we might choose to write them. Historically many different civilizations created numerous representations for these numbers, but their properties are universal. They are the same whether in Egyptian Hieroglyphics, Babylonian numerals, Chinese numbers, or our own Hindu–Arabic numbers. Are we seeing some hidden secrets of nature, or have we created something that only exists because we created it?

The type of number system we have is definitely a human creation. This does not automatically mean that the concept of a numerical quantity is a human creation as well. Thus, it does not necessarily follow, that Math is created, it could be discovered. Perhaps humans just create the means for talking about the concepts, but not necessarily create the concepts themselves (or do we?).

EXERCISES 5.1–5.4

ESSAYS AND DISCUSSION QUESTIONS:

1. Describe the difference between a positional and a non–positional number system.

2. Which civilization created the first positional number system? Also, describe that system, and its positive and negative features.

3. What were some of the unique features of the Mayan number system?

4. What impact if any did the abacus have on the advancement of a common number system? Support your answer.

5. What is the difference between a number concept and a number system? What, if anything, does this say about whether the number concept was either created or discovered?

PROBLEMS:

5.2 Non–Positional Number Systems

5.2.1 Egyptian Number System

Translate from Hindu–Arabic to Egyptian Hieroglyphics

1. 230
2. 1,349
3. 23,345
4. 1,000,111
5. 3,001,034
6. 367
7. 53,221
8. 643
9. 4,621
10. 103,273
11. 33,211
12. 99

Translate from Egyptian Hieroglyphs to Hindu–Arabic

13.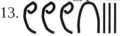

14.

15.

16.

17.

18.

19.

20.

21.

22.

23.

24.

Add the Egyptian Hieroglyphs

25. and

26. and

27. and

28. and

29. and

30. and

31. and

32. and

33. and

34. and

35. and

36. and

Subtract the Egyptian Hieroglyphs

37. from

38. from

39. from

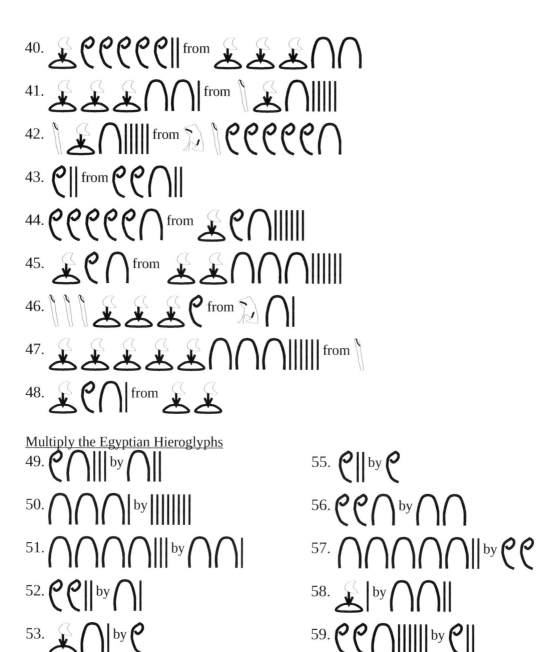

40. from

41. from

42. from

43. from

44. from

45. from

46. from

47. from

48. from

Multiply the Egyptian Hieroglyphs

49. by

50. by

51. by

52. by

53. by

54. by

55. by

56. by

57. by

58. by

59. by

60. by

Show How You Would Divide the Equivalent Egyptian Hieroglyphs Using Hindu–Arabic Numerals

61. $364 \div 4$

62. $128 \div 32$

63. $90 \div 15$

64. $273 \div 13$

65. $28 \div 7$

66. $858 \div 78$

67. $962 \div 74$

68. $767 \div 59$

69. $2,675 \div 107$

70. $520 \div 26$ 71. $1,204 \div 172$ 72. $465 \div 31$

5.2.2 Chinese Number System

Translate from Hindu–Arabic to Chinese Numerals

1. 1,349
2. 4,621
3. 99
4. 3,001
5. 367
6. 5,221
7. 643
8. 9,345
9. 273
10. 821
11. 1,111
12. 3,013

Translate from Chinese Numerals to Hindu–Arabic

13.

14.

15.

16.

17.

18.

19.

20.

21.

22.

23.

24.

5.2.3 Greek Numerals

Translate from Hindu–Arabic to Greek Numerals

1. 1,349
2. 4,621
3. 99
4. 3,001
5. 367
6. 5,221
7. 643
8. 9,345
9. 273
10. 821
11. 1,111
12. 3,013

<u>Translate from Greek Numerals to Hindu–Arabic</u>

13. π γ'

14. υ ο δ'

15. ,η φ ξ α'

16. ω η'

17. σ κ β'

18. χ ν'

19. ψ π θ'

20. λ ε'

21. τ ν ε'

22. ,α ω γ

23. ,Ϝ φ π ζ

24. ⟍ ξ ζ'

5.2.4 Roman Numerals

<u>Translate from Hindu–Arabic to Roman Numerals</u>

1. 1,349

2. 4,621

3. 99

4. 3,001

5. 367

6. 5,221

7. 643

8. 9,345

9. 273

10. 821

11. 1,111

12. 3,013

<u>Translate from Roman Numerals to Hindu–Arabic</u>

13. MCMIX

14. LIV

15. CDXVII

16. CCCXXXIII

17. MMXI

18. DXLIII

19. MMMCDXLIV

20. CCLXXVII

21. CMXXXVIII

22. LXXIX

23. DCCCXCIV

24. CCI

5.3 Positional Numbers

5.3.2 Babylonian Number System

<u>Translate from Babylonian Numerals to Hindu–Arabic</u>

1.

2.

3.

4.

5.

6.

7.

8.

9.

10.    11.    12.

5.3.3 Mayan Numerals

<u>Translate from Mayan Numerals to Hindu–Arabic</u>

1.

5.

9.

2.

6.

10.

3.

7.

11.

4.

8.

12.

References

1. Ifrah, G., (1985). (Originally published 1981).: *From One to Zero: A Universal History of Numbers*, R. R. Donnelley & Sons Company.

2. Ifrah, G., (2000). (Originally published 1994).: *The Universal History of Numbers: From Prehistory to the Invention of the Computer*, John Wiley & Sons, Inc.

3. Ifrah, G., (2001). (Originally published 1994).: *The Universal History of Computing: From the Abacus to the Quantum Computer*, John Wiley & Sons, Inc.

4. Menninger, K., (1992). (Originally published 1969).: *Number Words and Number Symbols: A Cultural History of Numbers*, Dover.

5. Smith, D.E., (1958). (Originally published 1923): *History of Mathematics, Volume I*, Dover.

6. Smith, D.E., (1958). (Originally published 1928): *History of Mathematics, Volume II*, Dover.

CHAPTER 6

Fractions & Decimals

A man is like a fraction whose numerator is what he is and whose denominator is what he thinks of himself. The larger the denominator, the smaller the fraction.

Leo Tolstoy

Decimals have a point.

Anonymous

6.1   Introduction

To the early Egyptians and the ancient Greeks all we really needed were the natural counting numbers. Whenever the concept of a part of a whole was needed, it was never thought of as a number, but as a ratio of natural numbers. Do we really need more numbers?

Where did fractions come from anyway, and what place do they occupy in our number system? Fractions came about from our need to consider parts of a whole. If you broke a whole into four equal parts and gave one part away, this would be one out of four parts. What remained was three out of four parts. To represent this quantity in an understandable way, early civilizations would talk about 1 in 4 or more succinctly as the ratio of 1 and 4, or 1/4. However, this was not thought of as a single number, but instead as a way to represent a part of a whole, using two counting numbers. To these early civilizations the only true numbers were the basic counting or natural numbers. It took a few thousand years before they became numbers in their own right.

6.2   Egyptian Fractions

To the early Egyptians a fraction was not a number, but rather a process of breaking a whole into parts that they could quantify using numbers. For most fractional quantities they used what we today call unit fractions. Using modern terminology, these are fractions where the numerator has a one only. (Now this is not totally accurate, for the Egyptians did have special symbols to represent some very common fractions, such as 1/2, 2/3 and 3/4, as we'll show below, but for the most part they used unit fractions.)

6.2.1 Writing Unit Fractions

To identify symbols as fractions the Egyptians would use a: , or simply , over the number. This fraction or part symbol was the symbol for "mouth" that in this context meant "part."

114

They would simply place this "part" symbol over a number and this meant, 1 in whatever number many parts were specified below the fraction symbol. We illustrate this with examples.

Examples:

Write the unit fraction in Egyptian Hieroglyph form:

1. $\dfrac{1}{3}$ becomes  or

2. $\dfrac{1}{10}$ becomes  or

3. $\dfrac{1}{100}$ becomes  or

6.2.2 Special Fraction Hieroglyphs

Sometimes the Egyptians would use unique symbols for some of the more common fractions, even if they were not unit fractions. The symbols for 1/2, 2/3, and 3/4, are shown below:

$\dfrac{1}{2}$ = , $\dfrac{2}{3}$ =  , and $\dfrac{3}{4}$ =

6.2.3 Writing Non-Unit Fractions

If the Egyptians wished to represent the fractions that were not unit fractions, such as 3/5, they could simply write the unit fraction for 1/5 three times, and since their non–positional system was additive, you just had to add the fractions together to find the total.

$\dfrac{3}{5}$ =

Now for some fractions, the string of symbols could become quite long. For example consider the fraction 7/10.

$\dfrac{7}{10}$ =

To reduce the needed writing, the Egyptians would sometimes try to find a smaller number of unit fractions, or special symbols, which together equaled an equivalent amount. They would frequently

try to find the fewest number of symbols needed to represent the fraction in question. This is quite challenging as we'll see below.

Thus, instead of writing seven one–tenth symbols for 7/10 as above, they instead might write:

$$\frac{7}{10} = \text{}$$

This is because we can think of 7/10 as follows: $\frac{7}{10} = \frac{5}{10} + \frac{2}{10} = \frac{1}{2} + \frac{1}{5}$. This form only requires the symbol for 1/2 and the symbol for 1/5 to represent its value, since when added together give the desired amount, 7/10.

As you can see from this simple example it could get quite complicated, and the answer may not be unique. We illustrate this through examples.

Examples:

1. Rewrite the fraction 23/24 in Egyptian Hieroglyphs.

    You would not want to write 1/24, 23 times, so we need to try and express the fraction in as few symbols as possible? In what follows we show three different ways to write 23/24 using each of the three special Egyptian hieroglyphs mentioned above:

    a. Here is one combination using the special fraction3/4:

    $$\frac{23}{24} = \frac{3}{4} + \frac{1}{6} + \frac{1}{24} = \text{}$$

    This works since 4 (the denominator of 3/4) is a factor of $6(4) = 24$.

    To form the special and unit fractions we multiply the 3 in 3/4 by the other factor in 24 above, or 6, to get 18, and rewrite the numerator term, 23, as the sum of 18 and 5.

    $$23 = 18+5$$

    Now break 5 down into the sum of factors of 24 that are smaller than 5 (4 and 1).

    $$23 = 18+4+1$$

    Using this we can rewrite 23/24 as:

$$\frac{23}{24} = \frac{18+4+1}{24} = \frac{18}{24} + \frac{4}{24} + \frac{1}{24}$$
$$= \frac{3(6)}{4(6)} + \frac{1(4)}{6(4)} + \frac{1}{24}$$
$$= \frac{3}{4} + \frac{1}{6} + \frac{1}{24}$$

The answer we provided above.

b. Another possible combination is:

$$\frac{23}{24} = \frac{2}{3} + \frac{1}{4} + \frac{1}{24} =$$

This works since three is also a factor of 24, $3(8) = 24$, and we can follow the same procedure from above.

To form the special and unit fractions we multiply the 2 in 2/3 by the other factor in 24 above, or 8, to get 16, and rewrite the numerator term, 23, as the sum of 16 and 7.

$$23 = 16+7$$

Now break 7 down into the sum of factors of 24 that are smaller than 8 (6, 4 and 1).

$$23 = 18+6+1$$

Using this we can rewrite 23/24 as:

$$\frac{23}{24} = \frac{16+6+1}{24} = \frac{16}{24} + \frac{6}{24} + \frac{1}{24}$$
$$= \frac{2(8)}{3(8)} + \frac{1(6)}{4(6)} + \frac{1}{24}$$
$$= \frac{2}{3} + \frac{1}{4} + \frac{1}{24}$$

The answer we provided above.

c. Another possibility uses 1/2.

$$\frac{23}{24} = \frac{1}{2} + \frac{1}{3} + \frac{1}{8} =$$

Can you see how we obtained it?

Can you think of other solutions? Perhaps some with four or more unit fractions. Perhaps some without the special hieroglyphs. Although it can be challenging, it can also be quite entertaining looking for other possible combinations.

2. Rewrite the fraction 9/16 in hieroglyphs.

First, look at the denominator and see which denominators of the special characters for, 1/2, 2/3, and 3/4, (2, 3, or 4), evenly divides the numerator of the fraction you are trying to rewrite. In this case both 2 and 4 evenly divide 16, but 3/4 is greater than 9/16, since

$$\frac{3}{4} = \frac{3}{4}\frac{4}{4} = \frac{(3)(4)}{(4)(4)} = \frac{12}{16},$$

so it is too large. We instead work with 2.

Two goes into sixteen eight times, so as in Example 1 above, we can rewrite 9/16 as:

$$\frac{9}{16} = \frac{8+1}{16} = \frac{8}{16} + \frac{1}{16} = \frac{1}{2} + \frac{1}{16}$$

This means we can write this using hieroglyphs as:

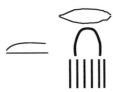

Four also goes into sixteen four times, so we can also rewrite 9/16 as:

$$\frac{9}{16} = \frac{4+4+1}{16} = \frac{4}{16} + \frac{4}{16} + \frac{1}{16} = \frac{1}{4} + \frac{1}{4} + \frac{1}{16}$$

This means we can write this using hieroglyphs as:

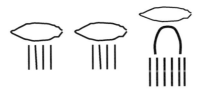

We could divide 8 into 16 as well and rewrite 9/16 as:

$$\frac{9}{16} = \frac{2+2+2+2+1}{16} = \frac{2}{16} + \frac{2}{16} + \frac{2}{16} + \frac{2}{16} + \frac{1}{16} = \frac{1}{8} + \frac{1}{8} + \frac{1}{8} + \frac{1}{8} + \frac{1}{16}$$

This means we could also write it using five unit fraction hieroglyphs if we wanted to, but will not do so here. The first solution has fewer hieroglyphs, so it would probably be more desirable, although there is nothing wrong with the other solutions.

Examples 1 and 2 above also establish a process that can be used.

a. First, look for which denominators of the special hieroglyphs go into the denominator of the fraction you are trying to rewrite.
b. Rewrite the numerator of the fraction in terms of a sum of the factors.
c. Simplify and write the hieroglyphs for the fraction.

d. If none of the special hieroglyphs denominators are a factor of the denominator of the fraction you are trying to rewrite, factor the denominator and rewrite the numerator in terms of a sum of combinations of these factors. Look for the fewest number of terms in the sum.
e. Use the fewest number of terms to rewrite the numerator, and reduce to unit fractions.
f. Write the hieroglyphs for the unit fractions.

We illustrate the second part in a final example (starting at step d. above).

3. Rewrite the fraction 24/35 in hieroglyphs.

Factor the denominator, $35 = (5)(7)(1)$. Rewrite 24 in terms of a sum of 5's, 7's, and 1's. You will have many possible combinations, you are looking for the fewest number of terms you can find.

$$24 = 5+5+5+5+1+1+1+1$$
$$24 = 7+7+7+1+1+1$$
$$24 = 5+5+5+7+1+1$$
$$24 = 5+5+7+7$$

The fewest number of terms is the last set, two fives and two sevens.

Rewrite the numerator using this combination of 5's and 7's.

$$\frac{24}{35} = \frac{5+5+7+7}{35} = \frac{5}{35} + \frac{5}{35} + \frac{7}{35} + \frac{7}{35} = \frac{1}{7} + \frac{1}{7} + \frac{1}{5} + \frac{1}{5}$$

Write the four unit hieroglyphs.

6.3   Greek Fractions

The Greeks primarily followed the Egyptians and used unit fractions, although they did not follow this as closely as the Egyptians did.

As we learned above, the Greeks would use an accent mark to distinguish a number from a word. To distinguish a natural or counting number from a fraction, they would use two accent marks.

Thus, they would write

$$\frac{1}{3} = \gamma''$$

The γ represents 3 and the double accent marks in the upper right means that γ is in the "denominator" and that this is the unit fraction of 1 over 3.

Any character string with a double accent mark was considered the "denominator" (using modern fraction terminology) term.

However, if they wanted to represent a fraction that was not a unit fraction, they would simply write the "numerator" term first, using a single accent mark identifying it as a number, and then they would write the "denominator" with a double accent mark right after it.

For example, to write the fraction 23/24, they would write:

$$\frac{23}{24} = \iota\gamma' \; \iota\delta''$$

where the $\iota\gamma'$ is the number 23 (which is the numerator), and the $\iota\delta''$ is the denominator of 24.

While it is relatively easy to write down a fraction using the Greek number system, it is extremely difficult to add, subtract, multiply or divide fractions using this number system. For this the Greeks would often resort to using an abacus that was developed by the Babylonians for these type of calculations.

6.4   Roman Fractions

Although the Romans used a decimal system for their whole numbers, for fractions the Romans used a base–12 or duodecimal system. Their system for fractions most probably came from their monetary as well as their weights and measure systems, where it was more common to compute 1/3 or 1/4 of a quantity, and this was much easier to do if your base unit allowed these to be computed easily. Since 12 was evenly divisible by 3 and 4, this served as a good base to use. We see this often where mathematics was used in two different ways with two different approaches in the same culture. We often have the practical day to day (business) math, as well as a "higher math" for the priests and scholars.

The base unit for Roman fractions was 1/12 which was called an *uncia*. The symbol they used was a simple dot •. Since they had an additive, non–positional number system, if they wanted to write 2/12 = 1/6, they would simply write two dots. The table below shows this and other values along with the associated symbols and names. They did have a special symbol for 6/12 = 1/2, and this was S (for semi, where we get the Latin word for half from), as well as some others shown below.

Fraction	Roman Numeral	Name
1/12	•	uncia
2/12 = 1/6	• •	sextans
3/12 = 1/4	• • •	quadrans
4/12 = 1/3	• • • •	triens
5/12	• • • • •	quincunx
6/12 = 1/2	S	semis
7/12	S •	septunx
8/12 = 2/3	S • •	bes
9/12 = 3/4	S • • •	dodrans
10/12 = 5/6	S • • • •	dextans
11/12	S • • • • •	deunx
12/12 = 1	I	as
Other Fractions		
1/24	Σ, or Ƨ, or Є, or Ɫ	semuncia
1/48	Ɔ	sicilicus
1/72	Ƨ	sextula
1/144	ƻ	dimidia
1/8	Σ•	sescunia
1/36	ƧƧ	duella

6.5 Babylonian Fractions

The Babylonian number system was different. Instead of inventing new symbols to represent their fractions, they used the same symbols they used for their whole numbers! You simply needed to know the position of the symbol to identify its place–value. For example, we already know that the symbol 𒁹 could represent a one, or a sixty, or a three–thousand–six–hundred or larger.

We can also follow the same pattern for fractions, but now we go in the opposite direction. Thus, it could also represent one sixtieth, or one thirty–six–hundreth, etc. Now instead of multiplying by increasing powers of 60 as we did in Chapter 5, we are dividing by increasing powers of 60. The place–values now range as follows:

PLACE–VALUES							
...	216,000	3,600	60	1	1/60	1/3600	1/216,000	...
...	$60 \times 60 \times 60 = 60^3$	$60 \times 60 = 60^2$	60^1	1	$\dfrac{1}{60} = 60^{-1}$	$\dfrac{1}{60 \times 60} = 60^{-2}$	$\dfrac{1}{60 \times 60 \times 60} = 60^{-3}$	...

To the left of one we multiply by factors of 60 and to the right of one we divide by multiples of 60.

The problem with the Babylonian notation is that they did not have what we would call a decimal point today, or more accurately what should be called a sexagesimal point for their number system. Thus, you would have to know from the problem being presented, where the whole number part ended and where the fractional part began.

As an example, consider the clay tablets below from around 1700 BCE. This is believed to be a students' homework assignment. Notice a rotated square with its diagonals is drawn on the artifact. In addition, numbers are placed along the diagonal and on one side of the square. We highlight the square and the decimal representation of the numbers in the second figure next to it.

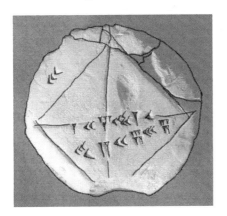 

Original image courtesy of Bill Casselman and the Yale Babylonian Collection. You can see all the original images at http://www.math.ubc.ca/~cass/euclid/ybc/ybc.html

After some analysis it was determined that the number 30 on the tablet is the length of the sides of the square and it should be interpreted as a whole number, and the numbers 1, 24, 51, and 10 are the Babylonian numbers that represent,

$$1 + \frac{24}{60} + \frac{51}{3600} + \frac{10}{216,000} = 1.41421296\overline{296}$$

Does this number look familiar?

In this number there is no indication that the first number, 1, is a whole number and that the numbers 24, 51, and 10, are to be interpreted as the fractional part of the number. This had to be figured out by the finders of the piece, just as the original readers would have done. It was expected to be known within the context of the problem. In this case the length of the diagonal of the square.

Now if we multiply the Babylonian approximation by 30 we obtain $42.426388\bar{8}$, which happens to be the approximate length of the diagonal written in our decimal fraction form.

The lower row of numbers 42, 25, and 35 on the tablet represent the number,

$$42 + \frac{25}{60} + \frac{35}{3600} = 42.426388\bar{8},$$

the approximate length of the diagonal (again, you have to know which is the whole and which is the fractional part of the number).

Now the actual value for $\sqrt{2}$ is given as: $\sqrt{2} = 1.414213562731...$ and $30\sqrt{2}$ is given as: $30\sqrt{2} = 42.4264068711929...$

Comparing the value to the Babylonian approximations we can see just how accurate they were. Unbelievable accuracy considering this was calculated over 4,500 years ago! Also, it is believed that this particular tablet was not written by an expert scribe and may have actually been a completed homework problem of a student at the time!

As you can see, the Babylonians didn't really have fractions such as 2/3 or 19/44, they instead used their place–value system to represent parts of a whole, much like our decimal fractions today (for additional details on decimal fractions see section 6.7 below). To represent the more common fractions the Babylonians would have look–up tables with exact, or if needed, approximate values for the fractions. This was especially true if the number did not divide evenly into 60 (such as 7 or 11). This is one reason why 60 was so useful for them. There are many numbers that divide into 60 without leaving a remainder. Can you name some of them?

Here are some examples of fractions and their base–60 representations and approximations that might have been used by the Babylonians. A word of caution though, it would have to be understood from the problem that these are all fractions, since they had no other way to explicitly identify this.

Fraction With its Decimal Equivalent	Babylonian Number
1/2 = 0.5	30/60=0.5 (exact)
1/3 = 0.333 ...	20/60 =0.333 … (exact, and does terminate)
1/4 = 0.25	15/60=0.25 (exact)

Fraction With its Decimal Equivalent	Babylonian Number
1/5 = 0.2	6/60 = 0.2 (exact)
1/6 = 0.1666 ...	10/60 = 0.1666…(exact, and does terminate)
1/8 = 0.125	7/60+30/3600 = 0.125 (exact)
2/3 = 0.666 ...	40/60=0.666...(exact, and does terminate)
1/7 = 0.1428571429 ...	8/60+34/3600+17/216,000=0.1428564815…
1/11 = 0.090909 ...	5/60+27/3600+16/216,000=0.0909074074...
1/59 = 0.016949152 ...	1/60+1/3600+1/216,000 = 0.0169490740 ...
1/51 = 0.019607843 ...	1/60+10/3600+36/216,000 = 0.019611 ...

The Babylonians didn't do arithmetic with fractions, they instead translated them into their number system and added, or subtracted much as we do with our decimal fractions today. Most students today would love to avoid fractions altogether, just like the Babylonians did!

Notice that in the Babylonian system the first seven fractions in the table above can be written exactly. However, when using our familiar decimal fractions we have problems with numbers such as 1/3, 1/6, and 2/3. All these numbers require us to write an unending pattern of numbers. The Babylonians also noticed this. They saw that some fractions, like 1/7 and 1/11, were different, since they did not divide into 60 evenly. Their number form gave them non–terminating numbers (just like we see in our own number system). That meant that the representation of these numbers would go on forever and could never be completed! This bothered the Babylonians and they were not quite sure how to interpret this strange phenomenon. It is the case of a new concept that is beginning to make its way into our number world, and we'll study this in greater detail in Chapter 11 below. It is the concept of the unending or the infinite.

6.6   Hindu–Arabic Fractions

The Hindu–Arabic number system is the number system we grew up learning. It is a base–10 positional number system. It is similar to the Babylonian system above. The only difference is that we use powers of ten instead of sixty. The use of dividing by powers of ten to represent the fractional part of a number, however, did not develop until the Arabs did so, centuries after the system was first introduced. It was originally not as sophisticated as the Babylonian system was in its day. Fractions were instead represented as ratios of numbers, and not simply as numerals in locations specifying the powers of ten you were dividing by.

The Indian mathematicians Brahmagupta (628 CE), and Bhaskara (1150 CE), wrote fractions much as we do today, but without the horizontal bar. This was introduced later by the Arabs around 1200 CE. We will not review the process of working with the fractions we were taught as children. We instead assume that you have a basic knowledge of the arithmetic of fractions. We realize that most people still have a great deal of difficulties working with fractions, but it is not the aim of this text to address that shortcoming. We instead focus more on the concept and idea.

The idea of integrating fractions into the positional number system did not occur until the 16th century, over 4000 years after the Babylonians did so. They were called decimal fractions, or decimals for short.

6.7   Decimals (Decimal Fractions)

In any number system doing arithmetic with fractions can be quite difficult, often requiring many rules and steps in order to find the answer. The Babylonians in 2000 BCE had the best system for doing calculations that involved fractions of a whole, but this was a base–60 or what is also called a sexagesimal number system we presented above in section 6.5. Recall that their system used whole parts of sixtieths (1/60), or three–thousand–six–hundreths (1/3600), etc. This allowed them to do calculations without fractions. In fact, as we saw above they even used it to estimate the length of $\sqrt{2}$ in and around 1,700 BCE.

When the world moved over to using the Hindu–Arabic decimal number system, the efficient system of the Babylonians was forgotten, and it wasn't until 1585 that a Flemish mathematician/engineer, Simon Stevin wrote a 36 page essay (booklet), titled De Thiende (*The Tenth*, and an English translation written in 1608 titled *DISME: The Art of Tenths*), creating decimal numbers, or as he might have said: *A process on how to perform all arithmetic using only whole numbers, without the need for fractions.* Actually 55 years prior to Stevin, Christoff Rudolff wrote a book on decimals titled *Exempel–Buchlin* in Augsburg Germany, but it never really caught on. Even though Rudolff was the original inventor of decimal fractions, it was Stevin's book that started the decimal–fractions transformation.

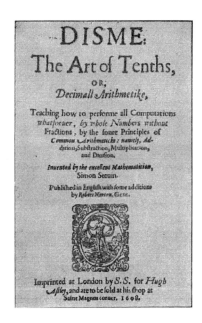

We should note that Stevin's original notation was very awkward by today's standard. We illustrate this with some examples below.

Example 1:

To write the decimal $\qquad$ 0.379,

Stevin's would have written $\qquad$ $3(^1)7(^2)9(^3)$

The superscript in parentheses identifies the power of 10 we divide by.

For example:

$$3(^1) \text{ means } 3 \times \frac{1}{10^1} = \frac{3}{10}$$

$$7(^2) \text{ means } 7 \times \frac{1}{10^2} = \frac{7}{100}$$

$$9(^3) \text{ means } 9 \times \frac{1}{10^3} = \frac{9}{1000}$$

Example 2:

The number 6.35 would be written $6(^0)3(^1)5(^2)$.

$6(^0)$ means $6 \times \frac{1}{10^0} = \frac{6}{1} = 6$, and $3(^1)$ means $3 \times \frac{1}{10} = \frac{3}{10}$, and $5(^2)$ means $5 \times \frac{1}{10^2} = \frac{5}{100}$

The notation was later refined into the form we use today by other mathematicians, most notably John Napier.

Christoff Rudolff's notation 55 years earlier was a bit easier to follow. Rudolff would have written 0.379 as |379, and 6.35 as 6|35. Instead of the decimal point he used a vertical line. This made arithmetic quite a bit easier to carry out.

With the invention of the decimal fractions, adding, subtracting, multiplying, and dividing fractions became a much easier task to perform. It seems that in the sixteenth century the west finally caught up with the Babylonians who lived 4,500 earlier!

EXERCISES 6.1–6.7

ESSAYS AND DISCUSSION QUESTIONS:

1. What are unit fractions and which civilizations used them?
2. Did the Greeks think of fractions as numbers? Explain your answer.
3. Did the Babylonians think of fractions as numbers? Explain your answer.
4. When did the Hindu–Arabic system of numbers introduce decimal fractions? Why did it take so long?
5. Which civilization first thought of fractions as numbers? Also, provide a possible reason why they did so.
6. Why did it take a long time for some cultures to think of fractions as numbers?

PROBLEMS:

6.2 Egyptian Fractions

6.2.1 Unit Fractions
Write the following as Egyptian unit fractions.

1. 1/5	4. 1/40	7. 1/267	10. 1/2
2. 1/13	5. 1/8	8. 1/75	11. 1/2,300
3. 1/120	6. 1/1,000	9. 1/32	12. 1/26

Write the following Egyptian unit fractions as Hindu–Arabic fractions.

13.

16.

19.

21.

14.

17.

15.

18.

20.

22.

23.  24.

6.2.3 <u>Non–Unit Fractions</u>
Write the following Hindu–Arabic fractions as Egyptian unit fractions (Note: you can have more than one correct answer).

1. 2/3	5. 3/4	9. 5/6	13. 6/11
2. 19/24	6. 13/16	10. 4/7	14. 29/36
3. 17/32	7. 9/14	11. 11/20	15. 11/15
4. 33/50	8. 7/12	12. 11/12	16. 7/15

6.3 Greek Fractions

Write the following as Greek unit fractions.

1. 1/5	4. 1/40	7. 1/267	10. 1/2
2. 1/13	5. 1/8	8. 1/75	11. 1/923
3. 1/120	6. 1/546	9. 1/32	12. 1/26

Write the following as Greek fractions.

13. 2/3	17. 3/4	21. 5/6	25. 6/11
14. 19/24	18. 13/16	22. 4/7	26. 29/36
15. 17/32	19. 9/14	23. 11/20	27. 11/15
16. 33/50	20. 7/12	24. 11/12	28. 7/15

6.4 Roman Fractions

Write the following as Roman unit fractions.

1. 1/6	4. 1/4	7. 1/144	10. 1/2
2. 1/8	5. 1/12	8. 1/72	11. 1/48
3. 1/36	6. 1/3	9. 1/24	

6.5 Babylonian Fractions

Write the equivalent Hindu–Arabic fraction. Assume that all symbols represent values to the right of the units' column, i.e., they are smaller than one. (Use a calculator to compute the decimal approximation).

1.

2.

3.

4.

5.

6.

7.

8.

9.

10.

11.

12.

References

1. Aaboe, A., (1964). Episodes from the Early History of Mathematics.

2. Adkins, L., Adkins, R., (2000). The Keys of Egypt: The Obsession to Decipher Egyptian Hieroglyphs. HarperCollins Publishers.

3. Allen, J.P. (1999). Middle Egyptian: An Introduction to the Language and Culture of Hieroglyphs. Cambridge University Press.

4. Blackwell, (1987). Language and Number. Blackwell, Oxford.

5. Blohm, H., Beer, S., and Suzuki, D., (1986). Pebbles to computers : the thread. Oxford University Press, Toronto.

6. Cajori, F., (1928). A history of mathematical notations. Two volumes. Chelsea, New York.

7. Calinger, R., (1999). A conceptual history of mathematics, Upper Straddle River, N. J.

8. Clawson, C.C., (1994). The mathematical traveler : exploring the grand history of numbers. Plenum Press, New York.

9. Closs, M.F. (editor), (1986). Native American mathematics. University of Texas Press, Austin.

10. Collier, Mark & Bill Manley (1998). How to Read Egyptian Hieroglyphs: a step-by-step guide to teach yourself. British Museum Press.

11. Conant, L.L., (1923). The Number Concept, its Origin and Development. Macmillan, New York.

12. Crump, T., (1990). The Anthropology of Numbers. Cambridge University Press, Cambridge.

13. Dantzig, T., (1930). Number: The Language of Science. Macmillan, New York.

14. De Villiers, M., (1923). The Numeral-Words, their Origin, Meaning, History and Lesson. Witherby, London.

15. Faulkner, R.O., (1962). Concise Dictionary of Middle Egyptian, Griffith Institute.

16. Flegg, G., editor (1989). Numbers Through the Ages. Macmillan.

17. Friberg, J., (1978). The third millenium roots of Babylonian mathematics. I. A method for the decipherment, through mathematical and metrological analysis, of proto-Sumerian and proto-Elamite semipictographic inscriptions, Department of Mathematics, University of Göteborg No. 9.

18. Gardiner, Sir Alan H., (1973). Egyptian Grammar: Being an Introduction to the Study of Hieroglyphs. The Griffith Institute.

19. Glaser, A., (1981). History of Binary and other Nondecimal Numeration, Tomash, Los Angeles.

20. Gvozdanovic, J., editor (1992). Indo-European Numerals. de Gruyter, Berlin-New York.

21. Hill, G.F., (1915). The Development of Arabic Numerals in Europe. Clarendon Press, Oxford.

22. Hurford, J. R., (1975). The Linguistic Theory of Numerals. Cambridge University Press, Cambridge.

23. Ifrah, G., (1985). From One to Zero: A Universal History of Numbers. Translated by Lowell Bair. Viking, New York.

24. Ifrah, G., (1998). A Universal History of Numbers : From prehistory to the invention of the computer, London.

25. Ifrah, G., (2000). The Universal History of Numbers: From Prehistory to the Invention of the Computer. Translated by David Bellos, E. F. Harding, Sophie Wood, Ian Monk. John Wiley & Sons.

26. Joseph G.G., (1991). The crest of the peacock, London.

27. Kamrin, Janice (2004). Ancient Egyptian Hieroglyphs: A Practical Guide. Harry N. Abrams, Inc. McDonald, Angela. Write Your Own Egyptian Hieroglyphs. Berkeley: University of California Press, 2007 (paperback).

28. Li, Y. and Shi Ran Du. (1987). Chinese Mathematics, a Concise History, Translated from the Chinese by John N. Crossley and Anthony W.-C. Lun. Clarendon Press, Oxford.

29. Maher, D., Makowski, W. J. and John F., (2001). Literary Evidence for Roman Arithmetic With Fractions, Classical Philology Vol. 96 #4 pp. 376-399.

30. McLeish, J., (1991). The story of numbers. Fawcett Columbine, New York

31. Menninger, K. (1957). Zahlwort und Ziffer. Eine Kulturgeschichte der Zahl. Two volumes. Vandenhoeck & Ruprecht, Göttingen. Translated as Number Words and Number Symbols: A Cultural History of Numbers. MIT Press, Cambridge, Mass., 1969.

32. Neugebauer, O. and Sachs, A., (1945). Mathematical Cuneiform Texts, New Haven, CT.

33. Schmandt-Besseart, D., (1992). Before Writing. Two volumes. Univ. Texas Press, Austin.

34. Smeltzer, D., (1959). Man and number. Emerson Books, New York.

35. Smith, D.E. and Ginsberg, J., (1937). Numbers and Numerals. Columbia University, New York.

36. Ulrich L., (1973). Chinese mathematics in the thirteenth century. MIT Press, Cambridge, Mass.

37. van der Waerden, B.L., (1954). Science Awakening, Groningen.

38. van der Waerden, B.L., (1983). Geometry and Algebra in Ancient Civilizations, New York.

39. Yoshio, M., (1974). The Development of Mathematics in China and Japan, 2nd edition, Chelsea Publ., New York, 1974. (1st ed. Leipzig, 1913.)

CHAPTER 7

Zero

I don't agree with mathematics; the sum total of zeros is a frightening figure.

Stanislaw J. Lec,

Black holes are where God divided by zero.

Steven Wright

If you divide by zero you die!

Anonymous

7.1 Introduction

Every grade school child is taught the number zero. We don't think anything of it. If we have nothing we represent that state with a 0. We can't understand why it might be such a big deal, but it is and was. Most of the early civilizations never had a zero, and for those that did, it took a while, and when they did use it, it caused many people a great deal of confusion (it still does so today).

> *The story of zero is an ancient one. Its roots stretch back to the dawn of mathematics, in the time thousands of years before the first civilization, long before humans could read or write. But as natural as zero seems to us today, for ancient peoples zeros was a foreign –and frightening –idea. An Eastern concept, born in the Fertile Crescent a few centuries before the birth of Christ, zero not only evoked images of a primal void, it also had dangerous mathematical properties. Within zero there is the power to shatter the framework of logic. It is hard to imagine being afraid of a number. Yet zero was inexorably linked with the void – with nothing. There was a primal fear of the void and chaos. There was also a fear of zero.*
> Zero – Excerpt from The Biography of a Dangerous Idea By Charles Seife

As the quote above suggests, the development of zero is a fascinating story. It is amazing that something as insignificant as nothing would have such a controversial history. Today, zero seems so simple and obvious. We don't question it at all, but let's think about this a little. Zero means nothing. Why do we need a symbol, which is something, to represent nothing? Also, is zero really a number like 1, 2 or 3, or is it just a concept, or better yet a place–holder? Why do we need it, and why and how did it come about? Did you know that we really celebrated the start of the new millennium (2000) a year too early due to a lack of zero in our number system, or that zero can be a dangerous quantity? These are all questions we will address in this chapter.

7.2 Why was zero needed?

If you used a non–positional number system like the Egyptians, Greeks, and early Chinese, you really had no need for zero. Why would you need a symbol, which is something, to represent nothing? You didn't, and worse, it could cause confusion. This is why the Egyptians, Greeks, and the early Chinese, never had a zero, or even developed it as a concept. Also, when civilizations started to count they didn't start at zero, they started at one (as we'll see this statement is true for all but one civilization). Thus, cultures with non–positional numbers didn't need the concept, so it never developed.

On the other hand, if you had a positional number system, like the Babylonians, the Mayans, and eventually the Indians, you needed the zero concept to avoid confusion in your number system. For example, if we had no way to represent an empty place–value we wouldn't be able to easily tell the difference between two and two–hundred. They would both be written with the symbol for two, but how would we know that we really meant 200 instead of 2, if we didn't have the zero symbol? Zero was needed as a place–holder in our number system.

For centuries, however, the Babylonians never used a zero, even though they really needed it. They would from time to time have the type of problem we outlined above, but they would get around this, most of the time, by counting on the fact that the user knew what was meant, based upon the problem they were working on. Since one number would be about 60 times larger, or more, than the other (one missing place–value in base–60). This would be similar to saying that the plane ticket cost me three fifty. You would assume that I meant three hundred and fifty dollars and not three dollars and fifty cents. However, if I said the bus downtown cost three fifty, you would assume the three dollars and fifty cent amount instead. This is how the Babylonians got around the problem for some problems, but it still caused issues. For certain problems, if the "zero" was in the middle of a number they might put a space between the numbers in the different place–value locations so that you knew that, that place–value was not occupied, and this would help.

For example:

Which in Babylonian mathematics, would mean two one's, no sixty's, and four three–thousand–six–hundred's. Or a total of fourteen–thousand–four–hundred and two. As opposed to the number,

Which represented four sixty's and two one's, or two–hundred and forty–two.

However, this approach was not foolproof. Are you sure that the numbers above don't have another zero or two after the last symbol? Or, perhaps they are fractions and there should be a zero or two before them. What if whoever wrote the number put a slightly larger or smaller space between the numbers? How can you be sure? Something eventually had to happen and that something was the creation of a symbol to occupy the empty place–value location.

In and around the second or third century BCE, the Babylonians began to use a symbol to identify an empty place–holder in their number system. Somebody decided to end the confusion and created a symbol to represent an empty place–value. Now, they did not think of this place–holder symbol as a number that represented a quantity of zero, but rather like a punctuation mark to help us distinguish two different numbers from another. This was the first use of zero, as we know it. Not as a number, but as a place–holder, or punctuation mark.

Since the Babylonians wrote on clay tablets they would simply add one or two slanted wedge shapes (the shape of the end of their writing tool for writing a unit symbol) to represent an empty place–value, i.e. ⟨ or ⟨⟨ .

This was the first use of zero, and its purpose was to clarify their number system. Thus we'd see

and know the value of the number to contain four 3600's, no (zero) 60's and two 1's, or fourteen–thousand four–hundred and forty–two, without having to make an educated guess.

However, we want to reinforce the point that zero was NOT thought of as a number by the Babylonians, but rather as a punctuation mark or an empty place–holder. The Babylonians never had a symbol for the number zero! They instead had a symbol to represent an empty place–holder to help solve a confusing problem in their number system. Zero was needed as the positional number system began to take hold. It began as a punctuation mark, and eventually transformed itself into an actual number, but the pathway was not simple nor direct.

7.3 Why was Zero a Difficult Concept to Understand?

The pathway for zero to achieve number status is long and torturous. Its eventual development is a story with many interesting twists and turns. At times it became a clash between religion and progress, or bold new ideas and strong beliefs. Many different cultures were involved in its development. It's a number unlike any other. Why is it so difficult to understand? Zero has a dual use –It's both a place–holder (punctuation mark), and also a bona fide number. However, zero doesn't act like other numbers when performing arithmetic. It goes by it's OWN rules.

About four hundred years after the Babylonians developed the zero place–holder concept, on the other side of the world, the Mayans were the first civilization to have the number zero. As we've shown in Chapter 5 the Mayans had a vigesimal (base–20) positional number system. However, the Mayans had a symbol for zero and they even used it in counting. They began their counting with zero, and then moved on to one, two, etc. Zero was not just a place–holder in their number system, but a number in its own right. We know that the Mayan's counted with zero since the first day of their month was the zeroth day. Also, when they started their calendar they also started with the zeroth year. Our calendar, however, never had a zero. This meant that when we celebrated the passing of the Millennium on December 31st, 1999. It was actually the wrong year! Why is this, you might ask? Let's demonstrate.

We use the Gregorian Calendar to calculate the passing of years. In 1592 Pope Gregory XIII ordered a change in how we added days to a year to correct for the fact that a year does not exactly equal 365.25 days (it is actually closer to the tropical year of 365.24219 days). This calendar, however, did not change the starting date which was tied to the year Christ was born (although further analysis has shown that value is incorrect too). The problem is that when the original Julian Calendar was made, and then eventually shifted to the "Christianized" version, a year was actually left out. Here is how:

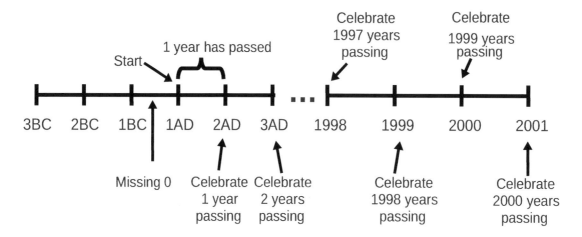

Notice that the number zero is missing. By starting our counting at 1AD we don't celebrate the passing of one year until the start of 2AD. Carrying this further, we wouldn't celebrate the passing of 2000 years until the beginning of 2001AD. This means the second millennium ended on December 31st 2000, not 1999!

Now while it is true that the Mayans had the number zero in and around 400 CE, we don't have any records of them defining it mathematically for multiplication and division. They did not advance the concept beyond counting and addition. It appears that their system of arithmetic wasn't as advanced as some other civilizations, and they were fairly isolated on the other side of the world. They did not have the benefits of the cultural interactions brought about by trade.

On the other side of the world, the concept of zero was advancing independently. To be fair, though, there is a large amount of uncertainty in who actually developed what concept first. Especially, since during the beginning of the common era in the East, when there was quite a bit of trading going on between India and China, as well as with the rest of the world. However, as best we can ascertain it looks like the Indian merchants may have brought back a Chinese abacus to India sometime in the sixth or the seventh century. Shortly after this period the Hindu's dropped their usage of numerals greater than 9 and adopted a positional number system. Eventually this Hindu number system made its way back to China, and the Chinese number system incorporated the zero concept developed by the Indians, into their number system as well, probably by adapting the Hindu system into their own. Some accurately datable historical evidence may be discovered to refute this, but for now we'll assume that this is the strongest possibility, although we will always keep our minds open. The generally accepted view is that the acceptance of zero as a number actually took place in India in the 7the century.

It is difficult to understand that as late as 1592 the western world still didn't have a zero, though. How could this be? The Mayans had it over eleven hundred years earlier, and the Indians had it nine hundred years previously. In western culture, however, the idea of zero took a long time to catch on. Even though Fibonacci brought the new Hindu–Arabic number system back with him to Europe in the thirteenth century, zero was not accepted as a part of that number system by the Europeans for several hundred years. In the east, on the other hand, the concept of nothingness was common in their culture, and was actually thought upon as a good thing. It was out of the void (sunya) that the universe came into being, and where we ultimately strive to return (nirvana).

The problem of accepting zero in the West actually started way back in Ancient Greece (600 BCE). To the Ancient Greeks zero was worse than meaningless, it upset the balance of the world. In Greece, numbers were always associated with shapes (Geometry), and there were no geometric shapes that needed the concept of zero. Also, in the Greek universe, created by Pythagoras, Aristotle, and Ptolemy, there is no void. There is no such thing as nothing. There is no zero!

> …it was not ignorance that led the Greeks to reject zero, nor was it the restrictive Greek number–shape system. It was philosophy. Zero conflicted with the fundamental beliefs of the West, for contained within zero are two ideas that were poisonous to Western doctrine. Indeed, these concepts would eventually destroy Aristotelian philosophy after its long reign. These dangerous ides are the void and the infinite. Charles Seife

The Greeks would eventually come to use the Babylonian zero (place–holder form) in calculations, but were not comfortable with it. They never used zero, or thought of it as a number, outside of their calculations. They always translated the Babylonian number system back into their number system, after the made their calculations. There was no zero in their system, so they were able to ignore it. Otherwise it would have caused them a philosophical problem or dilemma.

Also, in trying to understand the material world, various schools of thought emerged in Ancient Greece. On one side you had those that believed that matter, space, and time must be discrete. Meaning you could not break up a piece of matter indefinitely. Eventually you would end up with an indivisible piece of matter which they called an atom. Leucippus and his student Democritus 460 BCE were the main proponents of this belief, and were known as Atomist's. The problem with this view is that is had difficulties called paradoxes (a logical thought process leading to an absurd conclusion) associated with answering questions about motion (We'll look at this in detail later on in Chapter 11 on Infinity when we discuss Zeno's paradoxes.)

The alternative view was that matter, space, and time were continuous and that it was possible to subdivide them forever, never stopping. Aristotle was the major proponent of this perspective. The problem with this second view was that it too admitted paradoxes (also illuminated by Zeno.) Aristotle did not fully understand the subtleties of Zeno's paradoxes, so he simply discarded them as being too vague, and then attempted to clarify the concept of infinity, so that he could answer the objections of the Atomists.

The Atomist argument was simple. If we can divide the length of an object forever (continuously), then that means we would theoretically get to the point where the object we were dividing would have zero length. We could then do this to all the other subparts we divided from the original part, and that would mean that the whole, being the sum of all its parts, would be made up of things with zero length. But how could a whole be made by adding together things of zero size? If you add zero together, however many times you wish, you still end up with zero! Since this leads to a contradiction, then it must mean that our assumption of being able to divide forever is incorrect. Thus, we cannot divide things forever, and there must be some smallest indivisible part, which they called an atom.

If we believe in the concept of an atom then we would also believe that if an atom is moved in space, it must pass from one part of space to another, and where it moved from it must have left a finite "space" or more precisely it must leave a void in space behind it. The Atomist would then have to accept the concept of the void, as a natural result of believing in a discrete worldview.

On the other hand, Aristotle argued that space, matter, and time were all continuous and not discrete as the Atomists believed. He believed that dividing forever was **potentially** possible, but that it could not **actually** be done, so that you never arrived at an object with zero length. Thus, when an object moved in space it did so in a continuous way (without discrete jumps), and there were no finite voids left behind by the motion. It simply moved continuously through space which was also made up of another continuous substance called *pneuma* (Greek for breath, Later the term *aether* was adopted.) As a result, there was no void. If one accepted the void then one would have to accept the Atomist belief in the discrete nature of things, and Aristotle fought this with a passion.

Although the Atomist belief in some way provides a closer view to how we now perceive of matter in modern times, it lost favor to Aristotle's philosophy of the continuum in ancient times. Aristotelian philosophy eventually became the foundation of western thought, and one of its important conclusions, was that there is NO void. It even found its way into religion. Thomas Aquinas took the philosophy of Aristotle and adapted it to Christianity. In his work, *Summa Theologica* (1265–1274), Aquinas blended Aristotelian philosophy with Christianity, permanently joining the two and putting all other beliefs at odds with Christian teaching. By association, this meant that the concept of the void was against Christianity. Consequently, Atomism became associated with a materialistic and atheistic worldview.

During the Middle Ages in the west, Christian Theology dominated Europe. Christian Theology had adopted Aristotle's view of the universe, and adapted the Church's beliefs to this view. There was NO void! The void/chaos/darkness had existed, but God filled the void, tamed the chaos, and brought the darkness into light. The void brings you back to a "dark" time, which is associated with evil and the devil. There as no need for zero, and in fact, it only brings evil, darkness, and chaos back into the world, so it should be a**void**ed.

During this same period, Leonardo de Pisa (Fibonacci) visited the East (~1200 CE) and returned to Europe with the Hindu–Arabic number system, that contained both zero, and negative numbers. The use of the new number system eventually flourished in Europe due to its ease of use for calculations, but the ideas of zero and negative numbers took many centuries to take hold. People still feared zero for religious and superstitious reasons, and since it was closely associated with the infinite, it caused problems.

Zero had a tough hill to climb in the west, so it is understandable that it did not come from these cultures. The early western civilizations were able to avoid the concept of zero. The discovery/invention of zero would need to come from another source.

Perhaps no one has embraced nothing as strongly as the Indians who never had a fear of the infinite or of the void. Hinduism has embedded within it, a complex philosophy of nothingness, seeing everything in the world as arising from the pregnant void, known as Sunya. The ultimate goal of the Hindu was to free himself from the endless cycle of pain found in continual reincarnation and reconnect with the Nothingness that is the source and fundament of the All. For Indians, the void of Sunya was the very font of all potential; nothingness was liberation. No surprise then that it is from this sophisticated culture that we inherit the mathematical analog of nothing, zero. Like Sunya, zero is a kind of place holder, a symbol signifying a pregnant space where any other number might potentially reside. Charles Seife

Zero had an easier time being developed in the East. The Greek influence was not an issue in the East where algebra and not geometry was dominant. Also, in Hinduism, the predominant religion of India, the god Shiva was both creator and destroyer of the world. Shiva also was associated with nothingness. One aspect of the deity, Niskala Shiva, was literally the Shiva "without parts." He was the ultimate void, the supreme nothing; lifelessness incarnate. But out of the void, the universe was born, as was the infinite. Nothingness is what the world came from, and to achieve nothingness again became the ultimate goal of humankind. The void is actually associated with a good thing. In fact, Hinduism even has a goddess associated with the void. Aditi, is the Hindu Goddess that symbolizes the energy of the void; the sacred space that is essential to make room for new creation. Thus, it is not surprising that it was India that made the first major advance in treating zero as a number.

We know that from about 400 BCE to 300 CE the Indians used a non–positional number system made up from Kharosthi Numerals, which were similar to Roman Numerals. At some point they made a change from their non–positional to a positional number system. It is widely speculated that their exposure to the Babylonian number system, used for astronomical calculations by the Greeks, may have initiated this change. At that time the Babylonian system had the place–value symbol for the empty place already in use. One of the earliest Indian astronomers/mathematicians Aryabhata (476–550 CE) called the empty place value Kha (the Hindu word for emptiness). Also, it is possible that the use of the Chinese abacus may also have initiated the change. However it occurred, it appears that after the Indians were exposed to positional numbers, with the zero place–holder concept, it wasn't a far stretch before they not only accepted, but extended and embraced the new concept. Zero became another number just like 1, 2, 3, 4, etc., to the Indians, although it did behave differently.

The Indian mathematician Brahmagupta was the first person known to write about zero as a number. In particular, in 628 CE he wrote the book "Brahmasphutasiddhanta" (The Opening of the Universe). In this book Brahmagupta described properties of the number zero, roughly in the following way:

1. *When zero is added to a number or subtracted from a number, the number remains unchanged.*
2. *A number multiplied by zero becomes zero.*
3. *A number divided by zero, becomes a fraction with zero in the denominator.*
4. *Zero divided by zero is zero.*

The first two of these are correct. The last one (4) is definitely wrong, but the third one tells us nothing. It is, however, a major step forward in incorporating zero into the number system. Now, the place–value–holder role of zero of the Babylonian number system, which the Hindu's were exposed to, had been changed, so that zero was also a number, much like 1, 2, 3, 4, … (although as we'll soon see, it is not exactly like these other numbers).

In modern day mathematical notation, Brahmagupta established the following properties of zero:

1. $0 + 0 = 0$
2. $0 - 0 = 0$

and if a is any number then,

3. $a + 0 = a$
4. $a - 0 = a$
5. $a \times 0 = 0 \times a = 0$
6. $\dfrac{a}{0} = \dfrac{a}{0}$ A meaningless statement showing he really didn't understand this.
7. $\dfrac{0}{0} = 0$ An incorrect statement further showing his confusion.

Some Indian mathematicians after Brahmagupta tried to correct his mistakes and misconceptions, but some actually made it worse. One Hindu mathematician, Mahavira, two hundred years after Brahmagupta, said that statement 6 above should really be $\dfrac{a}{0} = a$, which is even further from the truth. This only goes to show that zero is a different sort of number. It doesn't always "behave" in predictable ways.

We see that zero was not a straight forward addition to the number system. This was because some number systems didn't need it, and the one's that did, couldn't comprehend it. Part of the problem is that zero has two uses. The first is just as a punctuation mark in a place–value number system (which is how the Babylonians used it), and the second is as a number itself (which is eventually how the Indians defined it). Perhaps the dual use of zero helped to add confusion to an already perplexing concept.

7.4   Is Zero Like Other Numbers?

Zero is not like other numbers.

Performing arithmetic with zero gives some unique results. Zero is the only number that can:

1. be added to or subtracted from itself and not change, its still zero
2. be added to or subtracted from any other number and not change the value of the other number
3. multiply any number of any size or sign and make that number vanish (become zero)
4. be divided by any number and remain the same (zero)
5. not be divided into any number, including itself

These five properties highlight the fact that zero is definitely different from the natural or counting numbers. It has caused some to question whether or not it is even a number (in the abstract, non–symbolic sense) at all! Some have made the same assertion about the number one (1), claiming that it is really just the creator of numbers, but not a number in its own right. The numbers one and zero play another unique role in arithmetic. One (1) is the multiplicative identity element,

$$1 \cdot a = a$$

and zero (0) is the additive identity element

$$a + 0 = a$$

Also, zero is the only number that is its own additive inverse.

$$0 + 0 = 0$$

Another difference is that any number divided by itself is one, except for zero, for which it is an undefined process.

We can see some further differences independent of arithmetic. For many mathematicians it is fairly obvious that zero is an even number, since by definition an even number is divisible by two, or a number that has two as a factor, i.e. a number n is even if:

$$n = 2m \text{, where } m \text{ is an integer}$$

But zero is a different sort of even number. Zero is the only even number that has ANY and all numbers as factors

$$0 = m \cdot 0$$

This means that it is divisible by any number.

Perhaps this is why many people pause before they answer the question about the evenness of zero.

Zero is also the only number that is neither positive nor negative.

It really seems to be a different sort of number and probably why early civilizations, as well as young students, find it difficult to comprehend as a number and as a concept. It is for these and other reasons that zero is a hard to accept, understand, and eventually embrace. It comes with many challenging and dangerous ideas.

7.5   A Dangerous Idea

We have shown that zero is unique and seems to march to a different beat than the other numbers. Why should we care?

We should care because zero is real and can be dangerous. The arithmetic of zero can make large quantities vanish, or small quantities explode. It can move mountains, bridges, or even stop a ship. Zero can be quite powerful, which is why it should be understood and reckoned with.

On September 21st, 1997 the engines of the USS Yorktown guided missile cruiser ship came to a sudden halt while performing maneuvers off the coast of Cape Charles Virginia. Apparently a zero was incorrectly entered into the Remote Database Management software, causing the entire propulsion system to fail. The ship was inoperable for hours. Luckily this did not happen in a critical battle situation. Otherwise, all aboard would most certainly have died.

Also, consider this. When something called the determinant of the stiffness matrix goes to zero, a structure will buckle. Just like the I 35W bridge collapse in Minneapolis, Minnesota on August 14th, 2007, where unfortunately 13 people lost their lives.

Earthquakes occur when the seismic–moment tensor has a zero determinant. We don't need to know exactly what this is, just that zero is related to the physical phenomenon in question. The damage caused by this zero can be catastrophic, as evidenced by the smoldering remains of the 1906 earthquake in San Francisco, where over 3,000 people lost their lives.

The collapse of the Tacoma Narrows Bridge in Washington State on November 7th, 1940 was caused by aerodynamic flutter. Aerodynamic flutter is caused by a lurking zero in the physics of the bridge, causing it to shake, rattle, and roll apart. Thankfully no one was hurt in this event.

Black holes, the things that swallow up stars and planets, are believed to have a zero called a singularity at their center. This is perhaps the most powerful and dangerous zero of them all. (Artist rendition of a black hole, courtesy of NASA)

These are just a few examples of the power and danger of zero. It is not quite like any other number. At the outset it seems so tame and tranquil, after all it is simply nothing. Who would think that all this destructive power could lie in nothing.

EXERCISES 7.1–7.5

ESSAYS AND DISCUSSION QUESTIONS:

1. Why was zero such a difficult concept to understand and develop?

2. Did Aristotle believe in zero? Why or why not?

3. Which cultures developed zero as a number and why?

4. What do we mean by the statement that zero has dual uses?

5. Did the Babylonians have zero as a number in their number system? Explain.

6. Who is Brahmagupta and what impact did he have on the development of zero?

7. Can you define arithmetic of natural numbers without a zero element? Explain.

8. Is zero a must have or a nice to have number? Explain.

9. Why is zero considered a dangerous idea?

References

1. Calinger, R., (1999). A conceptual history of mathematics, Upper Straddle River, N. J.

2. Grünbaum, A., (1967). *Modern Science and Zeno's Paradoxes,* Middletown: Connecticut Wesleyan University Press

3. Ifrah, G., (1987). From one to zero : A universal history of numbers, New York.

4. Ifrah, G., (1998). A universal history of numbers : From prehistory to the invention of the computer, London.

5. Joseph, G.G., (1991). The crest of the peacock, London.

6. Kaplan, R. & E., (1999). The Nothing that Is: A Natural History of Zero, Oxford Press.

7. Mukherjee, R., (1991). Discovery of zero and its impact on Indian mathematics, Calcutta.

8. Seife, C., (2000). *Zero: The Biography of a Dangerous Idea*, Penguin Books.

CHAPTER 8

Number Base Systems in General

In the binary system we count on our fists instead of our fingers.

Anonymous

There are 10 types of people in the world – Those who understand binary, and those who don't.

Anonymous

8.1 Introduction

In Chapter 5 we discussed three of the important early numbers systems: the Babylonian, the Mayan, and the Hindu-Arabic systems (a base–60, base–20, and a base–10 number system respectively.) In this chapter we take a more detailed look into number base systems in general. All positional number systems have two fundamental parts. The symbols used to represent the digits of the system, and the place–value associated with the location of these digits. The number of symbols required in any base, is equal to the base number, and the place–values are multiples of the base starting with an initial place–value of one. Using these two basic facts we can represent any number in any base number system

8.2 Base–2 or Binary Number System

A binary system is a system where we count in groups of two. Now this may seem rather pointless. Why count in groups of two, when we have what seems to be much more practical higher base system. The answer, quite simply, is that our electronic devices use the binary system. So we should be aware of what is happening behind the screens of all our gadgets. The reason electronic devices, such as computers and cell phones, use binary is because at each switch you can either have a voltage or not have a voltage applied. This is how they work. Having a voltage can be indicated by a value of 1, to mean on, and not having a voltage can be indicated by a value of 0, to mean off. Putting a series of switches together we have what is called a binary number. Once you have a binary number you can use this to create letters, sounds, and images. These are all the things we see and hear that addict us to these electronic toys. So let's demonstrate how it all works.

In a binary system we need two symbols to represent all possible numbers. These two symbols could be anything, but we'll use 1 and 0. The place–values for this system start with one and increase by a factor of two (2) as we move to the left. This means we multiply the previous place–value by two to get the next place–value. We show the place–values in the place–value chart below.

This chart is important as we will use this chart to translate a binary number into a decimal, and also translate a decimal number into a binary number.

Place Values											
...	1024	512	256	128	64	32	16	8	4	2	1
...	2^{10}	2^9	2^8	2^7	2^6	2^5	2^4	2^3	2^2	2^1	2^0

Notice, in addition to the place–value written as a base–10 number, we have also included the exponential form of that number, as well, in the chart above. Thus, we observe that to get the next place–value to the left, we simply add one to the exponent of the base term, 2.

Before we do the translations, let us first become familiar with binary numbers by learning how to count in binary.

The first binary number is 0. Next we have 1. This is no difference up to this point than the counting we do with our decimal numbers. After 1 we would have two, but we have a problem. We cannot use the symbol 2 in a binary number. We can only use a 0 and a 1. However, we remember that the place–value to the left of the first binary column is the two's position. This column tells us how many two's our number has. So, for the next number 2, we have 1 two, and 0 one's, so the next number is 10_2. Notice that we have written a subscript next to our number, This subscript will be used to identify the base our number is from. If we don't have a subscript, this means we should interpret it as a base–10 or decimal number.

What is the next number? Well, in base–10 it would be 3, but we can't write a 3, but we can break 3 into a 2 plus 1. Meaning we have 1 two and 1 one, or in binary, 11_2. Do you see the pattern yet? Let's try the next number, 4. Four is really two groups of two, but we can't write a 2 in the binary system, so we can't use that. However, if we look at the next group over from 2 in our place–value chart, we see that it is 4. Thus, our number has 1 four, 0 two's, and 0 one's or we can write it in binary as, 100_2. Let's put this all together in a table so you can see the pattern.

Counting in Binary							
Decimal	Binary		Decimal	Binary		Decimal	Binary
0	0_2		11	1011_2		22	10110_2
1	1_2		12	1100_2		23	10111_2
2	10_2		13	1101_2		24	11000_2
3	11_2		14	1110_2		25	11001_2
4	100_2		15	1111_2		26	11010_2
5	101_2		16	10000_2		27	11011_2

Counting in Binary							
Decimal	Binary		Decimal	Binary		Decimal	Binary
6	110_2		17	10001_2		28	11100_2
7	111_2		18	10010_2		29	11101_2
8	1000_2		19	10011_2		30	11110_2
9	1001_2		20	10100_2		31	11111_2
10	1010_2		21	10101_2		32	100000_2

The other thing that is unique with numbers from other bases, is how we read the number. We are so familiar and comfortable with our decimal system, that we read all numbers like they are decimals. For example when we see 10_2, we are tempted to say "ten in base two", but we have to resist because the word "ten" implies that this is a decimal number, which it isn't. The word ten doesn't exist in our base–2 world. Instead we must read this number as "one–zero." The next number 11_2 should be read as "one–one" and NOT as "eleven." In a similar way, 100_2 is "one–zero–zero" and NOT "one–hundred."

Look at the chart above again, and see if you understand where all the binary numbers come from, and if you can read them correctly.

Now let's look at translating binary numbers into decimal numbers.

8.2.1   Translating Binary Into Decimal

We will show how to find the decimal equivalent to a binary number using examples.

Example 1: Find the decimal equivalent to the binary number 1011010_2

To do so we rewrite our place value chart above and place our binary number with a single digit in each column as follows.

Binary Number	1	0	1	1	0	1	0
Place–Value as a Decimal Number	64	32	16	8	4	2	1
Binary Digit Times Place–Value	1×64	0×32	1×16	1×8	0×4	1×2	0×1
Values	64	0	16	8	0	2	0
Sum the Row Above	$64 + 0 + 16 + 8 + 0 + 2 + 0 = \mathbf{90}$						

In row three above, we multiply the contents in row one and two above, and place the value in row four. In row five we sum the results of row four to obtain our answer.

The answer is 90. We have found that the binary number $1011010_2 = 90$.

Example 2: Find the decimal equivalent to the binary number 11100001_2

To do so we rewrite our place value chart above and place our binary number with a single digit in each column as follows, and follow the same procedure from above.

Binary Number	1	1	1	0	0	0	0	1
Place–Value as a Decimal Number	128	64	32	16	8	4	2	1
Binary Digit Times Place–Value	1×128	1×64	1×32	0×16	0×8	0×4	0×2	1×1
Values	128	64	32	0	0	0	0	1
Sum the Row Above	$128 + 64 + 32 + 0 + 0 + 0 + 0 + 1 = \mathbf{225}$							

The answer is 225. We have found that the binary number $11100001_2 = 225$.

8.2.2 Translating Decimal Into Binary

The process of going from a decimal number into a binary number is a little more complicated, but fairly straightforward as the examples below illustrate. There is a multi–step process involved, and you must learn how to carry it out.

Example 3: Find the binary equivalent to the decimal number 328.

We begin by looking for the largest place–value that is less than 328. In looking at the place–value chart above we see that this number is 256, since 512 is too large and 128 is too small.

Now subtract 256 from 328 to obtain:
$$\begin{array}{r} 328 \\ -256 \\ \hline 72 \end{array}$$

Now we look at the remainder which is 72, and do the same with this value.

The largest place–value to go into 72 is 64, and subtract again, $72-64=8$.

Finally, the largest place–value to go into 8 is 8 and when we subtract we have a 0 remainder.

Thus we can rewrite 328 as: $328 = 256 + 64 + 8$

Now expand this showing all the place–values for a binary number using a zero coefficient for any missing values (we see why we need zero now), to obtain:

$$328 = 1\times256 + 0\times128 + 1\times64 + 0\times32 + 0\times16 + 1\times8 + 0\times4 + 0\times2 + 0\times1$$

Reading off the coefficients in front of the place–values gives the binary number 101001001_2.

Example 4: Find the binary equivalent to the decimal number 759.

We begin by looking for the largest place–value that is less than 759. In this case it is 512, since 1024 is too large and 256 is too small.

Now subtract 512 from 759 to obtain:
$$\begin{array}{r} 759 \\ -512 \\ \hline 247 \end{array}$$

Now we look at the remainder which is 247, and do the same with this value.

The largest place–value to go into 247 is 128, and subtract again,.
$$\begin{array}{r} 247 \\ -128 \\ \hline 119 \end{array}$$

Now we look at the remainder which is 119, and do the same with this value.

The largest place–value to go into 119 is 64, and subtract again, $119-64=55$

Again, the largest place–value to go into 55 is 32, and we subtract, $55-32=23$

Again, the largest place–value to go into 23 is 16, and we subtract, $23-16=7$

Again, the largest place–value to go into 7 is 4, and we subtract, $7-4=3$

Again, the largest place–value to go into 3 is 2, and we subtract, $3-2=1$

Finally, we have a remainder of 1.

Thus we can rewrite 759 in expanded form as:

$$1\times512 + 0\times256 + 1\times128 + 1\times64 + 1\times32 + 1\times16 + 0\times8 + 1\times4 + 1\times2 + 1\times1$$

Reading off the coefficients in front of the place–values gives the binary number 1011110111_2.

We don't have to end with base–2, we could use any number as our base. In what follows we'll explore this further.

8.3   Base–5 or Quinary Number System

A quinary system is a system where we count in groups of five. Some early societies actually counted in groups of five. Can you find a possible reason why? If you recall, the Mayan vigesimal system has a base–5 subsystem within it.

In a quinary (base–5) system we need five symbols to represent all possible numbers. We'll borrow these five symbols from our decimal system, using 0, 1, 2, 3, 4. Notice that the last symbol is always one less than the base. The place–values for this system start with one and increase by a factor of five. We show the place–values in the place–value chart below. We will use this chart to translate a quinary or base–5 number into a decimal, and also translate a decimal number into a quinary number.

Place Values							
...	15,0625	3125	625	125	25	5	1
...	5^6	5^5	5^4	5^3	5^2	5^1	5^0

Before we do the translations, however, let us become familiar with quinary numbers by learning how to count in quinary.

The first quinary number is 0. Next we have 1. In fact, there is no different between quinary and decimal counting up to 4. After 4 we would have five, but we have a problem. We cannot use the symbol 5. We can only use a 0, 1, 2, 3, or 4. However, we remember that the place–value to the left of the first quinary column is the five's position. This column tells us how many five's our number has. So, for the next number 5, we have 1 five, and 0 one's, so the next number is 10_5. Notice that we have again written a subscript next to our number, This subscript tells us this is a base–5 number.

What is the next number? Well, in base–10 it would be 6, but we can't write a 6, but we can break 6 into a 5 plus 1. Meaning we have 1 five and 1 one, or in quinary, 11_5. Do you see the pattern yet? Let's put this all together in a table so you can see the pattern.

Counting in Quinary							
Decimal	Quinary		Decimal	Quinary		Decimal	Quinary
0	0_5		11	21_5		22	42_5
1	1_5		12	22_5		23	43_5
2	2_5		13	23_5		24	44_5
3	3_5		14	24_5		25	100_5
4	4_5		15	30_5		26	101_5
5	10_5		16	31_5		27	102_5
6	11_5		17	32_5		28	103_5
7	12_5		18	33_5		29	104_5
8	13_5		19	34_5		30	110_5
9	14_5		20	40_5		31	111_5
10	20_5		21	41_5		32	112_5

Again, just like the binary numbers, we must read off the digits and not read as a decimal. Thus, 134_5 is to be read as "one–thee–four in base–5" and not as "one–hundred and thirty–four."

Look at the chart above again, and see if you understand where all the quinary numbers come from, and if you can read them correctly.

Now let's look at translating quinary numbers into decimal numbers.

8.3.1 Translating Quinary Into Decimal

Example 1: Find the decimal equivalent to the quinary number 423_5

To do so we rewrite our place value chart above and place our quinary number with a single digit in each column and follow the same process we used above for binary numbers.

Quinary Number	4	2	3
Place–Value as a Decimal Number	25	5	1
Quinary Digit Times Place–Value	4×25	2×5	3×1
Values	100	10	3
Sum the Row Above	$100 + 10 + 3 = \mathbf{113}$		

The answer is 113. We have found that the quinary number $423_5 = 113$.

Example 2: Find the decimal equivalent to the quinary number 32241_5

To do so we rewrite our place value chart above and place our quinary number with a single digit in each column as follows.

Quinary Number	3	2	2	4	1
Place–Value as a Decimal Number	625	125	25	5	1
Quinary Digit Times Place–Value	3×625	2×125	2×25	4×5	1×1
Values	1875	250	50	20	1
Sum the Row Above	$1875 + 250 + 50 + 20 + 1 = \mathbf{2196}$				

The answer is 2,196. We have found that the quinary number $32241_5 = 2,196$.

8.3.2 Translating Decimal Into Quinary

The process of going from a decimal number into a quinary number is similar to the binary calculations above. The only difference is that we now use the base–5 place–values, and instead of simple subtraction, we may have to divide.

Example 3: Find the quinary equivalent to the decimal number 867.

We begin by looking for the largest place–value that is less than 867. In this case it is 625, since 3,125 is too large and 125 is too small.

We notice that 625 can go into 867 at most one time so that we can simply subtract 625 from 867 to obtain:

$$\begin{array}{r} 827 \\ -625 \\ \hline 202 \end{array}$$

Now we look at the remainder which is 202, and do the same with this value.

The largest place–value to go into 202 is 125. Again 125 can go in at most one time so we can again subtract, $202 - 125 = 77$.

Now we look at the remainder which is 77, and do the same with this value.

The largest place–value to go into 77 is 25, but this remainder is different, because 25 can go into 77 three times. Thus we divide 25 into 77 to obtain:

$$\begin{array}{r} 3 \\ 25\overline{)\ 77} \\ -75 \\ \hline 2 \end{array}$$

Since the remainder 2 is less than 5 we are done.

Thus we can rewrite 867 as:

$$1 \times 625 + 1 \times 125 + 3 \times 25 + 0 \times 5 + 2 \times 1$$

Reading off the coefficients in front of the place–values gives the quinary number 11302_5.

Example 4: Find the quinary equivalent to the decimal number 596.

We begin by looking for the largest place–value that is less than 596. In this case it is 125, since 625 is too large and 25 is too small.

We now divide 125 into 596 and obtain:

$$\begin{array}{r} 3 \\ 125\overline{)596} \\ -375 \\ \hline 121 \end{array}$$

Now we look at the remainder which is 121, and do the same with this value.

The largest place–value to go into 121 is 25, and divide 25 into 121 to obtain:

$$\begin{array}{r} 4 \\ 25\overline{)121} \\ -100 \\ \hline 21 \end{array}$$

Now we look at the remainder which is 21, and do the same with this value.

The largest place–value to go into 21 is 5, and it goes in 4 times with a remainder of 1.

Thus we can rewrite 596 in expanded form as:

$$3 \times 125 + 4 \times 25 + 4 \times 5 + 1 \times 1$$

Reading off the coefficients in front of the place–values gives the quinary number 3441_5.

This is a bit more complicated than binary numbers, so we'll consider another example.

Example 5: Find the quinary equivalent to the decimal number 1,339.

We begin by looking for the largest place–value that is less than 1,339. In this case it is 625, since 3,125 is too large and 125 is too small.

We now divide 625 into 1,339 and obtain:

$$\begin{array}{r} 2 \\ 625\overline{)\ 1339} \\ -1250 \\ \hline 89 \end{array}$$

Now we look at the remainder which is 89, and do the same with this value.

The largest place–value to go into 89 is 25, and divide 25 into 89 to obtain:

$$\begin{array}{r} 3 \\ 25\overline{)\ 80} \\ -75 \\ \hline 14 \end{array}$$

Now we look at the remainder which is 14, and do the same with this value.

The largest place–value to go into 14 is 5, and it goes in 2 times with a remainder of 4.

Thus we can rewrite 1,339 in expanded form as:

$$2 \times 625 + 0 \times 125 + 3 \times 25 + 2 \times 5 + 4 \times 1$$

Reading off the coefficients in front of the place–values gives the quinary number 20324_5.

8.4   Base–8 or Octal Number System

An octal system is a system where we count in groups of eight. We sometimes see this number system when working with electronic devices, just as we did for binary numbers.

In an octal (base–8) system we need eight symbols to represent all possible numbers. We'll borrow these symbols from our decimal system, using 0, 1, 2, 3, 4, 5, 6, and 7. Again, notice that the last symbol is always one less than the base. The place–values for this system start with one and increase by a factor of eight. We show the place–values in the place–value chart below. We will use this chart to translate an octal or base–8 number into a decimal, and also translate a decimal number into an octal number.

Place Values					
...	4096	512	64	8	1
...	8^4	8^3	8^2	8^1	8^0

Before we do the translations, however, let us become familiar with octal numbers by learning how to count in octal.

The first octal number is 0 and there is no different between octal and decimal counting up to 7. After 7 we would have eight, but we have a problem. We cannot use the symbol 8. We can only use a 0, 1, 2, 3, 4, 5, 6, or 7. However, we again remember that the place–value to the left of the first octal column is the eight's position. This column tells us how many eight's our number has. So, for the next number 8, we have 1 eight, and 0 one's, so the next number is 10_8. Notice that we have again written a subscript next to our number, This subscript tells us this is a base–8 number.

What is the next number? Well, in base–10 it would be 9, but we can't write a 9, but we can break 9 into a 8 plus 1. Meaning we have 1 eight and 1 one, or in octal, 11_8 Do you see the pattern yet? Let's put this all together in a table so you can see the pattern.

Counting in Octal							
Decimal	Octal		Decimal	Octal		Decimal	Octal
0	0_8		11	13_8		22	26_8
1	1_8		12	14_8		23	27_8
2	2_8		13	15_8		24	30_8
3	3_8		14	16_8		25	31_8
4	4_8		15	17_8		26	32_8
5	5_8		16	20_8		27	33_8
6	6_8		17	21_8		28	34_8
7	7_8		18	22_8		29	35_8
8	10_8		19	23_8		30	36_8

Counting in Octal							
Decimal	**Octal**		**Decimal**	**Octal**		**Decimal**	**Octal**
9	11_8		20	24_8		31	37_8
10	12_8		21	25_8		32	40_8

Again, just like the binary and quinary numbers, we must read off the digits and not read as a decimal. Thus, 746_8 is to be read as "seven–four–six in base–8" and not as "seven–hundred and forty–six."

Look at the chart above again, and see if you understand where all the octal numbers come from, and if you can read them correctly.

Now let's look at translating octal numbers into decimal numbers.

8.4.1 Translating Octal Into Decimal

Example 1: Find the decimal equivalent to the octal number 715_8

To do so we rewrite our place value chart above and place our octal number with a single digit in each column as follows.

Octal Number	7	1	5
Place–Value as a Decimal Number	64	8	1
Octal Digit Times Place–Value	7×64	1×8	5×1
Values	448	8	5
Sum the Row Above	$448 + 8 + 5 = \textbf{461}$		

The answer is 461. We have found that the octal number $715_8 = 461$.

Example 2: Find the decimal equivalent to the Octal number 3476_8

To do so we rewrite our place value chart above and place our octal number with a single digit in each column as follows.

Octal Number	3	4	7	6
Place–Value as a Decimal Number	512	64	8	1

Octal Digit Times Place–Value	3×512	4×64	7×8	6×1
Values	1536	256	56	6
Sum the Row Above	$1536 + 256 + 56 + 6 = \mathbf{1854}$			

The answer is 1,854. We have found that the octal number $3476_8 = 1{,}864$.

8.4.2 Translating Decimal Into Octal

The process of going from a decimal number into an octal number is similar to the binary and quinary calculations above. The only difference is that we now use the base–8 place–values.

Example 3: Find the octal equivalent to the decimal number 746.

We begin by looking for the largest place–value that is less than 746. In this case it is 512, since 4,096 is too large and 64 is too small.

We notice that 512 can go into 746 at most one time, so that we can simply subtract 512 from 746 to obtain:

$$\begin{array}{r} 746 \\ -512 \\ \hline 234 \end{array}$$

Now we look at the remainder which is 234, and do the same with this value.

The largest place–value to go into 234 is 64. We divide 64 into 234 to obtain:

$$\begin{array}{r} 3 \\ 64\overline{)\ 234} \\ -192 \\ \hline 42 \end{array}$$

Now we look at the remainder which is 42, and do the same with this value.

The largest place–value to go into 42 is 8. We divide 8 into 42 to obtain:

$$\begin{array}{r} 5 \\ 8\overline{)\ 42} \\ -40 \\ \hline 2 \end{array}$$

Since the remainder 2 is less than 8 we are done.

Thus we can rewrite 746 as:

$$1\times512 + 3\times64 + 5\times8 + 2\times1$$

Reading off the coefficients in front of the place–values gives the octal number 1352_8.

Example 4: Find the octal equivalent to the decimal number 3,642.

We begin by looking for the largest place–value that is less than 3,642. In this case it is 512, since 4,096 is too large and 64 is too small.

We now divide 512 into 3,642 and obtain:

$$
\begin{array}{r}
7 \\
512\overline{)\ 3642} \\
-3584 \\
\hline
58
\end{array}
$$

Now we look at the remainder of 58, and do the same with this value.

The largest place–value to go into 58 is 8, and divide 8 into 58 to obtain:

$$
\begin{array}{r}
7 \\
8\overline{)\ 58} \\
-56 \\
\hline
2
\end{array}
$$

Now we look at the remainder which is 2, and since 2 is less than 8 we stop.

We rewrite 3,642 in expanded form as:

$$7\times512 + 0\times64 + 7\times8 + 2\times1$$

Reading off the coefficients in front of the place–values gives the octal number 7072_8.

8.5 Base–12 or Duodecimal Number System

A duodecimal system is a system where we count in groups of twelve. In a duodecimal (base–12) system we need twelve symbols to represent all possible numbers. Now we have a slight problem. The base–12 system needs more symbols than our base–10 system. Thus, we need to add two more symbols if we are to use the same symbols we use for base–10.

The two new symbols we add will be a T to stand for ten, and an E to stand for eleven (again our final number must be one less than the base). Our symbols in base 12 are: 0, 1, 2, 3, 4, 5, 6, 7, 8, 9, T, and E.

We show the place–values in the place–value chart below. We will use this chart to translate a duodecimal or base–12 number into a decimal, and also translate a decimal number into a duodecimal number.

Place Values					
...	20,736	1,728	144	12	1
...	12^4	12^3	12^2	12^1	12^0

Before we do the translations, however, let us become familiar with duodecimal numbers by again learning how to count in duodecimal, like we did in the other bases.

The first ten digits of the duodecimal number system are the same as the decimal system, i.e. 0, 1, 2, 3, 4, 5, 6, 7, 8, 9, the next two numbers symbols are T for ten, and E for eleven. The next number is twelve. We write twelve with a 1 in the twelve's position and a 0 in the one's position, or 10_{12}. Notice that we have again written a subscript next to our number, This subscript tells us this is a base–12 number.

What is the next number? Well, in base–10 it would be 13, but we can't write a 13, but we can break 13 into a 12 plus 1. Meaning we have 1 twelve and 1 one, or in duodecimal, 11_{12}. Do you see the pattern yet? Let's put this all together in a table so you can see the pattern.

Counting in Duodecimal							
Decimal	Duodecimal		Decimal	Duodecimal		Decimal	Duodecimal
0	0_{12}		11	E_{12}		22	$1T_{12}$
1	1_{12}		12	10_{12}		23	$1E_{12}$
2	2_{12}		13	11_{12}		24	20_{12}
3	3_{12}		14	12_{12}		25	21_{12}
4	4_{12}		15	13_{12}		26	22_{12}
5	5_{12}		16	14_{12}		27	23_{12}
6	6_{12}		17	15_{12}		28	24_{12}
7	7_{12}		18	16_{12}		29	25_{12}
8	8_{12}		19	17_{12}		30	26_{12}
9	9_{12}		20	18_{12}		31	27_{12}
10	T_{12}		21	19_{12}		32	28_{12}

Again, just like the binary numbers, we must read off the digits and not read as a decimal. Thus, 528_{12} is to be read as "five–two–eight in base–12" and not as "five–hundred and twenty–eight."

Look at the chart above again, and see if you understand where all the duodecimal numbers come from, and if you can read them correctly.

Now let's look at translating duodecimal numbers into decimal numbers.

8.5.1 Translating Duodecimal Into Decimal

Example 1: Find the decimal equivalent to the duodecimal number ET_{12} (The extraterrestrial)

To do so we rewrite our place value chart above and place our duodecimal number with a single digit in each column as follows. We also add another row to write the decimal equivalent of our duodecimal symbol, since it is not possible for this to be greater than nine.

Duodecimal Number	E	T
Decimal Value of the Digit	11	10
Place–Value as a Decimal Number	12	1
Duodecimal Digit Value Times Place–Value	11×12	10×1
Values	132	10
Sum the Row Above	$132 + 10 = \mathbf{142}$	

The answer is 142. We have found the duodecimal number for the "Extra–Terrestrial."
$$ET_{12} = 142.$$

Example 2: Find the decimal equivalent to the duodecimal number $4TE_{12}$

To do so we rewrite our place value chart above and place our duodecimal number with a single digit in each column as follows.

Duodecimal Number	4	T	E
Decimal Value of the Digit	4	10	11
Place–Value as a Decimal Number	144	12	1
Duodecimal Digit Value Times Place–Value	4×144	10×12	11×1
Values	576	120	11
Sum the Row Above	$576 + 120 + 11 = \mathbf{707}$		

The answer is 707. We have found that the duodecimal number $4TE_{12} = 707$.

Example 3: Find the decimal equivalent to the duodecimal number $8E93_{12}$

To do so we rewrite our place value chart above and place our duodecimal number with a single digit in each column as follows.

Duodecimal Number	8	E	9	3
Decimal Value of the Digit	8	11	9	3

Place–Value as a Decimal Number	1,728	144	12	1
Duodecimal Digit Value Times Place–Value	8×1728	11×144	9×12	3×1
Values	13,824	1,584	108	3
Sum the Row Above	13824 + 1584 + 108 + 3 = **15519**			

The answer is 15,519. We have found that the duodecimal number $8E93_{12}$ = 15,519.

8.5.2 Translating Decimal Into Duodecimal

The process of going from a decimal number into a duodecimal number is similar to the prior calculations. The only difference is that we now use the base–12 place–values.

Example 4: Find the duodecimal equivalent to the decimal number 129.

We begin by looking for the largest place–value that is less than 129. In this case it is 125, since 144 is too large and 1 is too small.

We now divide 12 into 129 and obtain:

$$\begin{array}{r} 10 \\ 12\overline{)\,129} \\ -120 \\ \hline 9 \end{array}$$

We have a divisor of 10 and a remainder 9. Since 9 is less than 12 we are done.

We can rewrite 129 in expanded form as:

$$10\times12 + 9\times1$$

Reading off the coefficients in front of the place–values, and replacing the 10 in front of the 12 by T, gives the duodecimal number $T9_{12}$.

Example 5: Find the duodecimal equivalent to the decimal number 6,333.

We begin by looking for the largest place–value that is less than 6,333. In this case it is 1,728, since 20,736 is too large and 144 is too small.

We divide 1,728 into 6,333 to obtain:

$$\begin{array}{r} 3 \\ 1728\overline{)\,6333} \\ -5184 \\ \hline 1149 \end{array}$$

Now we look at the remainder which of 1,149, and do the same with this value.

We divide 144 into 1,149 to obtain:

$$
\begin{array}{r}
7 \\
144 \overline{)\ 1149} \\
-1008 \\
\hline
141
\end{array}
$$

Now we look at the remainder which of 141, and do the same with this value.

We divide 12 into 141 to obtain:

$$
\begin{array}{r}
11 \\
12 \overline{)\ 141} \\
-12 \\
\hline
21 \\
-12 \\
\hline
9
\end{array}
$$

Now we look at the divisor which is 11 and the remainder of 9. Since our remainder is less than 12, we can stop

We can rewrite 6,333 as:

$$3 \times 1728 + 7 \times 144 + 11 \times 12 + 9 \times 1$$

Reading off the coefficients in front of the place–values, and replacing the 11 with E, gives the duodecimal number 37E91 $_{12}$.

8.6   Base–16 or Hexadecimal Number System

A hexadecimal system is a system where we count in groups of sixteen. Hexadecimal numbers are also used quite a bit in electronic devices. This is due to their unique relation to base–2 numbers, i.e., $2^4 = 16$. You may have noticed that when you computer crashes you might have seen a series of numbers and letters appear on the screen. These are actually hexadecimal numbers. The represent what was stored in various memory locations on your computer and can be used to troubleshoot your crashing problem.

In a hexadecimal (base–16) system we need sixteen symbols to represent all possible numbers. Now, just as in base–12, we have a slight problem. The base–16 system needs more symbols than our base–10 system. Thus, we need to add two more symbols if we are to use the same symbols we use for base–10. The six new symbols we add will be the capital letters of the alphabet from A to F.

Our symbols in base–16 are: 0, 1, 2, 3, 4, 5, 6, 7, 8, 9, A, B, C, D, E, and F.

Where A=10, B=11, C=12, D=13, E=14, and F=15.

We show the place–values in the place–value chart below. We will use this chart to translate a duodecimal or base–12 number into a decimal, and also translate a decimal number into a duodecimal number.

Place Values					
...	65,536	4,096	256	16	1
...	16^4	16^3	16^2	16^1	16^0

Before we do the translations let us, as we've always done, become familiar with hexadecimal numbers by learning how to count in hexadecimal.

The first ten digits of the hexadecimal number system are the same as the decimal system, i.e. 0, 1, 2, 3, 4, 5, 6, 7, 8, 9, the next six numbers symbols are A for ten, B for eleven C for twelve, D for thirteen, E for fourteen, and F for fifteen. The next number is sixteen. We write sixteen with a 1 in the sixteen's position and a 0 in the one's position, or 10_{16}. Notice that we have again written a subscript next to our number, This subscript tells us this is a base–16 number.

What is the next number? Well, in base–10 it would be 17. We can't write a 17, but we can break 17 into a 16 plus 1. Meaning we have 1 sixteen and 1 one, or in hexadecimal, 11_{16}. Do you see the pattern yet? Let's put this all together in a table so you can see the pattern.

Counting in Hexadecimal					
Decimal	Hexadecimal	Decimal	Hexadecimal	Decimal	Hexadecimal
0	0_{16}	11	B_{16}	22	16_{16}
1	1_{16}	12	C_{16}	23	17_{16}
2	2_{16}	13	D_{16}	24	18_{16}
3	3_{16}	14	E_{16}	25	19_{16}
4	4_{16}	15	F_{16}	26	$1A_{16}$
5	5_{16}	16	10_{16}	27	$1B_{16}$
6	6_{16}	17	11_{16}	28	$1C_{16}$
7	7_{16}	18	12_{16}	29	$1D_{16}$
8	8_{16}	19	13_{16}	30	$1E_{16}$
9	9_{16}	20	14_{16}	31	$1F_{16}$
10	A_{16}	21	15_{16}	32	20_{16}

Again, we must read off the digits and not read as a decimal. Thus, 528_{16} is to be read as "five–two–eight in base–16" and not as "five–hundred and twenty–eight."

Look at the chart above again, and see if you understand where all the hexadecimal numbers come from, and if you can read them correctly.

Now let's look at translating hexadecimal numbers into decimal numbers.

8.6.1 Translating Hexadecimal Into Decimal

Example 1: Find the decimal equivalent to the "horrible" hexadecimal number BAD_{16}

To do so we rewrite our place value chart above and place our hexadecimal number with a single digit in each column as follows. We also add another row to write the decimal equivalent of our hexadecimal symbol, since it is not possible for this to be greater than nine.

Hexadecimal Number	B	A	D
Decimal Value of the Digit	11	10	13
Place–Value as a Decimal Number	256	16	1
Hexadecimal Digit Value Times Place–Value	11×256	10×16	13×1
Values	2816	160	13
Sum the Row Above	$2816 + 160 + 13 = \mathbf{2989}$		

The answer is 2,989. We have found that a bad number in the hexadecimal system is equal to $BAD_{16} = 2,989$.

Example 2: Find the decimal equivalent to the hexadecimal number $9AC_{16}$

To do so we rewrite our place value chart above and place our hexadecimal number with a single digit in each column as follows.

Hexadecimal Number	9	A	C
Decimal Value of the Digit	9	10	12
Place–Value as a Decimal Number	256	16	1
Hexadecimal Digit Value Times Place–Value	9×256	10×16	12×1
Values	2304	160	12
Sum the Row Above	$2304 + 160 + 12 = \mathbf{2476}$		

The answer is 2,476. We have found that the hexadecimal number $9AC_{16} = 2,476$.

Example 3: Find the decimal equivalent to the hexadecimal number $9A7F_{16}$

To do so we rewrite our place value chart above and place our duodecimal number with a single digit in each column as follows.

Hexadecimal Number	9	A	7	F
Decimal Value of the Digit	9	10	7	15
Place–Value as a Decimal Number	4,096	256	16	1
Hexadecimal Digit Value Times Place–Value	9×4096	10×256	7×16	15×1
Values	36,864	2,560	112	15
Sum the Row Above	$36864 + 2560 + 112 + 15 = \mathbf{39551}$			

The answer is 39,551. We have found that the duodecimal number $9A7F_{16} = 39,551$.

8.6.2 Translating Decimal Into Hexadecimal

The process of going from a decimal number into a hexadecimal number is similar to the process above for the other bases. The only difference is that we now use the base–16 place–values.

Example 4: Find the hexadecimal equivalent to the decimal number 125.

We begin by looking for the largest place–value that is less than 125. In this case it is 16, since 256 is too large and 1 is too small.

We now divide 16 into 125 and obtain:

$$
\begin{array}{r}
7 \\
16 \overline{)\ 125} \\
-112 \\
\hline
13
\end{array}
$$

We have a divisor of 7 and a remainder 13. Since 13 is less than 16 we are done.

We can rewrite 125 in expanded form as:

$$7 \times 16 + 13 \times 1$$

Reading off the coefficients in front of the place–values, and replacing the 13 in front of the 1 by D, gives the hexadecimal number $7D_{16}$.

Example 5: Find the hexadecimal equivalent to the decimal number 2,750.

We begin by looking for the largest place–value that is less than 2,750. In this case it is 256, since 4,096 is too large and 16 is too small.

We divide 256 into 2,750 to obtain:

$$
\begin{array}{r}
10 \\
256 \overline{)\ 2750} \\
-2560 \\
\hline
190
\end{array}
$$

Now we look at the remainder of 190, and do the same with this value.

We divide 16 into 190 to obtain:

$$
\begin{array}{r}
11 \\
16\overline{)\ 190} \\
-16 \\
\hline
30 \\
-16 \\
\hline
14
\end{array}
$$

The remainder is14 which is less than 16 so we are done.

We can rewrite 2,750 as:

$$10 \times 256\ +\ 11 \times 16\ +\ 14 \times 1$$

Reading off the coefficients in front of the place–values, and replacing the 10 with A, 11 with B, and 14 with E, gives the hexadecimal number ABE_{16}.

EXERCISES 8.1–8.6

ESSAYS AND DISCUSSION QUESTIONS:

1. Why are binary numbers so important today in our society?
2. Describe how a base–20 system would work.
3. Is the base–10 (decimal) system better than the other systems? Why or why not?

PROBLEMS:

8.2.1 Translating Binary to Decimal

1. 10011_2	4. 111111_2	7. 111000_2	10. 10101010_2
2. 101110_2	5. 1000000_2	8. 10001_2	11. 11001100_2
3. 1011_2	6. 1110101_2	9. 111010_2	12. 100100_2

8.2.2 Translating Decimal to Binary

1. 36	4. 328	7. 1,024	10. 115
2. 24	5. 137	8. 78	11. 69
3. 129	6. 723	9. 449	12. 865

8.3.1 Translating Quinary to Decimal

1. 123_5
2. 44_5
3. 340_5
4. 14400_5
5. 2034_5
6. 3041_5
7. 301_5
8. 1004_5
9. 1010_5
10. 444_5
11. 3334_5
12. 2341_5

8.3.2 Translating Decimal to Quinary

1. 375
2. 56
3. 78
4. 629
5. 773
6. 124
7. 619
8. 378
9. 89
10. 187
11. 1,236
12. 1,801

8.4.1 Translating Octal to Decimal

1. 17_8
2. 44_8
3. 177_8
4. 534_8
5. 621_8
6. 1170_8
7. 4030_8
8. 476_8
9. 511_8
10. 125_8
11. 5061_8
12. 3531_8

8.4.2 Translating Decimal to Octal

1. 73
2. 24
3. 36
4. 189
5. 376
6. 502
7. 511
8. 1,267
9. 3,450
10. 63
11. 573
12. 6,321

8.5.1 Translating Duodecimal to Decimal

1. 69_{12}
2. $T10_{12}$
3. $T0E_{12}$
4. $89T_{12}$
5. $68E_{12}$
6. $11TE_{12}$
7. $75E3_{12}$
8. 4480_{12}
9. $296T_{12}$
10. $36E1_{12}$
11. E_{12}
12. 111_{12}

8.5.2 Translating Decimal to Duodecimal

1. 15
2. 35
3. 97
4. 111
5. 459
6. 734
7. 1,230
8. 3,035
9. 143
10. 179
11. 1,711
12. 1,693

8.6.1 Translating hexadecimal to Decimal

1. 46_{16}
2. 73_{16}
3. BAD_{16}
4. CAB_{16}
5. $DEAD_{16}$
6. $FEED_{16}$
7. $FA35_{16}$
8. $1AF0_{16}$
9. $BEAD_{16}$
10. $FADE_{16}$
11. $9B27_{16}$
12. $65F_{16}$

8.6.2 Translating Decimal to Hexadecimal

1. 58	4. 479	7. 9,321	10. 3987
2. 79	5. 52,421	8. 7,546	11. 61,453
3. 250	6. 29,327	9. 4095	12. 46,789

References

1. Atkinson, K., (1985). Elementary Numerical Analysis, John Wiley & Sons.

2. Dunham, W., (1994). The Mathematical Universe, John Wiley & Sons.

3. Ifrah, G., (1999). The Universal History of Numbers : From Prehistory to the Invention of the Computer, Wiley.

4. Mallory, J.P. and Adams, D.Q., (1997). Encyclopedia of Indo-European Culture, Fitzroy Dearborn Publishers, London and Chicago.

5. Nissen, H.J., Damerow, P., Englund, R., (1993). Archaic Bookkeeping, University of Chicago Press.

6. Ore, O., (1976). Number Theory and Its History, Dover Publications.

7. Paulos, J.A., (1992). Beyond Numeracy, Vintage Books.

8. Smith, D.E., (1911). Hindu Arabic Numerals, The Atneneum Press.

9. Schmandt-Besserat, D., (1992). How Writing Came About, University of Texas Press.

CHAPTER 9

Negative Numbers

I am shocked that quantities smaller than nothing are discussed, it is the root of aberration of human reason.

Francois–Charles Busset

As the sun eclipses the stars by his brilliancy, so the man of knowledge will eclipse the fame of others in assemblies of the people if he proposes algebraic problems, and still more if he solves them.

Brahmagupta

9.1   Introduction

A major hurdle in the abstraction of the number concept was the introduction of negative numbers. These were numbers that were so unlike any other numbers before them. The concept of number had been limited to a magnitude or a quantity in the eyes of mathematicians. Negative numbers had to break this mold. It would take a new level of conceptual abstraction to create the concept of a negative number, and this took over 2000 years to happen, and the pathway there was not very smooth!

First, you had two competing forums (or playgrounds) for the development of numbers in mathematics: geometry, and algebra. Second, outside of mathematics you also had the compelling forces of trade, commerce, and government. Finally, as was discussed in Chapter 7, humankind had come to an agreement that zero is the smallest quantity there is. Zero means nothing, and obviously nothing can be less than nothing. It was all these factors, sometimes pulling in opposite directions, that together caused the slow and painful journey towards the birth and acceptance of negative numbers.

However, the need for negative numbers arose out of necessity. When a merchant wanted to sell an item, and he and the buyer agreed on a price, but it was for more money than the buyer had, they would agree that the buyer owed the additional money. He had a debt to repay to the merchant. Thus, if a buyer had 10 units of money, and the price of the object was 15, then 15 from 10 would yield a quantity related to how much the seller still owed the buyer. Thus, there was a debt of 5 units of money. For many cultures negative quantities were not numbers in the mathematical sense, instead they were concepts that existed in the business or commerce world only.

Although negative numbers were born out of necessity, the mathematicians of the day didn't like or even accept them as numbers. This should sit well with a lot of students that find the idea of negative numbers to be confusing at best. If you are in this group, you should take some comfort in the fact that numerous highly regarded mathematicians felt the same way.

In this chapter we outline the historical development of negative numbers. We highlight the conceptual hurdles humans faced in their development, as well as focusing on the process humans use to create and adopt new levels of abstractions.

9.2 The History of Negative Numbers

Negative numbers have been used throughout many ancient civilizations, but they were typically used only in trade to keep track of debts on ledger sheets. They weren't really numbers in the traditional sense, no more than the early use of fractions were thought of as a single number, but instead were really a ratio of two numbers. It would take many centuries before negatives were accepted as numbers on their own.

The history of negative numbers is one of many stops and starts. The earliest users of the concept were the Chinese who somewhere between 100 BCE and 260 CE were solving simultaneous equations involving negative quantities. However, it seems that the Chinese hadn't quite made the mental leap to consider these negative quantities as numbers equivalent in stature to the basic positive counting numbers. The ancient Greeks, on the other hand, outright rejected negative numbers, calling them absurd. The Hindus, independent of the Chinese, were really the first civilization to use these negative quantities as numbers, and defined their properties starting in around 628 CE, again with Brahmagupta. The ancient Islamic nations adopted and preserved many of the Hindu achievements, but still associated negative quantities with real things, such as debts, and did not heartily embrace the number concept. The Europeans, during the Renaissance, finally began to take a serious look at negatives, but it took several centuries before all came on board and accepted them as numbers. In the US. Negative numbers have always been considered with a great deal of suspicion, and it was not until after our Civil War of the 1860's that negative numbers were finally accepted in the US.

The evolution of negative numbers illustrated the gradual transition from the process of subtraction to an object called a negative number. The process of subtracting, or taking a quantity away from another quantity, could also be thought of as adding a "negative" quantity to a positive quantity. Thus, subtraction, a process, could be thought of as adding a negative number, an object, to a positive number. This is a subtle, but important distinction that is sometimes forgotten about, or swept under the carpet.

Adding further to the confusion, was a complex tug of war between two mathematical disciplines, that of algebra and geometry, that impacted the development of negative numbers. Algebra needed the negative numbers to grow and prosper, geometry did not, and in fact it is a bit awkward even heretical to even consider negative quantities in geometry. What is a negative length anyway?

All of this development and transition was immersed in and stimulated by a commercial and cultural setting that needed negatives for the businesses and the governments of the societies to function properly. The development of negative numbers was happening independently throughout the world, largely driven by trade and commerce. The eastern cultures such as China, India, and Persia, were predominantly influenced by algebra and algebraic equations, which frequently would lead to solutions that required the concept of negative numbers. This is probably why they led the way in the development/discovery of negative numbers. To the Greeks, negative numbers were not needed to solve problems in geometry. And to the Greeks, mathematics was geometry. So negative numbers had no place in their mathematics. Thus, we had to rely on the east for this concept to develop. With this background we present a short historical overview of the contributions and at times the impairment of its development by the different cultures.

Eastern Cultures

China

The Chinese were the first civilization to use the concept of negative quantities. The earliest documented use was in the 3rd century CE with the publication of the *Nine Chapters on the Mathematical Art (Jiu zhang suan–shu)* in 263 by Liu Hui. This work was the compilation of all the known mathematics in China at the time. It is not known exactly when the concept of negative quantities were actually first introduced, as it could have been a few hundred years earlier than this published work, but we definitely know it was in use by this date. The chapters in this work were:

Chapter 1	Field Measurement
Chapter 2	Cereals (proportions)
Chapter 3	Distribution by Proportion
Chapter 4	What Width? (area and volume)
Chapter 5	Construction Calculations
Chapter 6	Fair Taxes
Chapter 7	Excess and Deficiency (linear equations)
Chapter 8	Rectangular Arrays (systems of equations)
Chapter 9	Gougu (Pythagorean theorem)

As the names of the chapters suggest, the Chinese were very practical where mathematics was concerned. It was a tool for calculating necessary things. The concept of "negatives" arose quite naturally for them as they needed to subtract larger quantities from smaller quantities. Now this did not confuse them, since it was simply a process to obtain a practical result. It seems, however, that the Chinese did not think of these negative quantities as standalone numbers. Instead they were the result of the subtraction process that would be reinterpreted in terms of positive numbers. They did not accept a negative number as a solution to a problem or equation. Instead, they would always restate a problem so the result was a positive quantity. This is why they often had to treat many different "cases" of what was essentially a single problem. The most common application where negatives arose was with debts and assets.

The concept of debts and assets arose quite naturally from their commerce system. They even developed a procedure for combining debts and assets using colored rods. A red rod meant an asset (today we use the color black for an asset) and a black rod signified a debt (today this is the color red). They would use these rods in a device similar to an abacus to determine the net value of the transaction. However, it doesn't appear that they thought of these rods as representing numbers, since they had a separate set of symbols for numbers, distinct from these. It was more or less the process of reconciling transactions, not of adding and subtracting numbers. This is probably why they had no process for multiplying negative numbers until about 1300 CE. They were not really numbers in their eyes. All things considered though, the Chinese certainly led the way in developing the concept of negative numbers, but it would take others to help complete it.

India

The Hindu's of India were one of the first civilizations to develop both the concept and a symbol for the number zero (called Shunya in Sanskrit, and Sifr in Arabic). At the same time they introduced negative quantities as standalone numbers to the world again through the work of Brahmagupta (598–665 CE) in 628 CE. Brahmagupta's work, *Brahmasphutasiddhanta* (The Opening of the Universe), was the first known mathematical treatment of negative numbers. In this work negatives were treated as actual numbers and not just as a concept or a process. He did, however, describe them in terms of debts (rina), and their properties in terms of assets/fortunes (dhana) and zero (shunya). For example, here is a translation of how he described the arithmetic of negative numbers.

1. *A debt minus zero is a debt.*
2. *The product or quotient of two debts is one fortune.*
3. *The product or quotient of a debt and a fortune is a debt.*
4. *The product or quotient of a fortune and a debt is a debt.*

These are identical to the properties we use today. The second one, however, is not without controversy, as well see later.

The development in India appears to be independent of what was happening in China. While China's focus was on the practical use of negative quantities, India took it to a higher level and created an even more abstract object called a negative number.

Even in India, though, the development of negative numbers did not always proceed smoothly. They would push the concept forward by solving algebraic equations, but sometimes it was difficult. In the 12 century CE (500 years after Brahmagupta), Bhaskara II found negative roots for quadratic equations, but he rejected them. In the problem he was solving he believed that a negative value is "in this case not to be taken, for it is inadequate; people do not approve of negative roots." So even in India there was doubt about a negative quantity being an actual number hundreds of years after they had defined them so.

The developments in India were carried over to the Persian empire, who valued and treasured education and knowledge.

Persia (Islamic/Arab culture)

During the height of the Persian empire, knowledge was highly revered. Mathematical knowledge from throughout the world was preserved in the great libraries of Persia. The Arab culture also pushed the development of algebra further. In fact the name, algebra, comes from the Persian mathematician Muhammed al–Khwarizmi (780–850 CE). Al–Khwarizmi was a scholar at the House of Wisdom in Baghdad, where he worked on translating Greek and Hindu scientific works. He wrote the first book on algebra, *Hisab al–jabr w'al–muqabala*, which is where we obtain the name algebra (al–jabr). He is the preeminent Arab mathematician, and we owe a great deal to him for his contributions to preserving and advancing the mathematics that went before him.

172

It was during the period of al–Khwarizmi that the Islamic world learned about negative numbers from translating Brahmagupta's works, and by 1000 CE Arab mathematicians were using negative numbers. However, they too only used them when talking about debts. A true negative number, it seems, was a hard concept to swallow. It was due to the actions of the Arab world that we in the west were able to obtain the mathematics of the east. For it was through Persia that Europe received its mathematics.

Western Cultures

Ancient Greece

The Greeks never introduced negative numbers because their numbers were based on geometry, and a negative length simply did not make any sense. Pythagoras only accepted whole positive numbers. Fractions, negative numbers, and zero were not considered numbers by him or his followers. Nowhere in the surviving literature from ancient Greece, or Babylon can we find reference to negative numbers. From Thales (624–546 BCE) to Diophantus (200–284 CE), not one Greek mathematician ever even considered the concept of a negative number.

Diophantus of Alexandria has been called the "Father of Algebra." He wrote a treatise, *Arithmatica*, consisting of 13 books which contained solutions to algebraic equations. Even though he worked extensively in solving algebraic equations, he never considered the concept of negative numbers. He referred to an equation that was equivalent to 4x + 20 = 0 (which has the solution of x = –5), saying that the equation was absurd.

Since the Greeks were giants in mathematics if they didn't have it, there must be a reason. The fact that they never conceived of the concept and even rejected the concept as absurd, probably hindered its acceptance.

Europe

In Europe, Fibonacci (Leonardo of Pisa) (1170–1250 CE), was the first to use negative numbers, bringing the concept back with him from the Arab world to Europe in his major work *Liber Abaci (Book of Abacus or Book of Calculation)*, published in 1202. He, however, only used them when working with debts, just like all the other previous cultures.

Nicolas Chuquet (1445–1500 CE (dates uncertain)), a French mathematician used negative numbers as exponents and referred to them as "absurd numbers." Gerolamo Cardano (1501–1576 CE), an Italian mathematician, amazingly did not accept negative numbers, although he is credited with helping to discover the new related form of number, the imaginary number (which we will discuss in Chapter 13.) Cardano described negative numbers as "fictions" and as being "as subtle as it is useless."

Rene Descartes (1596–1650 CE), the father of the Cartesian coordinate system, called negative solutions of algebraic equations "false solutions." Needless to say, his coordinate system, used to graph algebraic equations, originally had no negative numbers associated with it. It was only the first quadrant!

Leonhard Euler (1707–1783), a Swiss Mathematician, freely accepted and worked with negative numbers, but he could not provide a satisfactory reason why a negative times a negative is positive. He simply stated that since a negative times a positive is negative, that the product of two negatives must be different, so it had to be positive. For many others in the 18th century it was common practice to ignore any negative solution resulting from solving equations, on the assumption that they were meaningless. In 1759 the British mathematician Francis Maseres wrote that negative numbers "darken the very whole doctrines of the equations and make dark of the things which are in their nature excessively obvious and simple," concluding that negative numbers didn't exist.

Even Lazare Carnot (1753–1823), a prominent French mathematician openly questioned the existence of negative numbers. In 1803 he believed that to "obtain an isolated negative quantity, it would be necessary to cut off an effective quantity from zero, to remove something of nothing; impossible operation. How thus to conceive an isolated negative quantity?"

As you can see the road to acceptance for negative numbers was a long and torturous one. The reasons for this are varied, but certainly the fact that they seem to contradict our intuition is one of them. Another possible reason is related to the distinction between process and object.

It was difficult to accept negative numbers because negative numbers must first be accepted as a valid process arising from subtraction, and then the process must then be accepted as an object, a number that can stand by itself. It is the distinction between process and object that becomes a hurdle. A negative number is an object or thing, but subtraction is a process. Eventually the process and the thing became one, but not without a lot of confusion and some controversy.

Before the concept of a negative number can be accepted as a number in its own right, it must first become a process, a process related to subtraction or owing something to someone. It is not a number but a debt, that is –6 does not mean the number negative six, but rather the description of the process where a positive six will be subtracted from another positive number, when required. It does not stand alone, or so it appears. This is one of the reasons there was a hesitancy for accepting negatives as numbers by ALL cultures. They did not come from something concrete such as counting objects; they came from performing an act with two counting numbers. They seemed to defy intuition and common sense. They were at best confusing.

9.3 Why are Negative Numbers so Confusing?
– A Web of Inconsistent Explanations, Symbols, and Analogies

You have all taken a course in basic algebra, in which you learned the rules along with all the devices to try and remember how to do arithmetic with negative numbers. That is NOT our focus in this book. We want to know WHY, NOT HOW! So the purpose here is not to retest you on the properties of negative numbers, but rather to look at why understanding them is so confusing and difficult in the first place.

Now mathematically it isn't really a problem, because the negative numbers are defined to follow specific rules, which is why the properties are what they are. The problem is that most people need further justification outside of mathematics, so non–mathematical analogies are created to explain their properties. The analogies used, however, are not consistent. We often use one analogy for addition and subtraction, and another for multiplication and division, and they are not consistent with each other! No wonder students are confused.

We'll illustrate what we mean in what follows. Some common interpretations of negative numbers use the idea of debts, moving to the left on the number line, decreasing temperature values, moving backward in time, or some form of negation or opposition. The problem is that none of these analogies alone gives a sufficient and consistent explanation of negative numbers and all their properties as we'll show.

First, we consider positive numbers. Let's add two positive numbers together: $3+3=6$. We can interpret this as adding an asset to an asset and obtaining a greater asset, which makes perfect sense.

We then multiply two positive numbers: $3\times3=9$. An asset times an asset gives us an even larger (or multiple) asset. Again, something we wouldn't argue with.

Now consider the process of adding two negative numbers: $(-3)+(-3)=(-6)$. If you interpret the quantity (-3) as a debt, then a debt plus a debt yields a greater debt and the operation makes sense, and provides a useful analogy for explaining negative numbers.

Now, however, carry over the same interpretation to multiplying two negative numbers, such as, $(-3)\times(-3)=9$, and see what happens. Using the debt interpretation, we have a debt times a debt yields an even greater asset. What does that mean? It doesn't make any sense. Using the analogy from above for positive numbers, you would think you should have an even greater debt!

Perhaps you'll say this analogy is not appropriate, so let's use another. Using motion on the number line, we have the following explanation.

Considering positive numbers again, we add two positive numbers together, $3+3=6$, and interpret each positive number, as moving us to the right on the number line. So adding two positive numbers together means we have moved further to the right on the number line, which makes perfect sense.

We then multiply two positive numbers: $3\times3=9$. Moving to the right, times moving to the right, means we have moved a distance squared to the right, with no apparent contradictions.

Now consider the same analogy for adding two negative numbers: $(-3)+(-3)=(-6)$. If you interpret the quantity (-3) as moving to the left on the number line, then a motion to the left plus a motion to the left yields a greater motion to the left and the analogy makes sense.

Now, carry the same interpretation over to multiplying two negative numbers: $(-3)\times(-3)=9$. Using the same interpretation as above, we have a motion to the left times a motion to the left yields an even larger, squared motion, but to the right! How can that be? It again doesn't make any sense.

Now you might argue that the problem is in our interpretation, and that we haven't explained it correctly. The quantity (-3) really means two things at the same time. The 3 is the magnitude we move to the right (keeping with our earlier analogy for positive numbers) and the $-$ means a change in direction. (As you can see we are already making the explanation even more complex and convoluted, and a negative number is no longer a single thing, but is now defined as a process, however, we'll continue anyway.) The analogy now becomes, we start out facing to the right and then change direction and move three units to the left, but we now multiply this by (-3) which means we change direction to the right and move a multiple of three times the original distance we moved, or a total of nine units to the right (whew!). Okay, so it appears we have got it. Now let's make sure addition is fine with this analogy, and we're done.

We interpret $(-3)+(-3)=(-6)$ as facing right and changing direction and moving three units to the left. We then add another change in direction and motion so we now move three units to the left, and are back where we started. Thus, $(-3)+(-3)=0$. Something is not quite right here.

The only way around this is to use a different interpretation for the second (-3). Which now does not mean change direction and move three units in that direction, but to continue to move three units in the negative (left) direction. The problem is we have two interpretations for the same thing in the same analogy. Okay, maybe we should just stick with the second interpretation for both from the start, but this is what we already considered earlier!

This is what confused early mathematicians and continues to confuse pretty much all students who are first exposed to this concept. Eventually all accept that it must be true, but they really can't explain why. The others that can't "see" it, or don't readily accept it, are put down for their "lack of understanding," so they learn to keep quiet, and confused. It is a lot like convincing everyone that the emperor is wearing fine clothing, although he is completely naked. If everyone agrees to see something, all will follow. It takes a little innocent child, however, to see through the fiction, and ask the embarrassing question. Negative numbers do not make total sense, and cannot be explained away with one self–consistent analogy! This is especially true when someone asks the embarrassing questions.

Perhaps the problem is that we require a negative number times another negative number results in a positive quantity. How many times have you asked why a negative number times another negative number gives a positive number? You are pretty much just told that's the rule. It turns out that it is possible to define the rules of negative numbers, especially when it comes to multiplication, differently. In this case, multiplication with negative numbers is not commutative, though. Thus, a negative times a negative would be negative, and a positive time a negative would now be positive, while a negative times a positive would remain negative. It too is quite confusing and involved, but a bit more consistent from the analogy question.

In our number system there tends to be a bias towards the positive numbers. This comes about from our choice that the product of two negatives is positive (this is what some call an unsymmetric system, and why the analogies fail to work). Also, you can take the square root of a positive number, but not a negative number; or can you?

The truth is that negative numbers behave like they do based upon how we defined them, not due to our intuition. Their properties are at times counter–intuitive, and defy simple analogies. Here is another example of a problem with intuition and negatives.

We are all comfortable with the statement that negative numbers are less than or smaller than positive numbers. Certainly nobody would argue with the mathematical statement that

$$-1 < 1$$

People would also agree that if you take a number and divide it by a larger quantity you get a smaller quantity as a result. For example, if you divide one by two you get $\frac{1}{2}$, which is smaller than one. Also, if you take a larger quantity and divide by a quantity less than one you get a larger quantity. For example, $\frac{1}{1/2} = 2$. Thus, we'd agree on the statements that:

$$\frac{\text{smaller}}{\text{larger}} = \text{even smaller} \quad \text{and} \quad \frac{\text{larger}}{\text{smaller than one}} = \text{even larger}$$

Now, use 1 as the larger number and –1 as the smaller number and see what happens.

$$\frac{\text{smaller}}{\text{larger}} = \frac{-1}{1} = -1 \quad \text{and} \quad \frac{\text{larger}}{\text{smaller}} = \frac{1}{-1} = -1$$

Since they are both equal we have shown that

$$\frac{\text{smaller}}{\text{larger}} = \frac{\text{larger}}{\text{smaller}}$$

How can this be? This is a contradiction. The problem must be in our interpretation of negative numbers as being smaller than zero. Or for some, the belief that negative numbers are allowed to exist at all! This belief was held by many prominent mathematicians of their time, including John Wallis (1616–1703), and Antoine Arnauld (1612–1694) who both used this form of reasoning to try and debunk negative numbers.

If all this wasn't enough to confuse the average person (and not so average mathematician), we also have the fact that the symbol we use to indicate a number is negative, is the minus or negative sign (–). Why is this a problem? It is the same symbol we use for subtraction ($(5-3)$, the additive inverse (-3), as well as the negation of a quantity $[-(x+1)]$.) They all have similar, but actually slightly different meanings. The last two are essentially the same concept, it is the first that is actually different. The fact that the meanings are similar seems to call for a common symbol, but it does add fuel to the fire of confusion regarding negative numbers and how to understand and explain their properties.

It is not a coincidence that the widespread use and acceptance of negative numbers did not occur until the late 1800's, and this is true even for mathematicians. They are indeed very strange quantities and bring with them some overly confusing ideas and properties. The author of a handbook of mathematics (1843), Francois–Charles Busset, believed that the problems associated with teaching mathematics in France at the time, was due to the acceptance of negative numbers into mathematics. He was horrified that people believed that there were "quantities smaller than nothing," and that it was "the root of the aberration of human reason."

It might be hard to believe, but in the United States negative numbers were not fully accepted in mathematics until after our Civil War of 1863! They are indeed a very recent type of number, with a lot of conceptual baggage attached to them. We explore this further as we examine our arithmetic properties with them.

9.4   Working with Negative Numbers

Negative numbers first made their way into mathematics via the business world, as initially mathematicians denied their existence. They were not numbers, but instead were a way that merchants could keep tract of debts. In mathematics negative numbers seem to have come from the process of subtraction. However, as people began to subtract larger quantities from smaller quantities, they had to make sense out of the results. To be able to satisfy this apparent contradiction, negative numbers needed to be created (or were they already there to be discovered). Eventually mathematicians too could not resist the irresistible force and negative numbers were admitted into the number kingdom, but not without a long fight, as we have seen above.

We've already discussed the fact that the product of two negative numbers is a positive number is not at all intuitive. The Chinese never multiplied negative numbers, so it wasn't a real question for them. The Hindu's identified the product as being positive, but without the idea of proving or justifying it. One of the most prominent mathematicians of all times, Leonhard Euler, also used the fact without proof. He simply reasoned that since a positive times a negative is negative, that a negative times a negative must be different, so it must be positive.

The real and only reason that the product of two negative numbers is positive, is because this is how our system of numbers were defined. In mathematics something is defined using basic axioms, things we accept without proof, and rules to create new connected facts about numbers. It is through the axioms and rules that this new rule about the product of negatives comes about. It is something we can't see intuitively, but must learn it as an accepted fact.

We end this section by presenting a mathematical (not intuitive) justification of why it is so (We should point out that we will spend more time on defining what we mean by axioms, and these mathematical justifications or proofs, when we cover logic and reasoning later in the book.)

A mathematical justification can take a little getting used to. Here's how the process goes:

We begin with some fairly obvious facts:
1. Zero times anything equals zero.

2. Every number has exactly one additive inverse.
   Note: An additive inverse is a number that when added to the number it is the inverse of, gives zero as the sum. So the additive inverse of 2 is –2, since 2 + (–2) = 0, the additive inverse of 5 is –5, since 5+ (–5) = 0, and the additive inverse of –7 is 7, since (–7) + 7 = 0. The additive inverse of a is – a, since a + (–a) = 0. The additive inverse of (–b) is –(–b), since (–b) + [– (–b)] = 0.

We then add a basic rule on how to proceed:
3. We will require that all numbers obey the distributive law: $a(b+c) = ab+ac$

It is this new rule that makes us accept the new negative times negative equals positive, and this is how.
1. Consider the fact from (1) above: $a\cdot 0 = 0$
2. Combine this with step (2) from above, which tells us that:

$$b+(-b) = 0$$

We replace the zero in step (4) with $b+(-b)$ to obtain:

$$a[b+(-b)] = 0$$

Apply the distributive law from step (3):

$$a[b+(-b)] = ab+a(-b) = 0$$

or

$$ab+a(-b) = 0$$

3. Now add the additive inverse of $a(-b)$, which is $-[a(-b)]$, to both sides and obtain:

$$ab + a(-b) + [-a(-b)] = 0 + [-a(-b)]$$
$$ab + 0 = 0 + [-a(-b)]$$
$$ab = -a(-b)$$
$$ab = (-a)(-b)$$

4. Thus, if both a and b are positive numbers, then their product is also equal to the product of their additive inverses $(-a)$ and $(-b)$, which are both negative. Or, turning the last equation around:

$$(-a)(-b) = ab$$

The product of two negative numbers must be positive.

This was an informal process to provide mathematical justification for the rule. This is not a complete and rigorous mathematical proof. Such a proof is beyond the scope of this book.

Now this may have confused some of you, so we'll try and make this a little clearer by using fixed numbers in our calculations above.

1. $3(0) = 0$ Zero times any number is zero.

2. $2+(-2) = 0$ (-2) is the additive inverse of 2, so their sum is zero.

3. $3[2+(-2)] = 0$ Replace the zero in step (1) with $2+(-2)$ in step (2).

4. $(3)(2) + 3(-2) = 0$ Use the Distributive Property

5. $(3)(2) + 3(-2) + [-3(-2)] = -3(-2)$ Add the additive inverse of $3(-2)$, which is $-3(-2)$ to both sides of the equation

6. $(3)(2) = (-3)(-2)$ Simplify by canceling $3(-2) + [-3(-2)]$, and rewrite $-3(-2)$ as $(-3)(-2)$

Thus, we have shown that $(-3)(-2) = (3)(2)$, or the product of two negative numbers is positive.

It wasn't until 1815 that the Dutch mathematician Jacob de Gelder (1765–1848) published a treatise on negative numbers that one of the first proofs on the product of two negative numbers being positive was devised. Centuries after they were first contemplated by the early Chinese.

Summary

Overall, we can see that the concept of negative numbers was not an easy one to emerge. It is a very abstract concept. It is interesting to see how humans wrestle with abstraction, and eventually allow its use, provided the need is strong enough. It was almost as if the mathematics was forcing us to accept it. It is also reassuring that having a problem understanding negative numbers is quite common, even among great civilizations and cultures, as it should be. For those of us that had problems understanding and working with the process, we can be comforted by the fact that we are not alone, and that many great mathematicians had the same conceptual problems.

Extending our number system to include negative numbers was a major achievement in abstraction, but, as we'll see, more was still to come.

EXERCISES 9.1–9.4

ESSAYS AND DISCUSSION QUESTIONS:

1. What is the difference between viewing negative numbers as a process versus a real object?

2. Discuss why negative numbers are confusing.

3. Why were negative numbers a difficult concept to accept?

4. Name three aspects of negative numbers that make them difficult to understand, and explain why.

5. Explain why no single analogy works for explaining both addition and multiplication of negative numbers.

6. Give the basic arguments to mathematically justify that the product of two negatives is positive.

7. Similar to exercise 4 above, give a mathematical justification as to why the product of a positive and a negative number is negative.

8. Are negative numbers actual numbers or are they a process?

References

1. Aleksandrov, A., Kolmogorov, A., Lavrent'ev, M., (1999). Mathematics, its Content, Methods, and Meaning, Dover Publications, NY.

2. Beckers, D., (1999). Positive Thinking: Lacroix's theory on negative numbers in the Netherlands, Report No. 99, Department of Mathematics, University of Nijmegen.

3. Berggen, J.L., (1986). Episodes in the Mathematics of Medieval Islam, Springer-Verlag N.Y., And Berlin.

4. Boye, A. (2002). Papers on the history of science. Les Instituts de Recherche sur l'Enseignement des Mathématiques. Nantes, France, from http://nti.educa.rcanaria.es/penelope/uk_confboye.htm.

5. Darling, D., (2004). The Universal Book of Mathematics: From Abracadabra to Zeno's Paradoxes, John Wley & Sons, NJ.

6. Flegg, G., Hay, C., Moss, B., (1984). Nicola Chuquet, Renaissance Mathematician, Springer, NY.

7. Hazewinkel, M., (1987). Encyclopedia of Mathematics, Kluwer Academic Publishers, The Netherlands.

8. Henley, A.T., (1999). The History of Negative Numbers, PhD Thesis for South Bank University, London, UK.

9. Ifrah, G., (1998). The Universal History of Numbers . Harvill Press, London.

10. Li Y.and Du S., (1987). Chinese Mathematics: a Concise History, Oxford.

11. Menninger, K., (1969). Number Words and Number Symbols. M.I.T. Press Cambridge, Mass. andLondon.

12. Schubring, G., (2005). Conflicts Between Generalization, Rigor, and Intuition: Number Concepts Underlying the Development of Analysis in 17 - 19th Century France and Germany. Springer-Verlag N.Y.

CHAPTER 10

Exponential Numbers

The greatest shortcoming of the human race, is our inability to
understand the Exponential function.

Albert Bartlett

10.1 Introduction

Why would Albert Bartlett, a Physicist at Colorado State University, make such a strong statement as we have quoted above? What are exponential numbers, and why are they as important, as Bartlett would suggest? In this Chapter we will attempt to answer this question and more. It is a fascinating and important topic to discuss. It adds a new type of number to our system with some very important implications that affect our day to day life.

10.2 What is an Exponential Number?

The word exponential comes from two Latin words – expo meaning "out of" and ponere meaning "place." Thus exponential means "out–of–place." This is because the number called the exponent is placed above and to the right of another number called the base. The first use of an exponential number is credited to the English mathematician Michael Stifel in his 1544 book *Arithemetica Integra*.

The purpose of exponential numbers is to provide a concise way of representing numerous multiplications of the same number. For example, if we wanted to multiply seven factors of two in a row, we could write $2 \times 2 \times 2 \times 2 \times 2 \times 2 \times 2$, or we could write 2^7 instead. The number seven (7) is called the exponent, or the out–of–place number, and the number two (2) is called the base. The number 2^7 is read as "two to the seventh," and is known as an exponential number.

> **Definition:** An **exponential number** is any number that can be written as:
> $$(\text{base})^{(\text{exponent})} \quad \text{or} \quad a^n$$
> where a and n are real numbers, and a must be positive. a is called the **base** and n is called the **exponent**.

Exponential numbers allow us to write a series of successive multiplications by the same base number in a very concise way. Exponential numbers have some very interesting properties, as well as some surprising results.

Let's start with a simple illustration. Let us say you are given a challenge. You are to take a piece of paper and fold it over 100 times. Seems like a doable exercise at first glance. Now it may start to get difficult, as the paper begins to bind up on the edges as you fold, but suppose you are given a bye, and are allowed to bypass this difficulty and we can continue with it as a mental experiment. The real question we'd like to answer is, how tall will the stack of paper be after you go through all the mental folds? Stop for a moment and think about it. How high do you think? Would it be a couple of inches high, a couple of feet, higher than the ceiling in the room, taller than that? Well, let's work it out and see.

Suppose the paper is seven mils thick (a fairly average thickness for a sheet of paper). Seven mils is just seven thousandths of an inch or 0.007 inches. Okay, we know the height of one sheet of paper, now we need to know how many sheets of paper we will have after 100 folds. After one fold, we have two sheets, after two folds we double the amount, so now we have four sheets, if we fold the paper again, we will double it again, so we now have eight sheets, etc. Now we could continue this calculation for 100 folds, but let's see if there is a simpler way. It looks like every time we fold the paper we double the number of pages. Well there is a fairly simple mathematical expression we can use from basic algebra to write this out, and that is a number with an exponent. We suggest that folding a sheet of paper 100 times would give us 2^{100} sheets of paper.

Let's test this. If we fold it once we'd have 2^1 or 2 sheets of paper, fold it twice, and we'd have 2^2 or 4 sheets of paper. A third time would give us 2^3 or 8 sheets etc.. It seems fairly obvious then that after 100 folds we'd have 2^{100} sheets of paper. 2^{100} is an exponential number equal to the number of sheets of paper we obtain by folding a piece of paper 100 times.

Now if we want to find how high the final stack of paper will be, we just multiply the thickness of a single sheet of paper by the total number of sheets in our stack, or

$$0.007\,\text{inches} \cdot 2^{100}$$

Putting this into our calculator, we find that we have
$$8.87 \times 10^{27}\,\text{inches}$$

(Note: We are using Scientific Notation to represent the numerical value, which contains an exponential number, as it allows us to write our number more compactly. This is something you should already be familiar with).

Now this doesn't really mean much to us, so let's try putting it into more meaningful terms.

First, we know that there are 12 inches to a foot, so if we divide the expression above by 12, we'll know how many feet high it is. Doing this we find.

$$\frac{8.87 \times 10^{27}\,\text{inches}}{12\,\dfrac{\text{inches}}{\text{foot}}} = 7.39 \times 10^{26}\,\text{feet}$$

Again, this is still too complicated to understand. We know from conversion tables that there are 5,280 feet in a mile, so let's divide by 5,280 to see how many miles high it is.

$$\frac{7.39 \times 10^{26} \, \text{feet}}{5,280 \, \dfrac{\text{feet}}{\text{mile}}} = 1.40 \times 10^{23} \, \text{miles}$$

This still is not understandable. Let's change the game totally then. We know that light can travel at the rate of 186,000 miles per second. So if we divide the last number by 186,000 we'll get how many seconds it would take a beam of light to travel from the bottom to the top of this stack, as it travels at 186,000 miles per second.

$$\frac{1.40 \times 10^{23} \, \text{miles}}{186,000 \, \dfrac{\text{miles}}{\text{seconds}}} = 7.53 \times 10^{17} \, \text{seconds}$$

This is still too large to comprehend. Divide this by 3,600, since there are 3,600 seconds per hour, and the result will tell us how many hours it would take light to travel to the top.

$$\frac{7.53 \times 10^{17} \, \text{seconds}}{3,600 \, \dfrac{\text{seconds}}{\text{hour}}} = 2.09 \times 10^{14} \, \text{hours}$$

Again, let's keep going. Now there are 24 hours in a day. So dividing by 24 gives us the number of days.

$$\frac{2.09 \times 10^{14} \, \text{hours}}{24 \, \dfrac{\text{hours}}{\text{day}}} = 8.71 \times 10^{12} \, \text{days}$$

Continuing further, there are approximately 365.25 days in a year, so:

$$\frac{8.71 \times 10^{12} \, \text{days}}{365.25 \, \dfrac{\text{days}}{\text{year}}} = 23.8 \times 10^{9} \, \text{years} = 23.8 \, \text{billion years}$$

Finally something we can make sense out of (sort of). It seems it would take light nearly 24 billion years to reach the top of the stack. That means it would be 24 billion light years tall! Now, the diameter of the known universe is only about 24 billion light years. That means the stack of paper would stretch clear across the known universe!

How could something so simple as 2^{100} give us something so impossible to believe? That is the mystery and the power of an exponential number, and what we'll be discussing in this chapter

We can also visualize exponential numbers and see how they "grow" with increasing exponent value. We do this by graphing this exponential number, versus a changing n value for a fixed "a" value.

For example, consider the exponential number 2^n. If we look at how this number grows by calculating its value for increasing exponent values, we get the table of values:

$n-$ value	2^n
0	$2^0=1$
1	$2^1=2$
2	$2^2=4$
3	$2^3=8$
4	$2^4=16$
5	$2^5=32$
6	$2^6=64$
7	$2^7=128$
8	$2^8=256$

We can obtain a picture of what is happening by graphing the 2^n value versus the corresponding n value. We have shown this in the graph below. We plotted the values in the table above and then connected the "dots" with a smooth curve to get a more complete picture. Notice the shape of the graph. This is the characteristic shape of how exponential numbers grow. This is called exponential or **geometric** growth.

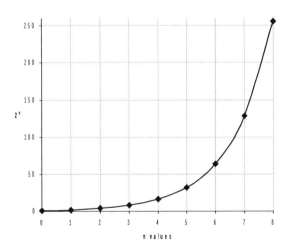

We can contrast this with what we call linear of **arithmetic** growth. Arithmetic growth is the type of growth humans tend to be "programmed" to expect, which is why exponential growth surprises us.

Definitions:

> **Exponential** or **geometric growth** is characterized by the expression a^n, where the number a (also called the base) is a fixed positive number, and the number n is an arbitrary positive number. The number a^n will grow, become larger in value, as the exponent n increases. For exponential growth n must be a positive number.

> **Linear** or **arithmetic growth** is characterized by the expression $a\,n$, where the coefficient a is a positive constant and the number n, which is also positive, is allowed to increase. For linear growth both a and n must both be positive.

Examples of both exponential (geometric) and linear (arithmetic) growth are shown in the table below. In addition we have also graphed both sets of numbers on the same graph below, to the right of the table, to highlight the significant differences.

Number	Exponential/ Geometric	Linear/ Arithmetic
n	2^n	$2\,n$
0	$2^0 = 1$	$2 \times 0 = 0$
1	$2^1 = 2$	$2 \times 1 = 2$
2	$2^2 = 4$	$2 \times 2 = 4$
3	$2^3 = 8$	$2 \times 3 = 6$
4	$2^4 = 16$	$2 \times 4 = 8$
5	$2^5 = 32$	$2 \times 5 = 10$
6	$2^6 = 64$	$2 \times 6 = 12$
7	$2^7 = 128$	$2 \times 7 = 14$
8	$2^8 = 256$	$2 \times 8 = 16$

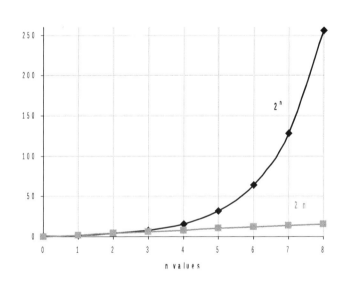

Notice how slowly the arithmetic number grows when compared to the geometric number. Arithmetic growth is also called straight–line growth, since its growth pattern is just a straight line as shown in the graph above. The slope of that line is a. We show this for several different values of a in the graph below.

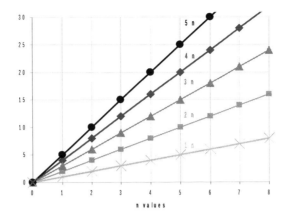

The graph shows five (5) straight lines, each corresponding to a different value of a, from $a = 1$ to $a = 5$. This shows the fundamental characteristic of arithmetic growth, and how we recognize it graphically; it's a straight line.

Exponential (geometric) growth, on the other hand, is quite different. Notice that it is not identified with a straight line, but rather it is a curve that as we move along the horizontal axis the curve continues to get steeper and steeper. The arithmetic growth graph has the same steepness throughout the graph for any particular value of a. It has a constant rate of growth. The exponential growth, is NOT constant.

> **Arithmetic (linear) growth** does NOT depend on the current amount present to determine the future amount, it only depends upon the constant growth amount.

> **Geometric (exponential) growth** depends upon the current amount present to determine the future amount.

A simple tool to identify arithmetic and geometric growth.

If we are able to plot the data, we are better able to distinguish between arithmetic and geometric growth. There is, however, a simple technique we can use without having to plot the data.

If we look at the data from the table above for the geometric and arithmetic data and subtract the previous value from the current value for different n's, we see a pattern.

	Geometric	Difference (current minus previous)		Arithmetic	Difference (current minus previous)
n	2^n			$2n$	
0	$2^0=1$	No previous		$2\times0=0$	No previous
1	$2^1=2$	$2-1=1$		$2\times1=2$	$2-0=2$

	Geometric	Difference (current minus previous)	Arithmetic	Difference (current minus previous)
n	2^n		$2n$	
2	$2^2=4$	$4-2=2$	$2\times2=4$	$4-2=2$
3	$2^3=8$	$8-4=4$	$2\times3=6$	$6-4=2$
4	$2^4=16$	$16-8=8$	$2\times4=8$	$8-6=2$
5	$2^5=32$	$32-16=16$	$2\times5=10$	$10-8=2$
6	$2^6=64$	$64-32=32$	$2\times6=12$	$12-10=2$
7	$2^7=128$	$128-64=64$	$2\times7=14$	$14-12=2$
8	$2^8=256$	$256-128=128$	$2\times8=16$	$16-14=2$

If we look at the difference column for the two sets of number–differences, the first thing we see is that for the arithmetic growth the difference is constant. In this case it is always equal to two (2). Notice that this is the same as the a value used to compute these values. This is no accident, it will always work out to be this value if the growth is arithmetic.

If we look at the difference column for the geometric growth we also see something interesting. We get exactly the same numbers in the previous column, only now they are shifted down a row. This is also no coincidence. This is true of all geometric growth. Also, the first number after the one (1) in the column equals two (2), the a value (base) for the geometric growth. Again, this is no accident. It will always equal this value.

Thus, we have a simple test:

If we subtract previous values from current values in a set of numbers and:

a. the value is constant throughout, then it is **arithmetic growth**, and the difference value is equal to a.

b. the number is NOT constant, but continues to increase exactly like the original set of numbers, only shifted down a row, then it is **geometric growth**, and the a value (base) is equal to the first number after the one (1) in the list (column).

Examples: Determine if the sets of numbers are arithmetic or geometric growth.

1.

n	Value
1	2
2	4
3	6

n	Value
4	8
5	10
6	12
7	12
8	18

We begin by subtracting the former value in the table from the next value, i.e.,

n	Value	Difference
1	2	No Previous
2	4	$4 - 2 = 2$
3	6	$6 - 4 = 2$
4	8	$8 - 6 = 2$
5	10	$10 - 8 = 2$
6	12	$12 - 10 = 2$
7	14	$14 - 12 = 2$
8	16	$16 - 14 = 2$

We notice that the value in the third column equals the same value of 2, so this is arithmetic growth.

2.

n	Value
1	1
2	2
3	4
4	8
5	16
6	32
7	64
8	128

We begin by subtracting the former value in the table from the next value, i.e.,

n	Value	Difference
1	1	No previous
2	2	$2 - 1 = 1$
3	4	$4 - 2 = 2$
4	8	$8 - 4 = 4$
5	16	$16 - 8 = 8$
6	32	$32 - 16 = 16$
7	64	$64 - 32 = 32$
8	128	$128 - 64 = 64$

We notice that the value in the third column is not constant. It equals the same value in the second column, only shifted down one row. This is an example of geometric/exponential growth.

Why should we care about any of this? Are exponential numbers really that important?

10.3 Why are Exponential Numbers Important?

The examples above highlight some interesting features of exponential numbers. What the examples don't tell us, though, is that populations tend to follow an exponential growth curve, and this could be a major problem for us. In 1798 Thomas Robert Malthus (1766–1834), a reverend and economic scholar, wrote an essay titled, *An Essay on the Principle of Population*, that made some dire predictions concerning overpopulation of the human race. Malthus believed that unrestrained population growth would end progress towards a utopian society, and that immediate steps should be taken to avoid the consequences. He made the case that populations grow exponentially, but that the resources to support those populations do not. He believed that the resources needed to

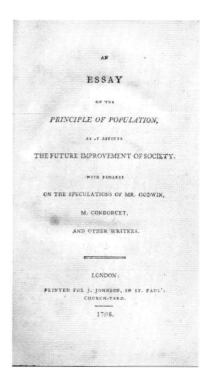

support the population grow arithmetically. Consequentially, there will come a time when there are too many people for the available resources. Malthus' work was very influential, and had a significant impact on people such as Charles Darwin (1809–1888), an English naturalist who came up with the evolutionary theory of natural selection. Malthus' work is still cited and debated to this day, over 200 years later.

Malthus was correct on one account. Populations do grow exponentially. This can be seen by looking at the world population from the beginning of the common era to the present day. From the graph below, you can see that over time the world population follows the characteristic exponential shape.

The graph is not totally smooth, as it has a few bumps in the road. There are a few dips in the graph from time to time, especially during the occurrence of the Black Death and the Plague during the middle ages, as shown in the inserted graph below. Even with these changes, though, the world population growth still managed to recover quickly and resume its exponential growth pattern.

The exponential growth of the world's population is essentially the reason that Albert Bartlett made the bold assertion we presented at the beginning of this chapter regarding the "greatest shortcoming of the human race."

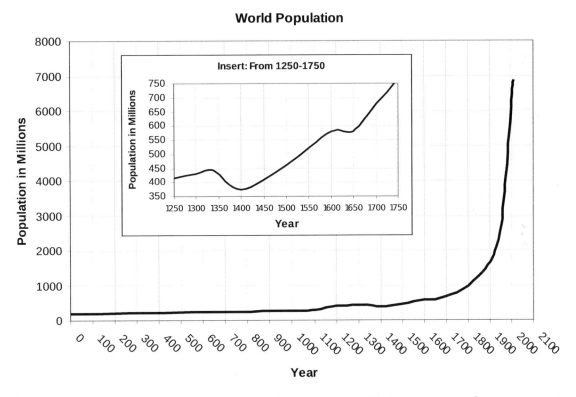

Like Malthus, and numerous others, Bartlett is concerned with how our population is growing unchecked. You can log onto YouTube on the web and see several videos of Professor Bartlett presenting his case for his grand claim. We highly recommend that you look at his videos. They are quite informative.

192

Bartlett (see picture to the left) makes the compelling case that if the world population increases at the rate of 1.3% per year, in 780 years the human population will "reach a density of one person per square meter on the dry region of the earth." Furthermore, in 2400 years "the mass of the world's population will equal the mass of the earth!"

Bartlett presents numerous compelling arguments, suggesting that population growth needs to be addressed without delay. Bartlett further asserts that zero population growth is going to happen, whether we like it or not, and that either we choose how to obtain it, or nature will choose for us. Bartlett presents both options, which we show in the table below.

Table of Options	
Increase Population	**Decrease Population**
Procreation	Abstention
Motherhood	Contraception/Abortion
Large Families	Small Families
Immigration	Stopping Immigration
Medicine	Disease
Public Health	War
Sanitation	Murder/Violence
Peace	Famine
Law & Order	Accidents
Scientific Agriculture	Pollution
Accident prevention	Smoking
Clean Air	
Ignorance of the Problem	

Things we choose from the left make the problem worse. All those are "American as apple pie," and even though they are the core of our value system, they have to be re–examined if we are to achieve zero population growth. If we don't make the tough choices, nature will make them for us, according to Bartlett.

Bartlett suggests that, given the list above, nobody would advocate for things in the right column. "Do you want more war, famine, murder, or disease?" We recognize that these are all undesirable things. Items in the left column are considered good, but they only make the problem worse. Everything from the left column makes the problem worse, and only the negative things from the right column help. Bartlett says that if we don't choose from the list, nature will choose for us, since eventually we will have to have zero population growth.

Another advocate for curbing population growth is Paul Ehrlich (1932–present) (see picture to the left), a biology professor in population studies from Stanford University, who in 1968 wrote a book titled, *The Population Bomb*. The book was essentially a modernized restatement of Malthus' earlier work.

If you listen to Bartlett, Ehrlich, and others, the evidence seems strong and irrefutable. Continued population growth is a problem that needs to be addressed ASAP, otherwise we will face some pretty dire consequences. Or is it?

As we have frequently seen, there are always two sides to every issue. While there are many who support Malthus' original argument there are others who strongly disagree. They don't deny that the world population is growing at an exponential rate. They instead question whether or not our resources are growing arithmetically, and whether or not we can support the growing population. Some in fact think exponential growth is actually necessary, if we as a species are going to survive for an extended period–of–time. We take up both sides of this fascinating debate in the next section.

10.4 Sustainability Versus Uninhibited Growth

A fundamental question is whether or not we can have sustainable growth. For Thomas Malthus, Paul Ehrlich, Albert Bartlett and others, sustainable growth is an oxymoron (two terms that contradict each other). Many believe that you cannot sustain growth permanently. Eventually, what is growing will overtake the limits of its environment, and growth will not be possible. Every environment has a limit called its **carrying capacity**, and populations above this are not possible. It is also a fact that ANY percentage increase in growth will eventually double in size.

The time it takes the population to double is called the doubling time. The doubling time, measured in years, can be approximated using the formula

$$\text{Doubling Time in Years} = \frac{70}{\text{Yearly Growth Rate as a Percentage}}$$

(This formula is derived using something called logarithms which is beyond the scope of this book, so we ask the reader to accept the validity of the formula without proof.)

Thus, to approximate the doubling time you simply compute 70 divided by the yearly percentage growth rate and the result is how many years it will take the population to double in size.

Examples:

1.   What is the doubling time if the yearly rate of growth is 7%?

$$\frac{70}{7\%} = 10 \text{ years}$$

2.   What is the doubling time if the yearly rate of growth is 2%?

$$\frac{70}{2\%} = 35 \text{ years}$$

We can also use this formula to find a particular growth rate, given a predefined doubling time in years, as the next example illustrates.

3.   What is the yearly rate of growth if the doubling time is 50 years?

$$50 \text{ years} = \frac{70}{n\%}$$

We can rewrite this equation using basic algebra as:

$$n\% = \frac{70}{50 \text{ years}} = 1.4\%$$

All this seems straightforward and obvious. Uncontrolled population growth is a problem, so why all the questions. The questions arise because the carrying capacity limit is true for most species, but humans have a unique ability to alter their environment. Thus, the carrying capacity of the earth for humans, years ago, is not the same as it is today. Furthermore, it will likely be different in another thousand years. This is due to the technological improvements humans make to their environment. When humans were hunter gatherers the carrying capacity of the earth was estimated to be about 100 million. Today, it is estimated as high as 40 billion. This is probably the extreme limit at present, and could not be realized unless the standard of living for the top 25% of the richer world's population was reduced significantly. New discoveries and innovations, however, are constantly changing the picture.

This leads us to another distinct voice in the discussion; an economics professor by the name of Julian Simon (1932–1998). Simon says (couldn't resist saying it) that overpopulation is not the problem, but rather the solution. This sounds outrageous when we look at the graph of the population growth presented earlier. However, Simon is not misguided or irrational. He was once a believer in the Malthusian doctrine, but changed his view after he started to look more carefully at the data. He was once part of the doomsayers, warning of the drastic effects of over population, but eventually devoted the last part of his life to being what has been described as a "doomslayer."

Simon looked at all the historical data dating back as far as we have accurate data for, and found that the resources we need to live on have not been getting less scarce, but are in fact more plentiful and relatively less expensive today. According to Simon, resources do not increase arithmetically as

Malthus has suggested, but grow geometrically, along with the population. Also, the cost of these resources has become less expensive over time, and not more (adjusting for inflation). In general, each passing decade provides us with an easier and less expensive style of living. This is because of the ability of human creativity and ingenuity to discover less expensive ways of obtaining the resources we need, or in finding new types of resources to accomplish the same thing as other resources, but more efficiently and effectively. According to Simon, we should not try and reduce population growth, but encourage it, because it is the collective innovation of humanity as a whole that allows us to survive and prosper. There are a series of video interviews with Julian Simon on YouTube that we recommend the reader watch. Simon is able to present his own views in his own words very effectively.

Interestingly enough, in 1980 there was a famous wager between Julian Simon, and Paul Ehrlich. The purpose of the wager was to challenge and test the predictions of their opposing views. Simon allowed Ehrlich to pick 5 natural resources and the bet by Simon was that in 10 years time, these resources would be less expensive and more abundant than they were in 1980, adjusting for inflation. When the data was examined in 1990, Simon had won the bet! Thus, there may be more to the story than Malthus, Ehrlich and Bartlett are presenting.

Another interesting piece of data is that it appears that humans are already reducing their population growth rate without imposing any catastrophic changes in how we live, and without having nature to intervene. Even over the period of one year we see the changes across the world in population growth rates presented below.

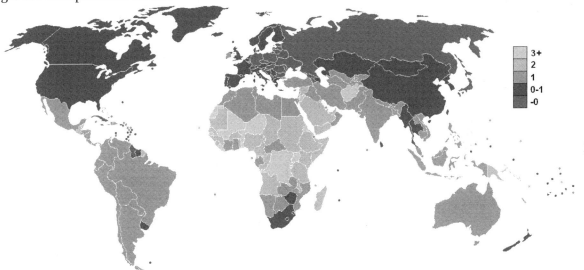

Estimated World Population Growth Rates (in percent) in 2005

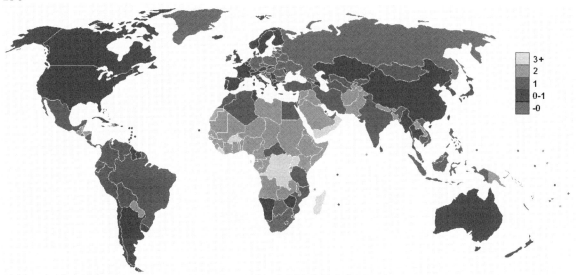

Estimated World Population Growth Rates (in percent) in 2006

What a difference a year makes! If you look at the two world maps above, you can see some significant changes in the growth rates of many countries, in just a single year. This is especially true in South America and Africa. The growth rates of countries in these regions is significantly less in 2006 that it was in 2005. Again, these are still just estimates, but the trend looks promising.

If we then account for the net world population growth rate by including data from all the countries dating back from 1950, another interesting trend is discovered. The world population growth rate on average is declining, as shown in the graph below. This chart also shows growth rate projections for the future.

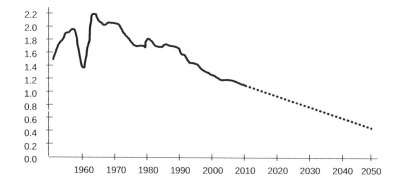

What we see here is the power of mathematics at work. Mathematics allows us to quantify and visualize the patterns to better understand our world. Mathematics can often provide an objective look at what is really happening.

Does that mean that the mathematics is telling us that we don't have to be concerned about population growth – not necessarily. The mathematics of exponential numbers has highlighted two very important aspects. First, when analyzing the data using the tools of mathematics, we are better at sorting through the actual versus the speculated impact. Second, and as equally important, we see populations growing exponentially. Why is this so important? If you look at an exponential graph of population growth, you see that if the horizontal axis represents time, and the vertical axis is population, the time it takes for populations to grow the same amount is rapidly being reduced as we advance in time. This means our response time to be able to make corrective changes is also decreasing rapidly. Consequently, since we cannot guarantee that we know everything about how the world works and what the total impact of human population growth will have on it, we have to watch it carefully, because we will have little or no time to respond when and if there is a problem. Could global warming be such a consequence – only time and mathematics will tell?

Humans do not naturally see things exponentially, which is why we must rely on mathematics to guide us in our understanding of hidden patterns. This is why Ehrlich, Bartlett and others are so concerned. They see our limited ability to correct things, if we have to. On the other hand, Simon and others don't want us to overreact and stifle the positive growth and innovation we've experienced since the beginnings of humanity on this planet. It is a hard choice either way.

As with most things in life, this is probably not an either or question, but rather one of careful scrutiny and balance.

10.5 Other Uses of Exponential Numbers

Exponential numbers have many other uses beyond population growth. We see this every day in the impact on our wallets. Our bank accounts, and investments grow exponentially. Our cost to live, also grows in an exponential way. In this section we look more closely at the impact of exponential numbers on finance. At the very end, though, we will also return to the problem of population growth with some actual problems. To do any of this we first need to do some mathematics.

We begin by considering how an investment can grow. Again we can have arithmetic or geometric growth. Arithmetic growth is also called simple growth, and is modeled using something called simple interest, when we are talking about finances.

Simple Interest

Simple interest does not depend on the current amount present, it only depends upon the initial amount invested. Furthermore, simple interest only adds a fixed amount during the interest period, and that amount does not change.

Let' see how a formula for simple interest can be developed.

Assume that you have $1,000 in an account and that you have agreed to receive 2% simple interest on the initial investment, to be added on a yearly basis. This is how your investment will grow. Each year you will receive a fixed amount of 2% of $1,000 or

$$2\% \text{ of } \$1,000 = 0.02 \times \$1,000 = \$20$$

After the first year has ended, you will now have $1,020 in the account. At the end of year two you will have $1,040 in the account, etc. Your account will grow by the fixed amount of the yearly interest rate times the initial investment.

If we let "r" represent the yearly interest rate written as a decimal, and not as a percentage, and "P" represent the amount of the initial investment, also called the principal, our yearly interest earnings I can be given with the formula:

$$I = r \times P \quad \text{or simply as} \quad I = rP$$

Thus, the amount, A, we have in the account after t years is given as:

$$A = P + It = P + Prt = P(1 + rt)$$

Giving us our **Simple Interest Formula**:

$$\boxed{A = P(1 + rt)}$$

Thus, after 10 years with the problem above, you would have a total of:

$$A = \$1{,}000(1 + 0.02 \cdot 10) = \$1{,}000(1.2) = \$1{,}200$$

Consider some additional examples on how the Simple Interest Formula can be used.

Examples:
1. You invest $10,000 at a simple yearly interest rate of 1.3%. How much will you have in the account after 10 years? What is the total amount of interest you earned?

   To find the amount A, you need to first find P, r and t. P is the principal or the initial investment amount of $10,000, r is the yearly interest of 1.3%, which we have to write as a decimal, so r = 0.013, and t is the time in years, which is 10.

   P = $10,000, r = 0.013, and t = 10. Substituting these into the formula above we have:

   $$A = \$10{,}000(1 + 0.013 \cdot 10) = \$10{,}000(1 + 0.13) = \$10{,}000(1.13) = \$11{,}300$$

   To find the amount of the interest earned, we simply subtract the original investment, P, from the final amount, A.

   $$I = A - P = \$11{,}300 - \$10{,}000 = \$1{,}300$$

In practice, simple interest is not often used for investments. It is, however, used sometimes for short–term loans. Here are examples of two such loans.

2. You borrow $3,000 at a simple **monthly** interest rate of 2%. How much will you owe after 8 months?

   To find the amount A, you need to first find P, r and t. P is the principal or the initial loan amount of $3,000, r is the monthly interest of 2%, which we have to write as a decimal, so r = 0.02, and t is the time in months (if r is in months, t is in months), which is 8.

P = $3,000, r = 0.02, and t = 8. Substituting these into the formula above we have:

$$A = \$3,000(1+0.02 \cdot 8) = \$3,000(1+0.16) = \$3,000(1.16) = \$3,480$$

3. You borrow $4,500 at a simple yearly interest rate of 12%. How much will you owe after 9 months?

    To find the amount A, you need to first find P, r and t. P is the principal or the initial loan amount of $4,500, r is the yearly interest of 12%, which we have to write as a decimal, so r = 0.12, and t is the time in years (if r is in years, t should be written in years), which is 9/12 = 0.75 years.

    P = $4,500, r = 0.12, and t = 0.75. Substituting these into the formula above we have:

    $$A = \$4,500(1+0.12 \cdot 0.75) = \$4,500(1+0.09) = \$4,500(1.09) = \$4,905$$

You can also use it to find out the simple interest rate you are paying, given an initial and final amount.

4. You borrow $5,400 and agree to pay back $6,000, after a year. What is the simple yearly interest rate you have agreed to pay?

    This requires that we solve the simple interest formula for our interest rate r, which we do and obtain:

    $$r = \frac{A-P}{Pt}$$

    Using this formula we can find r by evaluating the formula for A = $6,000, P = $5,400, and t = 1 year.

    $$r = \frac{\$6,000-\$5,400}{\$5,400 \cdot 1} = \frac{\$600}{\$5,400} = 0.1\overline{1}$$

    Writing this as a percent we obtain: $r \approx 11.11\%$, or approximately 11% if rounded to the nearest percentage.

Simple interest is rarely used, except in private and informal exchanges of money. Instead most money transactions involve something called compound interest.

Compound interest

Compound interest calculates a final amount by calculating the interest on the current amount during the interest accruing period. This current amount includes the prior interest earned along with the original principal amount. A growth amount calculated on the current amount is the definition of geometric (exponential) growth.

We now derive a formula for this type of growth. We'll use the same example we used above for deriving the simple interest formula. We invest $1,000 at 2% yearly interest, but now we must choose a compounding period. Let's start with a yearly compounding period, and we'll later change this.

Thus, we start with $1,000 and after 1 year we have $1,020. However, during the next year we earn 2% of $1,020, which is now $20.40 that must be added to the previous years amount. This gives us $1,040.40 after 2 years.

Let's put this in a table to organize it a bit better:

Years from loan beginning	Amount at beginning of compounding period $A_{beginning}$	Interest Earned $I = A_{beginning} \cdot r$	Amount at end of compounding period $A_{end} = A_{beginning} + I$
0	$1,000.00	$1,000.00·(0.02) = $20.00	$1,000.00+$20.00 = $1,020.00
1	$1,020.00	$1,020.00·(0.02) = $20.40	$1,020.00+$20.40 = $1,040.40
2	$1,040.40	$1,040.40·(0.02) = $20.81 *	$1,040.40+$20.81 = $1,061.81
3	$1,061.81	$1,061.81·(0.02) = $21.36	$1,061.81+$21.36 = $1,083.17
4	$1,083.17	$1,083.17·(0.02) = $21.66	$1,083.17+$21.66 = $1,104.83
5	$1,104.83	$1,104.83·(0.02) = $22.10	$1,104.83+$22.10 = $1,126.93
6	$1,126.93	$1,126.93·(0.02) = $22.54	$1,126.93+$22.54 = $1,149.47
7	$1,149.47	$1,149.47·(0.02) = $22.99	$1,149.47+$22.99 = $1,172.46
8	$1,172.46	$1,172.46·(0.02) = $23.45	$1,172.46+$23.45 = $1,195.91
9	$1,195.91	$1,195.91·(0.02) = $23.92	$1,195.91+$23.92 = $1,219.83
10	$1,219.83		

* At this point in the table calculations we are rounding all our answers to the nearest penny. Thus, the answer will not be exact, but will be approximate. The answer closer to the exact value is actually $1,218.99, as we'll show later with a formula we develop. The difference comes about from rounding between steps, which the formula does not do.

The difference between the compounded value in the table and the simple interest value from our earlier problem is:

$$\$1,219.83 - \$1,200 = \$19.83$$

This doesn't seem like much in this example, but once we derive a simple formula to use and calculate other examples, we'll see a rather dramatic difference for certain examples.

We now want to take the process we have written in the table above and duplicate it for any principal value, P, and interest rate, r.

After one compounding period (one year for this example) the results is exactly the same as the simple interest case or:

$$A_1 = P+I = P+Pr = P(1+r)$$

Now, for the next compounding we compute the interest on A_1 which is equal to $rA_1 = rP(1+r)$. Add this amount to A_1 to get A_2:

$$A_2 = A_1+I_1 = P(1+r)+rP(1+r) \quad \text{Factor out a (1+r) term}$$
$$= (1+r)(P+rP) = (1+r)P(1+r) \quad \text{Factor a P out from the second term}$$
$$= P(1+r)(1+r) = P(1+r)^2 \quad \text{Simplify}$$

Notice we have the exponent 2 appearing in our calculation. If we compound again we have:

$$A_3 = A_2+I_2 = P(1+r)^2+rP(1+r)^2 = P(1+r)^2(1+r) = P(1+r)^3$$

Continuing the process we see that after m compoundings we would have,

$$A_m = P(1+r)^m$$

Again we see our exponential number with the exponent m.

If we were to compound n times per year, then this formula would equal the amount we'd have after

$$\frac{m}{n} = t \text{ years}$$

Solving this formula for m we have, m = nt. The interest rate for each compounding is r/n, which when we substitute into the equation for A_m, and drop the subscript m for the A, we obtain the

Compound Interest Formula:

$$\boxed{A = P\left(1+\frac{r}{n}\right)^{nt}}$$

Where:

$$P = \text{Initial Amount (Principal)}$$
$$r = \text{Yearly Interest Rate as a decimal}$$
$$n = \text{Number of Compoundings per Year}$$
$$t = \text{the Number of Years}$$
$$A = \text{The Final Amount}$$

Instead of using the table above we can now simply substitute the required numbers into the formula and compute the final amount, A, as follows:

$$P = \$1,000$$

$$r = 0.02$$
$$n = 1$$
$$t = 10$$

$$A = P\left(1+\frac{r}{n}\right)^{nt} = \$1{,}000\left(1+\frac{0.02}{1}\right)^{1\cdot10} = \$1{,}000\,(1.02)^{10} = \$1{,}000\cdot(1.218994\ldots) = \$1{,}218.99$$

The new variable we have in this formula is the number of compoundings per year, n. This requires some additional explanation. The parameter n can either be provided directly as a simple counting number, or it can be indicated indirectly using the following terminology.

If we say the interest is compounded,

yearly or **annually**,	then n =1,	or one compounding per year
semi–annually,	then n = 2,	or two compoundings per year
quarterly,	then n = 4,	or four compoundings per year
monthly,	then n = 12,	or twelve compoundings per year
weekly,	then n = 52,	or fifty–two compoundings per year
daily,	then n = 365,	or three–hundred and sixty–five compoundings per year

Consider the following examples on how to calculate with this formula.

Example 1:

You invest \$25,000 at a yearly interest rate of 6.3% compounded monthly. How much will you have in the account after 10 years? How much would you have earned using simple interest?

Solution: To find the amount A, you need to first find P, r, n and t.

P is the principal or the initial loan amount, so P = \$25,000,
r is the yearly interest of 6.3%, which we have to write as a decimal, so   r = 0.063,
n is the number of compoundings per year, which is 12 for being compounded monthly, and
t is the number of years, so t = 10.

Substituting these values into the Compounded Interest Formula above we have:

$$A = P\left(1+\frac{r}{n}\right)^{nt}$$

$$= \$25{,}000\cdot\left(1+\frac{0.063}{12}\right)^{12\cdot10} = \$25{,}000\cdot(1+0.00525)^{12\cdot10}$$

$$= \$25{,}000\cdot(1.00525)^{120} = \$25{,}000\cdot(1.87451885\ldots)$$

$$= \$46{,}862.97$$

Make sure you follow your order of operations when doing this calculation. To get the correct answer you must do the computations in this order. We are working from the inner parentheses out.

Process	Example Above
1. Compute r/n,	$0.063/12 = 0.00525$
2. Add 1 to the result in step–1 (1),	1.00525
3. Multiply n times t,	12 times 10 = 120
4. Raise the results in step–two (2) to the value in step–three (3),	$(1.00525)^{120} = 1.874518854...$
5. Multiply the result in step–four (4) by P,	$25,000 times 1.874518854... to get the final result, $46,862.97.

Using the simple interest formula we find that the amount, A, we would have after 10 years is:

$$A = P(1+rt) = \$25,000\left[1+(0.063)\cdot 10\right] = \$25,000\cdot(1.63) = \$40.750.00$$

Just by having your interest compounded monthly you earned an additional $6,112.97 !

What is happening here is that you are earning interest on your interest. That is the very definition of compounding.

Example 2:

You invest $15,000 at a yearly interest rate of 3.9% compounded quarterly. How much will you have in the account after 25 years? How much would you have earned using simple interest?

Solution: To find the amount A, you need to first find P, r, n and t.

P is the principal or the initial loan amount, so P = $15,000,
r is the yearly interest of 3.9%, which we have to write as a decimal, so   r = 0.039,
n is the number of compoundings per year, which is 4 for being compounded quarterly, and
t is the number of years, so t = 25.

Substituting these values into the Compounded Interest Formula above we have:

$$A = P\left(1+\frac{r}{n}\right)^{nt}$$

$$= \$15,000\cdot\left(1+\frac{0.039}{4}\right)^{4\cdot(25)} = \$15,000\cdot(1+0.00975)^{4\cdot(25)}$$

$$= \$15,000\cdot(1.00975)^{100} = \$15,000\cdot(2.638676712...)$$

$$= \$39,580.15$$

Using the simple interest formula from above, we find that the amount, A, we would have after 25 years is:

$$A = P(1+rt) = \$15,000\left[1+(0.039)\cdot25\right] = \$15,000\cdot(1.975) = \$29,625.00$$

Compounding quarterly would earn your nearly $10,000 more !

Other things that grow exponentially are the value of homes or the cost of health care. The same formula we derived for the compounding of an investment, also applied to those two examples. We illustrate the process using two examples.

Example 3:

You can purchase a starter home today on Long Island in some areas for $225,000. If the value of the home appreciates at 7% per year, how much money would you need to purchase that same home in 15 years?

Solution: To find the amount A, you need to first find P, r, n and t.

P is the current value of the home, so P = $225,000,
r is the yearly appreciation rate of 7%, which we have to write as a decimal, so   r = 0.07,
n is the number of compoundings per year, which for appreciation will be assumed to be once yearly, and
t is the number of years, so t = 15.

Substituting these values into the Compounded Interest Formula above we have an estimate for the value of the home 15 years from now:

$$A = P\left(1+\frac{r}{n}\right)^{nt}$$

$$= \$225,000\cdot\left(1+\frac{0.07}{1}\right)^{1\cdot(15)} = \$225,000\cdot(1.07)^{1\cdot(15)}$$

$$= \$225,000\cdot(1.07)^{15} = \$225,000\cdot(2.759031541...)$$

$$= \$620,782.10$$

The cost of health care has been growing at a rate much greater than inflation. Healthcare costs have risen at values at greater than 10% in recent years. Consider the following example.

Example 4:

An average cost of an inpatient surgical procedure in 2003 was $40,000. Between 2003 and 2007 costs rose from 25% – 41% with an average yearly percent increase of 8.25%. If the average percent increase stays the same what will the cost of an average inpatient surgery be in 2020?

Solution: To find the amount A, you need to first find P, r, n and t.

P is the 2007 value of the surgery, so P = $40,000,

r is the yearly appreciation rate of 8.25%, which we have to write as a decimal, so
r = 0.0825,

n is the number of compoundings per year, which for inflation will be assumed to be once yearly, and

t is the number of years, so t = 2020– 2007 = 13 years.

Substituting these values into the Compounded Interest Formula above, we have an estimate for the cost of inpatient surgery in 2020 as:

$$A = P\left(1+\frac{r}{n}\right)^{nt}$$

$$= \$40,000\cdot\left(1+\frac{0.0825}{1}\right)^{1\cdot(13)} = \$40,000\cdot(1.0825)^{1\cdot(13)}$$

$$= \$40,000\cdot(1.0825)^{13} = \$40,000\cdot(2.802610633...)$$

$$= \$112,104.43$$

The average cost for inpatient surgery will nearly triple in 13 years ! This is why many are so concerned over the rising costs of health care.

Continuous Compounding

In all the previous examples we have always managed to compound a finite number of times. An obvious question to ask is – what happens if instead of every day, hour, minute, second, or microsecond, we instead compound on a "continuous" basis? The answer is rather surprising, and is best understood by considering a simple made–up example.

Let's consider that we are starting with a Principal value of 1, and we are curious how much this increases in a single year, as the number of compoundings increases without end. For this problem, the compounding formula would be given as:

$$A = P\left(1+\frac{r}{n}\right)^{nt} = \left(1+\frac{r}{n}\right)^{nt}$$

Let us now make a substitution, and replace n with the value r times m. This changes the formula to the form:

$$A = \left(1+\frac{r}{mr}\right)^{mrt} = \left(1+\frac{1}{m}\right)^{mrt} = \left[(1+\frac{1}{m})^m\right]^{rt}$$

We will no focus on the term: $\left(1+\dfrac{1}{m}\right)^m$ and ask what happens as m becomes larger and larger?

What happens to the value of expression above? It is easier if we put the value of this expression, for different values of increasing m, into a table to see if we can find a pattern.

m	$\left(1+\dfrac{1}{m}\right)^m$
1	2.000000000
10	2.593742460 …
100	2.704813829 …
1,000	2.716923932 …
10,000	2.718145927 …
100,000	2.718268237 …
1,000,000	2.718280469 …
1,000,000,000	2.718281827 …

If you notice, as we let m get larger and larger (this is equivalent to letting n increase without bounds), the value of the expression $\left(1+\dfrac{1}{m}\right)^m$, appears to get closer and closer to a specific fixed number, which is approximately 2.71828182. It turns out that this is a very special irrational number (we'll learn more about irrational numbers shortly) like π. It was discovered by a Swiss mathematician Leonard Euler (1707–1783) (see picture below), pronounced "oiler", and is called, e, the natural base.

$$e \approx 2.71828182\ldots$$

Consequently, as m increases without bounds, and $\left(1+\dfrac{1}{m}\right)^m$ gets closer to 2.71828182 … = e, we can replace it with e in the expression $\left[\left(1+\dfrac{1}{m}\right)^m\right]^{rt}$ and obtain:

$$e^{rt}$$

This gives a new formula for **Continuous Compounding**:

$$\boxed{A = Pe^{rt}}$$

NOTE: To do the exercises at the end of this chapter you will need a calculator that has an "*e*" button on it, or if you have a calculator with either a ∧, y^x, or x^y button, you can use 2.71828182 for the "*e*" value and use either of these keys on the calculator and get approximately the same result. Ask your instructor for help here.

You may run into the number *e* in other courses or in a variety of different settings. It is a very important number in mathematics.

In this lesson, we will use the above formula whenever continuous compounding is called for. It is often used as the replacement formula for population growth and for inflation problems. This is because it is more consistent to model population growth and price changes as continuous processes, rather than assuming it increases at one time at the end of the year. As we'll show, the answers using either formula are not drastically different.

If we are estimating population growth, we typically rewrite the equation by replacing A with P (for population), and P with P_0 (for initial population, at time t = 0), to obtain the

Population Growth Formula

$$P = P_0 e^{rt}$$

If we are estimating inflationary growth, we typically rewrite the equation by replacing P with A and P_0 with A_0 (for initial amount, at time t = 0), to obtain the

Inflationary Growth Formula

$$A = A_0 e^{rt}$$

Examples:

1. The current world population is about 6.8 Billion people. If the net growth rate (birth rate minus death rate) is 5%, what will the population be in 20 years? Assume that the population grows continuously.

    Solution: To find the amount P, you need to first find P_0, r, and t.

    P_0 is the current world population, so P_0 = 6.8B. We will use B for billion and note that the final answer will come out in billions. We don't have to write 6,800,000,000,
    r is the yearly growth rate of 5%, which we have to write as a decimal, so r = 0.05,
    t is the number of years, so t = 20 years.

    Substituting these values into the Population Growth Formula above, we have an estimate for the world's population in 20 years as:

    $$P = P_0 e^{rt}$$
    $$= 6.8B\, e^{(0.05)\cdot 20} = 6.8B\, e^1 \approx 18.48B$$

Such a large increase in only 20 years is why Bartlett and others are concerned. Population growth could easily get out of hand and cause tremendous problems for the world.

2. An average cost of an inpatient surgical procedure in 2003 was $40,000. Between 2003 and 2007 costs rose from 25%–41% with an average yearly percent increase of 8.25%. If the average percent increase stays the same what will the cost of an average inpatient surgery be in 2020?

   Solution: To find the amount A, you need to first find P, r, n and t.

   P_0 is the 2007 value of the surgery, so P_0 = $40,000,
   r is the yearly appreciation rate of 8.25%, which we have to write as a decimal, so
   r = 0.0825,
   t is the number of years, so t = 2020– 2007 = 13 years.

   Substituting these values into the Compounded Interest Formula above, we have an estimate for the cost of inpatient surgery in 2020 as:

   $$P = P_0 e^{rt}$$
   $$= \$40,000\, e^{(0.0825)\cdot(13)} = \$40,000\, e^{1.0725}$$
   $$= \$116,907.08$$

   If we compare this with Example 4 above, we can see a difference of about $4,800. This is only a projection, and it is reasonably close, although the continuous compounding is a more consistent measure.

As you can see exponential numbers are all around us. The fact that humans cannot think exponentially causes problems, and frequently surprises us when we examine outcomes similar to the exercises above. Mathematics tells us we need to watch out for and monitor these numbers, since they have such a large impact on our lives.

To increase your understanding consider the exercises below. The goal is to properly make you aware through concrete examples of the power and surprising nature of these numbers.

EXERCISES 10.1–10.5

ESSAYS AND DISCUSSION QUESTIONS:

1. What is an exponential number?

2. Albert Bartlett said that, "The greatest shortcoming of humankind is our inability to understand the exponential function." While Julian Simon believed the opposite. Explain both views and then give and support your own opinion.

3. Who is Julian Simon and what is his opinion on population growth?

4. Who is Malthus, and what did he do that was so important?

5. Do you believe population growth is a problem? Explain why or why not.

6. What is the difference between arithmetic and geometric (exponential) growth?

7. What is the doubling formula, and what does it mean?

8. What is the difference between simple and compounded growth?

9. What is continuous compounded growth, and what unique number does this lead to?

10. Compare and contrast the beliefs of Bartlett and Simon, and then give your own opinion.

PROBLEMS:

1. Determine if the sets of numbers below represent arithmetic or geometric growth.

a)	b)	c)	d)	e)
2	2	1	2	1
4	4	5	22	3
6	8	9	42	9
8	16	13	62	27
10	32	17	82	81
12	64	21	102	243
14	128	25	122	729
16	256	29	142	2187

2. Find the approximate time it takes a population to double if the growth rate is 7%.

3. Find the approximate time it takes a population to double if the growth rate is 15%.

4. Find the approximate time it takes a population to double if the growth rate is 3.5%.

5. Find the approximate time it takes a population to double if the growth rate is 2.5%.

6. Find the approximate time it takes a population to double if the growth rate is 12%.

7. Find the approximate growth rate if it takes a population 15 years to double.

8. Find the approximate growth rate if it takes a population 100 years to double.

9. Find the approximate growth rate if it takes a population 95 years to double.

10. Find the approximate growth rate if it takes a population 35 years to double.

11. You borrow $2,000 for 6 months at 2.5% simple monthly interest. How much interest did you pay? What is the total amount you must pay on the loan?

12. You borrow $5,000 for 9 months at 10.9% simple yearly interest. How much interest did you pay? What is the total amount you must pay on the loan?

13. Your borrow $3,000 for 6 months at 13.5% simple yearly interest. How much interest did you pay? What is the total amount you must pay on the loan?

14. You borrow $1,500 for 8 months and agree to pay $250 in interest charges (simple interest). What yearly interest rate are you paying?

15. If you invested $10,000 at 7.5% simple yearly interest for 10 years, how much money would you have at the end?

16. If you invested $25,000 at 6 1/3 % simple interest for 20 years, how much money would you have at the end?

17. If you invested $10,000 at 7.5% interest compounded monthly for 10 years, how much money would you have at the end?

18. If you invested $25,000 at 6 1/3% interest compounded weekly for 20 years, how much money would you have at the end?

19. On Long Island you can purchase a starter home today for about $275,000. If the home appreciates at a rate of 7 ¼% per year, How much will a starter home be worth in 25 years when your children will be looking to purchase a home? If you want to give them 15% down on their now home, how much do you need to save to do that?

20. On Long Island you can purchase a starter home today for about $275,000. If the home appreciates at a rate of 9 ¾ % per year, How much will a starter home be worth in 30 years when your children will be looking to purchase a home? If you want to give them 20% down on their now home, how much do you need to save to do that?

21. The current world population is approximately 6.66 billion people. If the population were to grow at a rate of 7% per year, what would be the population in the year 2051 (roughly your retirement age)? Consider the current year to be 2011.

22. Long Island (including NY City) has a population of approximately 7.536 million people. If the population were to grow at 10% per year, how many people would be on the island in 2026?

23. In 2000 Suffolk County had a population of approximately 1.42 million people. The total land area of Suffolk County is 2,373 square miles. How many people per square mile were there in 2000? If the population were to grow at a rate of 10% per year, what will the population on the island be in the year 2045? How many people would there be per square mile?

References

1. Bartlett, A.A., (1993). The Arithmetic of Growth, Methods of Calculation. Population & Environment, 14(4), 359-387.

2. Bartlett, A.A., (1994). Reflections on Sustainability, Population Growth and the Environment. Population and Environment, 16(1), 5-35. Reprinted in: Renewable Resources Journal, 15(4), 1997, 6-23.

3. Bartlett, A.A., (1997) Is There a Population Problem< Wild Earth, 7(3), 88-90.

4. Bartlett, A.A., (1998). Malthus Marginalized. The Social Contract, 8(3), 239-251.

5. Bartlett, A.A., (1986). Forgotten Fundamentals of the Energy Crisis. Dir. and Perf., University of Colorado at Boulder.

6. Bartlett, A.A., (1998). Arithmetic, Population, and Energy.

7. Bartlett, A.A., (2004). The Essential Exponential! For the Future of Our Planet, University of Nebraska.

8. Drexler, K.E., (1986). Engines Of Creation, (see chapter on Limits to Growth).

9. Ehrlich, P.R., (1968). The Population Bomb. New York, Ballantine Press.

10. Hardin, G., (1999). The Ostrich Factor: Our Population Myopia.

11. Hawken, P., (1993). The Ecology of Commerce: A Declaration of Sustainability. New York: Harper Collins.

12. Ikle, F.C., (1994). Our Perpetual Growth Utopia. Natinal Review, 46, March 7, 36-44.

13. Lomborg, B., (2001). The Skeptical Environmentalist.

14. Maddox, J., (1972). The Doomsday Syndrome.

15. Maor, Eli, (1994), e The Story of a Number, Princeton University Press, Princeton, New Jersey.

16. Miller, T.G. (1996). Living in the Environment. New York: Wadsworth Publishing Company.

17. Moore, S., (2000). It's Getting Better All the Time : 100 Greatest Trends of the Last 100 Years by Stephen Moore, Julian Lincoln Simon manuscript finished posthumously by Stephen Moore.

18. Simon, J.L., The State of Humanity: Steadily Improving, Cato Policy Report, 1995 (based on the introduction to his 1995 book, The State Of Humanity)

19. Simon, J. L., (1981). The Ultimate Resource (Princeton: PUP)

20. Simon, J.L., (1996). The Ultimate Resource II.

21. Simon, J. L., (1984). The Resourceful Earth: A Response to "Global 2000" (1984), Julian Simon & Herman Kahn, eds

22. Simon, J. L., (1999). The Economic Consequences of Immigration into the United States, University of Michigan.

23. Simon, J. L., Effort, Opportunity, and Wealth: Some Economics of the Human Spirit

24. Simon, J. L., (1999). The Hoodwinking of a Nation, Transaction Publishers, Newark, N.J.

25. Simon, J. L., (1994). Scarcity or Abundance? A Debate on the Environment, with Norman Myers, WW Norton.

26. Simon, J. L., (1998). Economics of Population: Key Modern Writings, Transaction Publishers, Newark, N.J.

27. Simon, J. L., (1995). The State of Humanity, Blackwell Publishers.

28. Simon, J.L., (1980). Reply, Science, 19 December, 1980b, Vol. 210, No. 4476, pp. 1296-1308.

29. Simon, J. L., (1990). The Phony Farmland Scare, Washington Journalism Review, May, 1990, pp. 27-33.

30. Simon, J.L., (1986). Disappearing Species, Deforestation, and Data, New Scientist, 15 May, 1986, pp. 60-63.

31. Simon, J.L., (1986). Theory of Population and Economic Growth, New York: Basil Blackwell.

32. Simon, J.L., and Wildavsky, A,. (1984). On Species Loss, the Absence of Data, and Risks to Humanity, in The Resourceful Earth, Julian L. Simon and Herman Kahn, eds., 1984, pp. 171-183.

CHAPTER 11

Irrational Numbers

(Irrationals), a miracle of analysis, a monster of the ideal world,
almost an amphibian between being and not being.

Gottfried Wilhelm Leibniz

11.1 Introduction

Even before their was zero, the concept of a totally incomprehensible form of number was discovered. At the time, the followers of Pythagoras who discovered it, tried to put the genie back in the bottle, but to no avail. These new, irrational numbers, were here to stay. Even their given name makes them sound absurd, and not logical. In this chapter we'll investigate why, and hopefully along the way give you'll get a greater appreciation for why we have them, and also what this means in terms of our notion of reality.

11.2 What do you Mean, I can't Measure This Length?

During the sixth century BCE, Pythagoras and the society he founded, the Pythagoreans, were hard at work trying to discover the secrets of the universe. Together they looked towards mathematics to explain the patterns in the world. To them number was everything, and number to them comprised only the natural counting numbers. With these numbers you could unleash the harmony of music, carry out commercial trade, plot the seasons and cycles of the year, predict your future fate, and listen to the harmony of the planets and sun as they moved about in celestial grandeur. Numbers must be everywhere they thought, and they sought them out with a religious passion.

Pythagoras had visited Egypt and was intrigued by a little mathematical tool that they used to create right angles. The Egyptians found that if they created a triangle where one side was three units long and the other side was four units long, that if the longer side was five units long one of the angles of the triangle would always be ninety degrees. This was a great way to build things accurately. The other more amazing thing was that if you took the squares of the two shorter sides and added them together, that their sum was equal to the square of the longer side! Again, the power, mystery, and beauty of numbers at work. But wait, there were other triplets of numbers for which this worked. It was also true for five, twelve, and thirteen, or even eight, fifteen, and seventeen, the list seemed endless. Undoubtedly a mystery was locked in these numbers. There was a basic rule that if one of the angles was ninety degrees (also called a right angle), that if you took the length squared of each of the shorter sides of the triangle and summed the result together, it would always give the length of the longer side squared. In fact, inspired by Thales, they actually proved this result. The power they must have felt, sharing the knowledge of the Gods. Today we write this as:

$$a^2 + b^2 = c^2$$

The followers of Pythagoras set out to study this amazing discovery. One follower by the name of Hippasus, is said to have let the sides a and b equal one, and tried to find a value for c. Now to the Pythagoreans, the answer had to be either a counting number or a ratio of two counting numbers, because that was all there was. Try as he did, Hippasus couldn't find such a pair of numbers, that when divided and squared was exactly equal to the square of the longer side. This could not be, the answer must lie in the mathematics. So he then put the power of proof to work again, but this time to his dismay, a strange and dangerous new idea emerged. There was no such pair of numbers and there could not be, and he could prove it!

For those that are curious, a possible informal version of a proof might go as follows:

Let us assume that the length of this unknown longer side can be written as a ratio of two natural numbers, and furthermore that these two natural numbers have no factors in common. That is, the ratio is in reduced form.

We'll call these unknown numbers n and m, and write $c = \dfrac{n}{m}$. Now if we square c we obtain:

$$(1) \quad c^2 = \frac{n^2}{m^2},$$

By the Pythagorean Theorem this must be equal to the sides of the other two squared and summed, or

$$1^2 + 1^2 = 1 + 1 = 2$$

This implies that

$$(2) \quad \frac{n^2}{m^2} = c^2 = 2 \ \text{ or } \ \frac{n^2}{m^2} = 2$$

If we now multiply (2) by m^2 we have that

$$(3) \quad n^2 = 2m^2$$

We now know something about this number n. Since n^2 has 2 as a factor, n^2 must be an even number. This implies that n must also be an even number. Why? The only way to get an even number is to multiply two even numbers or an even and an odd number. Since the numbers are the same (they are both n), they must both be even.

Okay, since n is even it means we can write it as some undetermined natural number times two, or

$$(4) \quad n = 2p$$

It follows by squaring both sides of (4) that $n^2 = 4p^2$.

Now, let's use this in the second equation for (2) above and replace n^2

$$(5) \quad \frac{n^2}{m^2} = \frac{4p^2}{m^2} = 2 \quad \text{or} \quad \frac{4p^2}{m^2} = 2$$

Divide both sides of the second equation in (5) by 2 and again multiply the result by m^2 to obtain: $m^2 = 2p^2$.

Now this means that m must also be even, by the same argument we used to show that n was even earlier.

Now this is a problem At the very beginning we said that n and m had no factors in common, but if they are both even then they have a factor of 2 common to both. This is a contradiction! This must mean our initial assumption of being able to write $c = \frac{n}{m}$ is wrong! Meaning, there is no such number!

This meant one could NOT measure the length of the longer side. Since the square of its length is not measurable, the length itself could not be. It was deemed an incommensurable (not measurable) and inconceivable number. A frightening hole in the universe of the Pythagoreans. Something their counting numbers could not explain. Hippasus told Pythagoras, and he and the other Pythogoreans were sworn to secrecy. They did not want such heretical information to leave the sect. Their order depended upon it. However, Hippasus was too excited to hold it in and talked about it freely. I mean it was a monumental discovery, how could you keep quiet. Pythagoras, however, could not have his order shaken, so as legend has it, Hippasus was given a nice boat trip that he never returned from. However, the secret was already out.

11.3 What is an Irrational Number Anyway, and Why Should I Care?

We now know that there is a new type of number that cannot be written as a fraction (ratio of numbers). In our basic algebra course we are taught the various sets of numbers and how they are related. We are told about the irrational numbers, as a matter of fact. The teacher points out that we are already aware of them. They talk about π and $\sqrt{2}$ and we nod affirmatively, since we've heard about them for most of our lives. We don't question their existence. In fact, it all seems a bit boring to us. Why all this fuss about numbers, anyway.

We also learn a very important and easy way to distinguish a rational number from an irrational number. It turns out that if we were to write the decimal equivalent of a rational number, it will either terminate (give an unending string of zeros after the last nonzero digit), or there will be a repeating pattern of digits. An irrational number on the other hand, has a non–terminating, non–repeating decimal equivalent.

Examples of rational and irrational numbers and their decimal equivalents.

Rationals	**Irrationals**
$\frac{1}{2}=0.5$	$\sqrt{3}=1.732050808\ldots$
$\frac{2}{3}=0.3333\overline{3}$	$2\sqrt{2}=2.828427125\ldots$
$\frac{16}{25}=0.64$	$\sqrt{(\sqrt{5})}=1.495348781\ldots$
$\frac{4}{33}=0.1212\overline{12}$	$2\pi=6.283185307\ldots$
$\frac{37}{77}=0.480519\overline{480519}$	$\pi^{2}=9.869604401\ldots$

Okay, this seems fairly easy to understand. We have both rational and irrational numbers. The rationals have either terminating or repeating decimal forms, while the irrationals neither terminate nor do they repeat.

We are then told that the irrational numbers "complete" the number line. Most of us never knew there were any holes in it to begin with. It seems that we could always find any point we needed on the line without difficulty, simply using our rationals or decimals. The mathematicians tell us that, yes there were gaps in the number line, and this we'll talk about later. Furthermore, they tell us that the size of the missing irrational number gaps is infinitely greater than the number of rationals on that line. This explanation is a bit confusing, so we really just ignore it for now, and again simply move on.

We are then told that if we group all the irrational numbers together and combine them with all the rational numbers, the combined set makes up the set of real numbers. When we first hear this we take this as a given and move on without thinking about it or questioning it. After all, you can't really question math, it is what it is, right? Maybe it is, maybe it is not. Irrational numbers, however, should give us pause. We should question them, because they are quite different from any number we've yet experienced. What gives them any level of credibility at all is their apparent connection to geometry. The Greeks believed this. To the Ancient Greeks, numbers pretty much took their meaning from geometry. This is why the length of the side of a triangle, in their eyes, must be a number. Every other length was a number so surely this length also had to be one as well. This length, however, could not be measured.

So why should we care? Isn't is just something that just mathematicians are curious about? We should care because it points out a possible problem in how we perceive the world. One of the most fundamental perceptions we have is that matter in the universe is thought to be separated from other matter by space, and that we can measure this separation by calculating the distance between things. Mathematics, however, with the introduction of irrational numbers seems to be telling us that some distances can be measured, but if you move a small amount away, an amount so small we can't even hope to measure it, we go from a measurable distance to an unmeasurable distance. This apparent contradiction begins to highlight just how little we understand about the world and how it all works. Irrational numbers seem to hold one of the keys to understanding reality, and this is why we should care about them.

11.4 Some Special Irrational Numbers

Irrational numbers are all around us. They seem as real as any counting number, and most times even more useful and critical. In this section we look at three important irrational numbers and show how they came about.

11.4.1 Euler's Number

In Chapter 10 we saw that the number e was related to continuous compounded growth. This number is called Euler's number, and was also identified as the natural base. What we didn't show however, is that this number is also an irrational number. It's decimal representation doesn't terminate, nor does it repeat a pattern of digits. All irrational numbers come about by carrying out some process that goes on and on forever. The number e comes from examining the expression

$$\left(1+\frac{1}{m}\right)^m$$

and observing what numerical value this takes on as m gets larger and larger. We saw in Chapter 10, that this appeared to approach a fixed number, but we couldn't find it exactly. The number we approached we called e. If we continued to find more and more of the digits of e we would not see a repeated pattern, nor would the process ever terminate (finding all zeros after a final nonzero digit). This is why we have adopted a special symbol, e, to represent this number.

11.4.2 π

Another number we are all familiar with is π. We all have learned that π is irrational, although some of us have come to the misguided belief that 3.14 is equal to π. What we were actually taught is not that they are equal, but that you can use 3.14 in replace of π and get a good approximate answer to the problem we might be solving.

$$\pi \neq 3.14$$

The real $\pi = 3.1415926535\ 8979323846\ \ldots$, and requires an unending process to define it.

π was known by the ancient Greeks, and Egyptians. However, they didn't know of it as an irrational number, nor was it known by the π symbol we use today. Instead it was known from the geometric observation that the circumference of any circle divided by its diameter, was always the same number. This was a number that they could not find its exact value, which caused quite a bit of interest and eventually in the 1700's it was given the current name of π. How could something so basic and fundamental not have an exact answer?

Initially, during the 1600's, π was used as a symbol to represent the circumference or periphery of a circle. The in 1706 the English writer William Jones used π to represent the ratio of the circumference to the diameter of a circle. Then in 1737 Euler adopted the notation and that sealed the deal (as Euler went, so went mathematics).

The first person to do some rigorous estimation of π was Archimedes (287–212 BCE). Various approximations for π existed in Ancient Egypt, Babylon, and Greece prior to Archimedes. The earliest estimates date from around 1900 BCE. The Egyptians would use the equivalent of $\frac{256}{81} \approx 3.16$, and the Babylonians used the equivalent of $\frac{25}{8} = 3.125$. Both these approximations were more than adequate for their basic calculations. Even today using either 3.14 or $\frac{22}{7} \approx 3.14$, gives sufficiently accurate results for many applications.

Archimedes, however, took it a step further. Instead of looking for an approximation for π, he looked at trying to find out what the actual number equaled. Archimedes was definitely a mathematician before his time, which is why he is always on any list of the greatest mathematicians of all time. Archimedes created the method of exhaustion to estimate π. Again, this is another unending process. The idea goes as follows: We know that π is the ratio of the circumference to the diameter of a circle. If we choose a circle with a diameter, $D = 1$, then the circumference of this circle $C = \pi$.

$$\pi = \frac{C}{D} = \frac{C}{1} = C$$

So if we start with a circle with a diameter equal to one, then the circumference is equal to π. We then use regular polygons, whose perimeters we can find exactly, to estimate π as accurately as we desire. If we use inscribed polygons (polygons inside the circle) the length of its perimeter will be less than π, and using circumscribed polygons (the circle inside the polygon), the length of the perimeter will be greater than π.

Starting with a square, a polygon of four sides, we continue to double the number of sides of the polygon each time (see the picture below.) We see that after a very small number of doublings (4 to 8 to 16), the perimeter of the inscribed and circumscribed polygons gets closer to the length of the circumference of the circle, which for this circle is equal to π.

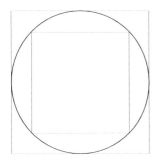

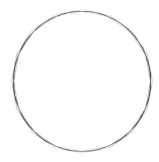

Four–Sided Polygons   Eight–Sided Polygons   Sixteen–Sided Polygons

Archimedes used the equivalent of a 96–sided polygon to come up with the estimate of π to be between $3\frac{10}{71}\approx 3.1408$ (the perimeter of the inner polygon), and $3\frac{1}{7}\approx 3.1429$ (the perimeter of the outer polygon.) If you take the average of these two values, the estimate is less than 1/100 of a percent off the actual value!

Theoretically, you could keep doubling the numbers of sides in the polygons, and both perimeters would get closer and closer to the actual value of π. This, however, is an unending process – the defining feature of an irrational number.

Since Archimedes time many other techniques have been discovered for calculating the value of π and as of today π has been calculated to over 5 trillion digits! Some people, obviously, have a lot of extra time on their hands.

11.4.3 $\sqrt{2}$

The last number we consider is the first irrational number discovered, $\sqrt{2}$. Approximations to $\sqrt{2}$ have existed since the time of the Babylonians in 1800 BCE. If you look back to section 6.5 on Babylonian Fractions in the text, we show a clay tablet with an inscription that is actually giving an approximation to $\sqrt{2}$. This is our first evidence of this number. The value the Babylonians used in this tablet (see Section 6.5 for a picture of the tablet) was:

$$1+\frac{24}{60}+\frac{51}{60^2}+\frac{10}{60^3}=1.41421\overline{296}$$

At the time the Babylonians did not have a concept of irrational numbers, so to them it was just a number to calculate with.

It wasn't for another 1300 years until the Pythagoreans discovered the truth about $\sqrt{2}$, that we had any idea about irrational numbers. The fact that $\sqrt{2}$ could not be written as a ratio of two natural numbers was proven by the Pythagoreans in about 500 BCE. The Greeks called this an incommensurable or inexpressible magnitude (or number). For the first time someone noticed a number that was important for calculations that we could not assign an exact value to. We couldn't even measure its length exactly. This was a very strange concept indeed, especially to the Pythagoreans who had put all their faith in numbers. You could take two sides of a right triangle, both equal to one unit, and you could NOT measure the length of the hypotenuse using any comparative tool.

11.5 Do Irrational Numbers Really Exist?

Although irrational numbers are common place throughout mathematics, we cannot point to anything in the real world that we can say is equivalent to an irrational number. Only mathematical objects are irrational. We cannot measure an irrational quantity, since to do so requires an infinite process of finding a more and more accurate measurement. Although irrational numbers are all around us in geometry (π, $\sqrt{2}$, etc.), it is strange that something so fundamental cannot be measured exactly. Not only can we not measure it, but we can't even find its precise value mathematically.

This raises a fundamental question – Do irrational numbers really exist? We can't measure them, we can't write their exact value, so how can they be "real"?

On the one hand, you could say that they do not exist, except in a mathematical sense. Meaning that mathematicians have created them as part of a mathematical system that is not connected to the real world, so we shouldn't expect them to be real. This, however, would seem to imply that all of mathematics is in the same category, and it is only a lucky coincidence that mathematics does fit the physical world quite well, even though mathematicians never intended it to. The fact that number patterns appear in nature that have the same physical relationships between them in nature as they do in their mathematical description isn't really something special. It is all a fortuitous accident that seems to keep repeating itself every time a new mathematical connection to the universe is pointed out.

On the other hand, maybe there is a deeper connection between mathematics and the universe. Perhaps it is not just coincidence that the very essence of how the world works is mathematical by nature, and that as Plato believed, we continue to discover new mathematical truths about the world that are already there waiting for us to find, precisely because it is mathematical. However, if this is the case, then what do irrational numbers tell us about the universe? A mathematical object that doesn't seem to exist in the physical world, but without it some basic calculations relating to area and length would not be possible. Plato might simply acknowledge that this is further proof of his world of ideals and that mathematics is the perfect world, and the physical world we know is just an imperfect copy. A world where irrationals do not exist. This might seem to be the case when we start to look more closely at space, time, and matter, and try to divide them indefinitely.

Thus, we can't really say if irrational numbers exist or not. It may seem rather unlikely, but just because we've never seen one, does not mean there isn't one. You can't prove a negative. Either way you look at it though, it does make you think. You begin to appreciate just how important mathematics is in our understanding of reality and the universe. It is amazing that all these new and confusing ideas all started from simply counting the objects around us.

11.6 Irrationals and Beyond

A new view of mathematics began to emerge with the insights of Karl Weierstrass (1815–1897), along with Augustin Cauchy (1789–1857), Bernhard Bolzano (1781–1848), and Georg Cantor, when they established a new modern interpretation of irrational numbers. It essentially separates mathematics from the real physical world. Mathematicians would no longer think or real numbers geometrically. Intuition in mathematics was cast aside and there was a reliance on definitions.

Irrational numbers are not a simple concept. Strangely and amazingly there are four types of irrational numbers. They range from the confusing to the unbelievable.

There are **algebraic**, irrationals such as $\sqrt{2}$, $\sqrt{3}$, $\sqrt[3]{2}$, ... These are the irrational numbers that come to us from the equations we saw in our algebra class, such as:

$$x^2 - 2 = 0, \quad x^2 - 3 = 0, \quad x^3 - 2 = 0, \dots$$

These algebraic irrationals also come in two different forms. Some of these are what we call **constructable**, meaning we can draw their lengths using only a straight edge and a compass (the mechanical device used to draw circles) using only a finite number of steps. Others, however, are **nonconstructable**, meaning we cannot draw (construct) their lengths. The $\sqrt{2}$ is constructable while $\sqrt[3]{2}$ is not.

In addition to the algebraic irrationals, we also have **transcendentals**. Transcendentals do not have an associated algebraic equation we can use to define them. Instead they require a process with an infinite number of steps to create them. Numbers such as π and e, are transcendental numbers, we saw in Section 11.4 above how we were able to define these numbers through an infinite process. It turns out that there are more transcendental numbers than algebraic numbers.

This is where things now get really strange, for in addition to algebraic and transcendentals we also have **uncomputable** irrationals. These are numbers that do not have a decimal representation we can compute, but we know that they exist. Wow, that is really strange! An example of such a number comes to us from the field of computer science. It is called Chaitin's constant, Ω. This is a number we can define in terms of a process, but we cannot compute it due to a degree of randomness involved in its definition. A more detailed explanation can be found elsewhere for the interested reader, but it is a bit complex and requires some knowledge about computers, which we don't want to get in to in this book. Surprisingly there are more uncomputable numbers than transcendental numbers.

Finally, we have the most bizarre of the irrationals. These are the **undefinable** irrational numbers. It turns out that almost all irrational numbers fit into this category. Numbers that defy even a definition for us to find them. These are numbers that we know exist, but we can't even begin to tell you how to find them. We can't even begin to talk about them!

Irrational numbers truly are a bizarre type of number. They cause us to rethink everything we know about the number concept, and to accept an unbelievable sort of abstraction. If only our number concept ended here, we would still be dumbfounded and confused, but it doesn't. It gets even more abstract and fascinating.

EXERCISES 11.1–11.6

ESSAYS AND DISCUSSION QUESTIONS:

1. What are irrational numbers?

2. Why are $\sqrt{2}$, π, and e, irrational numbers?

3. Are irrational numbers real? State your opinion and explain.

4. Explain Archimedes approach to finding the value of π.

5. Understand and explain the proof of why $\sqrt{2}$ is irrational.

6. Discuss and explain the different types of irrational numbers?

7. What do we mean by saying that we can't measure the length $\sqrt{2}$?

8. Are irrational numbers a creation of humankind, or are they discovered? Explain.

References

1. Beckmann, P., (1970). A History of Pi, 4th. ed., Golem Press.

2. Borwein, J.M. and P. B., (1987). Pi and the AGM, Wiley-Interscience.

3. Boyer, C.B. and Merzbach, U., (1991). A History of Mathematics, 2nd. ed., Wiley, 1991.

4. Cajori, F., (1980). A History of Mathematics, 3rd. ed., Chelsea, 1980.

5. Dörrie, H., (1965). 100 Great Problems of Elementary Mathematics: Their History and Solution, translated by D. Antin, Dover.

6. Von Fritz, K., (1945). The Discovery of Incommensurability by Hippasus of Metapontum, The Annals of Mathematics.

7. Flegg, G., (2002). Numbers: Their History and Meaning. Mineola, NY: Dover Publications.

8. Galovich, S., (1989). Introduction to Mathematical Structures. San Diego: Harcourt Brace Jovanovich.

9. Hardy, G. H. and Wright, E. M. An Introduction to the Theory of Numbers, 5th ed. Oxford, England: Clarendon Press, 1979.

10. Higgins, P.M., (2008). Number Story: From Counting to Cryptography. London: Copernicus.

11. Ifrah, G., (2000). The Universal History of Numbers: From Prehistory to the Invention of the Computer. Translated by David Bellos et al. New York: Wiley.

12. Kline, M., (1990). Mathematical Thought from Ancient to Modern Times. New York: Oxford University Press.

13. McLeish, J., (1994). The Story of Numbers: How Mathematics Has Shaped Civilization. New York: Fawcett Columbine.

14. Menninger, Kl,. (1992). Number Words and Number Symbols: A Cultural History of Numbers. New York: Dover Publications.

15. Nagell, T., (1951). Irrational Numbers, and Irrationality of the numbers e and pi. §12-13 in Introduction to Number Theory. New York: Wiley, pp. 38-40.

16. Nivens, I., (1965). Irrational Numbers, Mathematical Association of America, also 1985, 2005.

17. Nivens, I.M., (1956). Irrational Numbers. New York: Wiley.

18. Nivens, I.M., (1961). Numbers: Rational and Irrational. New York: Random House.

19. Pappas, T., (1989). Irrational Numbers & the Pythagoras Theorem, The Joy of Mathematics. San Carlos, CA: Wide World Publ./Tetra, pp. 98-99.

20. Wells, D. G., (1998). The Penguin Dictionary of Curious and Interesting Numbers. Rev. ed. London, UK: Penguin Books.

CHAPTER 12

Infinity

To see a world in a grain of sand,
And a heaven in a wild flower,
Hold infinity in the palm of your hand,
And eternity in an hour.

William Blake, Auguries of Innocence

12.1 Introduction

This is without a doubt the most intriguing concept we have considered so far. Yes, infinity is really a concept, it isn't quite a number, although we can sometimes treat it as if it is a number, with properties similar to ordinary numbers to help us try and understand it better.

In Chapter 10 we looked at exponential numbers, and saw how large they can be, and how we humans have trouble thinking exponentially. Infinity, however, puts these numbers to shame. When compared to infinity, exponential numbers don't even make the chart, no matter how big they are. Thus, if humans have difficulty trying to grasp exponential numbers, understanding infinity is a whole other ball game. It is quite frankly impossible. How can a finite being even begin to comprehend the infinite. We can, however, get a glimpse into its shocking nature.

In one way infinity is the inverted reflection of zero. Zero and infinity are forever linked. Perhaps this is what makes zero so much more interesting than the other numbers, and a bit more dangerous as well. The danger in zero it seems, is that it can give rise to the infinite. Or as the Hindus believed it is out of the void (zero) that the infinite universe is created.

In the last chapter we looked at irrational numbers, and saw how these strange new numbers are related to the concept of infinity, since they required an infinite number of decimal digits to represent them. What we intend to do in this chapter is to see if we can throw some light on this immensely intriguing but elusive concept. We will add quite a bit of new mathematics in our attempt, and also present some seemingly nonsensical results. What we hope to accomplish, though, is to open up your mind to the wonder of it all, and to make you think.

12.2 What is Infinity?

Infinity has a long history. Ever since humans began looking towards the nighttime sky and wondering if what they saw went on forever, infinity has been around. It is what caused the great debates in Ancient Greece between Aristotle, Democritus, and Zeno – Can anything be divided forever? Infinity both inspires and frightens us, impacting metaphysical discussions of reality, and practical calculations in geometry and calculus.

Look for a definition on infinity, and usually you'll find a variation of the following:

"endless, boundless, or unlimited extent of time, space, or quantity"

"an indefinitely large number or amount"

"an idea that something never ends"

We can define it as a process or entity that is not completable. Since we are finite beings, it is impossible for us to understand or define it any better, or so it seems.

The problem is, that once we wish to pose questions about the world, the concept of infinity must be addressed or understood to some extent, otherwise we will not obtain the insight we seek into our own reality.

It is in ancient Greece that we have chosen to begin the story of infinity. The Greeks posed many questions to try and understand the world. One question that arose was – is there an end to the universe in either time or space?

For example, the early Greeks posed some basic questions about matter. Could you divide a piece of matter indefinitely? Is there a smallest increment of time, or can time also be divided indefinitely? What about space – can we divide it forever? These are fundamental questions about our world. They go right to the heart of reality – so what are the answers? As we'll see we don't have many answers, only more questions, as well as some unbelievable consequences and events come from our questioning.

Aristotle believed that infinity was an unrealizable concept. It did have some meaning when you tried to understand the divisibility of matter, space and time, but you could never actually complete the process of dividing, since its completion would be a contradiction in terms. A process is by definition infinite, if it cannot be completed.

To Aristotle there were two different ways to look at infinity. The first is as a Potential infinity. This is something that has to potential to go on forever, but can never be completed. The other way to view infinity is if was an Actual infinity. An actual infinity is a completed infinity. This is what Aristotle saw as a contradiction. Something can't be infinite if it has been completed, so an actual infinity is an oxymoron. It just can't happen.

Aristotle put this argument forward to address a group of Greek philosophers of his day that were trying to make the case that you could not continue to divide matter indefinitely. Democritus and others believed that eventually you would come to a point where matter could no longer be divided and this smallest indivisible part was called an atom. It is the same name we use today to identify the basic building block of all matter.

Aristotle's view on infinity dominated western thought for over 2,000 years. His philosophy and beliefs on the infinity were even adopted by the Catholic church. Infinity was only a potential conclusion. The only actual infinity was God. Anyone that suggested otherwise was a heretic and preaching blasphemous ideas.

Towards the end of the 16th century, in the midst of the Renaissance, some people started to openly question the concept of infinity, and the role of God in its existence or non–existence. It was a dangerous time to question anything, and especially something so important for one of the most powerful religions in the world, the Catholic church.

In 1584 Giordano Bruno (1548–1600) published a work in Italy titled *De l'Infinito Universo et Mondi (On the Infinite Universe and Worlds)*. Bruno was an Italian Dominican Friar, philosopher, amateur mathematician and astronomer. He was also a contemporary of Galileo, although the two never met.

Aside from the work on the infinite universe, Bruno published many other works questioning the beliefs of the Catholic church on God and the universe. His heretical views did not go unnoticed.

Bruno was imprisoned in 1592 by the Roman Inquisition, and tortured for 8 years to try and get him to recant his beliefs about God and the universe. Bruno questioned the beliefs of Aristotle on the infinite and the void. He also questioned the belief in a personal god as opposed to a cosmological God of a Copernican universe. Copernicus (1473–1543) was a Polish born astronomer that first proposed that the sun, not the earth was at the center of the solar system (at that time the universe). Some in the Catholic church opposed this view and those that believed otherwise were branded as heretics.

Bruno did not recant his beliefs, and after 8 long years of torture he was tried, convicted and then burned at the stake for heresy in 1600. The exact role infinity played in all of this is not known, for Bruno expressed many beliefs contrary to Catholic doctrine. However, his views on the infinite world probably played a part in forming his other beliefs. What we do know is that sixteen years after Bruno was executed, Galileo Galilei (1564 –1642) was summoned to the same place by the same people for his heretical views on the universe.

Like Bruno, Galileo was also intrigued by the idea of the infinite and of the universe. Unlike Bruno, Galileo was a true scientist. He is considered the first modern scientist, and the first to discover the pivotal role that mathematics plays in understanding how the universe works. It was Galileo who made the bold statement:

Philosophy is written in this grand book – I mean the Universe – which stands continually open to our gaze, but it cannot be understood unless one first learns to comprehend the language and interpret the characters in which it is written. It is written in the language of mathematics, and its characters are triangles, circles

and other geometrical figures, without which it is humanly impossible to understand a single word of it.

Galileo forged a strong connection between mathematics and the physical world. He had taken the mathematical connection of numbers to the physical world of Pythagoras, to a new level. In this new mathematical/physical world, infinity held a special place. Galileo starts to provide us with another look at the infinite, beyond the common Aristotelian view of the day. It intrigues him, and helps to shape his view of the world and our place in it.

Here is an extended excerpt about infinity from Galileo's, *Dialogue on Two New Sciences*, 1638. Galileo conveyed his ideas through a dialogue between three men, two philosophers and one layman. One was named Simplicio who was a simple minded philosopher (representing the views of Aristotle). Another was the philosopher Salviati, who represented Galileo and the Copernican world view. The last was Sagredo, an intelligent layman that begins the debate being neutral between the two philosophical perspectives.

The three are discussing two line segments or unequal length, and trying to determine if one has more points than the other. The discussion quickly expands to notions of infinity.

> **Simplicio:** *Here a difficulty presents itself which appears to me insoluble. Since it is clear that we may have one line segment longer than another, each containing an infinite number of points, we are forced to admit that, within one and the same class, we may have something greater than infinity, because the infinity of points in the long line segment is greater than the infinity of points in the short line segment. This assigning to an infinite quantity a value greater than infinity is quite beyond my comprehension.*

> **Salviati:** *This is one of the difficulties which arise when we attempt, with our finite minds, to discuss the infinite, assigning to it those properties which we give to the finite and limited; but this I think is wrong, for we cannot speak of infinite quantities as being the one greater or less than or equal to another. To prove this I have in mind an argument, which, for the sake of clearness, I shall put in the form of questions to Simplicio who raised this difficulty.*

> *I take it for granted that you know which of the numbers are squares and which are not.*

> **Simplicio:** *I am quite aware that a squared number is one which results from the multiplication of another number by itself: thus 4, 9, etc., are squared numbers which come from multiplying 2, 3, etc. by themselves.*

> **Salviati:** *Very well; and you also know that just as the products are called squares the factors are called roots; while on the other hand those numbers which do not consist of two equal factors are not squares. Therefore if I assert that all numbers, including both squares and non–squares, are more than the squares alone, I shall speak the truth, shall I not?*

> **Simplicio:** *Most certainly.*

Salviati: If I should ask further how many squares there are, one might reply truly that there are as many as the corresponding number of roots, since every square has its own root and every root its own square, while no square has more than one root and no root more than one square.

Simplicio: Precisely so.

Salviati: But if I inquire how many roots there are, it cannot be denied that there are as many as there are numbers because every number is a root of some square. This being granted we must say that there are as many squares as there are numbers because they are just as numerous as their roots, and all the numbers are roots. Yet at the outset we said there are many more numbers than squares, since the larger portion of them are not squares. Not only so, but the proportionate number of squares diminishes as we pass to larger numbers. Thus up to 100 we have 10 squares, that is, the squares constitute 1/10 part of all the numbers; up to 10,000 we find only 1/100th part to be squares; and up to a million only l/1000th part; on the other hand in an infinite number, if one could conceive of such a thing, he would be forced to admit that there are as many squares as there are numbers all taken together.

Sagredo: What then must one conclude under these circumstances?

Salviati: So far as I see we can only infer that the totality of all numbers is infinite, that the number of squares is infinite, and that the number of their roots is infinite; neither is the number of squares less than the totality of all numbers, nor the latter greater than the former; and finally the attributes "equal," "greuter," and "less" are not applicable to infinite, but only to finite, quantities. When, therefore, Simplicio introduces several lines of different lengths and asks me how it is possible that the longer ones do not contain more points than the shorter, I answer him that one line does not contain more or less or just as many points as another, but that each line contains an infinite number. Or if I had replied to him that the points in one line segment were equal in number to the squares; in another, greater than the totality of numbers; and in the little one, as many as the number of cubes, might I not, indeed, have satisfied him by thus placing more points in one line than in another and yet maintaining an infinite number in each? So much for the first difficulty.

Sagredo: Pray, stop a moment and let me add to what has already been said an idea which just occurs to me. If the preceding be true, it seems to me impossible to say that one infinite number is greater than another. ...

This excerpt of the translation of Galileo's book was taken from: Sherman Stein's *Mathematics: The Man–made Universe,* 2nd edition, Freeman, 1962; pp. 314–315.

This discussion proposes and supports the idea that there is only one infinity. This is why infinite groups of object may appear to be larger than other infinite groups, but this is not possible. It only comes about because our finite minds, using our intuition, cannot begin to grasp the concept. It is where we will start a deeper discussion about infinity below.

Like Bruno, Galileo also fell into disfavor with the Catholic Church. Galileo's inquisitiveness caused him to question many of the strongly held beliefs by some in the church. In particular, whether the earth or sun was at the center of our world. Galileo believing it to be the sun and many in the church leadership saying it was the earth. For his public writings Galileo was called before the Inquisition twice. The first time was at the urging of Father Tommaso Caccini, who called for Galileo to be brought before the Inquisition for his writings. Caccini eventually had the view that the earth moved around the sun declared as heresy, but concerning Galileo, the Inquisition took no action against him, but surely he must have been aware that there were some who would be watching him closely thereafter.

Sixteen years later, in 1632, Galileo published his *Dialogues on the Two Chief Systems of the World*. In this work he challenged the earth centered view of the universe, which was generally accepted by the Catholic Church. He almost immediately found himself in trouble with the Church once more. He was again called before the Inquisition, and on September 23 1632, Galileo was put on trial for the second time. This time, however, he was found guilty and was forced to renounce his beliefs in the Copernican world view, that claimed the earth revolved about the sun, and not the sun revolving around the earth. He was initially condemned to life in prison, but the following day the sentence was changed to house arrest. For the remainder of his life Galileo lived under house arrest. His book was also banned by the Catholic Church and only in 1992 did the Church finally take back its condemnation of Galileo, although many in the Church leadership at the time still questioned the need to do so.

Galileo had taken a glimpse into the infinite and it had a tremendous impact on his views and his life. He recognized the unique aspect of the infinite regarding our understanding of the world, and it was a central piece of his investigations. Galileo also had a major influence on those that followed him.

Another notable mathematician that impacted the concept of infinity, although not in a major way, was John Wallis (1616–1703). John Wallis was an English mathematician, partly responsible for the development of modern calculus, but he is also credited with introducing the symbol, ∞, for infinity that we use today.

Nobody made any headway in advancing our understanding of the infinite until Georg Cantor (1845–1918) (see the picture of a young Cantor to the left) a Russian mathematician provided new insight that would change the concept forever. To help in his study of the infinite, Cantor created Set Theory in 1874 as a useful comparative tool. It was through set theory that the veil on infinity could begin to be lifted. In the following sections we highlight Cantor's significant contributions.

We should comment here that Cantor suffered a great deal due to his involvement with and his great achievements related to the idea of infinity. At times it is too great for his phenomenal, but unstable mind, and he struggles to remain out of sanatoriums due to his mental illness. To the left above and the right below we provide two pictures of Cantor: as a young man and then later in life.

12.3   Set Theory – A Tool to Study the Infinite

Set theory was originally created by Georg Cantor in 1874 to study the concept of the infinite. It is a framework by which we can organize, compare, and study collections of objects. These collections can be finite in number, or more to the point of Cantor, they can also be infinite. Set theory was a monumental achievement in mathematics, as it served to unify how mathematics was presented and talked about.

12.3.1   Why study set theory?

In this book we do not study set theory because it will be needed later in life, or because you will need it as a prerequisite in another course. Instead the study of set theory gives us an opportunity to examine how we search for truth. It is more of a philosophical question rather than one of mathematical content at this level.

From the beginning we have seen a progression to more and more abstract concepts. Starting from the development of natural numbers, zero, negative and rational numbers, exponential numbers, irrational numbers, and now infinity. Mathematics has been used to better understand and describe the world in which we live. For some it is the language of the universe, or for others it is the creation of a supreme being and to try and understand this is a part of who we are. A truly educated person needs to be aware of what we have achieved in this discipline we call mathematics; set theory is one of those grand achievements.

Finally, we study set theory because it allows us to see just how Cantor saw infinity. It is a wonderful and powerful tool that he created for seeing beyond the finite and to infinity and beyond.

12.3.2   Basics of set theory

We begin the introduction of set theory with some fundamental definitions and notations.

A **set** is a collection of objects.

The objects can be anything – numbers, letters, people, colors, animals, beliefs, cars, investments, etc.

An **element** of a set, is an individual object in the set.

Sets will be usually represented using braces, { }, with their elements contained between these pair of braces. We will adopt the convention of using upper case letters at the beginning of the alphabet to denote a set.

For example, the set of all natural numbers from one to nine inclusive would be written as:

$$A = \{1, 2, 3, 4, 5, 6, 7, 8, 9\}$$

The set of the first six letters of the alphabet could be written as:

$$B = \{a, b, c, d, e, f\}$$

Although we will typically represent a set using braces and a listing of its elements, we should point out that a set can be represented in one of three forms.

1. The **roster form,** where each element is shown between a pair of braces with commas between the elements. This is how we have presented them above, and is the most common form for representing sets.

2. The **descriptive form,** where a verbal description of the set is given without any symbolism.

   For example: the set of all whole numbers between zero and eleven, or the set of all students in this class taller than five feet six inches, or the set of all non–republicans in this class.

3. The **set–builder notation form,** where we essentially combine the two previous forms into one form.

   For example: the set of all whole numbers between zero and eleven from above could be shown as either:

   $$\{x \mid x \text{ is a whole number greater than 0 and less than 11}\}$$

   or

   $$\{x \mid x \text{ is a counting number greater than or equal to 1 and less than or equal to 10}\}$$

We read set–builder notation as follows:

When we see the brace, {, we say "the set of all elements." The "x" tells us the name we will use for these elements, in this case, "x." The vertical line is read as "such that," and then we have the description.

So the first example above is read as: "The set of all elements x, such that x is a whole number greater than zero and less than eleven."

The next example is read as: "The set of all elements x, such that x is a counting number greater than or equal to one and less than or equal to ten."

232

Set theory can be used to provide a very precise means of talking about collections of objects, along with a formalized way to discuss, modify, and manipulate these collection of elements. To show how this is done, we need to introduce some additional definitions, relationships, and concepts.

> A **finite set** is a set with a finite number of elements.

Examples
1. The set of primary colors
2. $\{x|\ x$ is an integer and $-5 \leq x \leq 5\ \}$
3. {Mary, Sue, Jane, Leslie, Danise, Kim}
4. The set of all planets in our Solar System.

> An **infinite set** is a set with an infinite number of elements.

Examples
1. $\{x|\ x \in \mathbb{R}$, and $-5 \leq x \leq 5\ \}$
2. The set of prime numbers.
3. $\{2, 4, 6, 8, \ldots\}$

   Notice that only numbers can make up a truly infinite set. Nothing else works, although we sometimes include non–infinite sets in this category, because their numbers are too large to count or write out. For example, some people would classify these sets as infinite, even though technically they are not.
4. The set of all stars in the Universe.
5. The set of all possible baseball teams that could be made from anyone in the world between the ages of 15 and 50.

> The **cardinality** of a set is the size of the set. It is a measure of how many elements the set has, and may be written as, $N(A)$, $n(A)$, $|A|$, or $\bar{\bar{A}}$. We will use the symbolic form of $n(A)$ to represent the cardinality of a set.

Examples
1. $A = \{x|\ x \in \mathbb{R}$, and $-5 \leq x \leq 5\ \}$, $n(A) = 11$
2. $A = \{$Mary, Sue, Jane, Leslie, Danise, Kim$\}$, $n(A) = 6$
3. $B =$ The set of all the planets in our Solar System, $n(B) = 9$
4. $C = \{2, 4, 6, 8, \ldots\}$, $n(A) = \infty$ (for now this is how we'll count infinite sets)

> The **universal set** is the set containing all the elements under consideration. If an element is not in the universal set, then it does not exist. We use the letter, U, to represent the universal set.

Examples
1. $U =$ the set of all car manufactures in the world.
2. $U =$ the set of all men in the United States.

3. U = the set of all galaxies in the universe.
4. U = {x| $x \in \mathbb{R}$ }
5. U = { ... −2, −1, 0 ,1, 2, ... }

> A set is said to be empty if it has no elements, and is called the **empty set** and is identified either by { }, or ∅ (Note: this is NOT a way to represent an empty set, { ∅ }. This is a set containing the empty set and is NOT empty.)

Examples
1. The set of all humans greater than 16 feet tall.
2. The set of all odd numbers divisible by 2.
3. The set of women over 100 that gave birth to their 100th child.

Sets and their Relationships

1. Membership

Membership in a set will be denoted using the symbol, $\in$.

Considering the two sets, A and B

$$A = \{ 1, 2, 3, 4, 5, 6, 7, 8, 9 \}, \quad B = \{a, b, c, d, e, f\},$$

we could say "d" is an element (or a member) of set B symbolically by writing:

$$d \in B$$

Not being a member can be denoted with the same symbol $\in$, but with a line through it, meaning not. Thus, to say 11 is not a member of set A symbolically, we would write:

$$11 \notin A$$

2. Union

> The **union** of two sets A and B, is a new set C, such that all the elements in both A and B are in C, but repeated elements are only counted once.

The union of A and B is written as:

$$A \cup B$$

Example: Given: A = {1, 2, 3, 4, 5}, and B = {4, 5, 6, 7, 8, 9}, then

$$A \cup B = \{1, 2, 3, 4, 5, 6, 7, 8, 9\}$$

Notice the elements 4 and 5 are only counted once in the union.

3. Intersection

The **intersection** of two sets A and B, is a new set C, such the elements in C are in both sets A and B at the same time.

The intersection of A and B is written as:

$$A \cap B$$

Example: Given: A = {1, 2, 3, 4, 5}, and B = {4, 5, 6, 7, 8, 9}, then

$$A \cap B = \{4, 5\}$$

4. Complement

The **complement of a set**, written as $\bar{A}$ or A', is the set of all the elements that are not in A, but are in the Universal set, U.

Example: Given: A = {1, 3, 5, 7, 9}, and U = {1, 2, 3, 4, 5, 6, 7, 8, 9}, then

$$\bar{A} = \{2, 4, 6, 8\}$$

5. Equal Sets

Two sets are **equal**, if they both have the same elements, and is written as A = B.

Example: Given: A = { 1, 2, 3, 4, 5}, and B = {1, 2, 3, 4, 5}, then

$$A = B$$

Since all the elements in A are contained in B, and all elements contained in B are in A.

6. Difference

The **difference** of two sets A and B, written as A − B, is a set whose elements are the elements of set A with all the elements of set B that are also in set A, removed.

Example: Given: A = {1, 2, 3, 4, 5, 6, 7, 8, 9} and B = { 1, 2, 3, 4, 5}, then

$$A - B = \{6, 7, 8, 9\}$$

Example: Given: A = {1, 2, 3, 4, 5, 6, 7} and B = { 5, 6, 7, 8, 9}, then

$$A - B = \{1, 2, 3, 4\}$$

7. Disjoint Sets

> Sets with no elements in common are called **disjoint sets**.

Example: A = a set of cows and, B = a set of birds. A and B are disjoint sets.

Example: A = a set of numbers greater than 10 and C = a set of numbers less than 5. A and C are disjoint sets.

8. Subset

> A **subset** is a set whose elements are contained in another set. If A is contained in B we say A is a subset of B and write this as $A \subseteq B$.

Example: Given: A = { 1, 2, 3, 4, 5}, and B = {1, 2, 3, 4, 5, 6, 7, 8, 9}, then

$$A \subseteq B$$

Since all the elements in A are contained in B.

9. Proper Subset

> A set A is a **proper subset** of set B, if there is at least one element in B that is not in A. It is written as $A \subset B$

Example: Given: A = { 1, 2, 3, 4, 5}, and B = {1, 2, 3, 4, 5, 6}, then

$$A \subset B$$

Since all the elements in A are contained in B, but there is at least one element in B, the number 6, that is not in A.

10. <u>Cartesian Product</u>

An **ordered pair** (a,b) is a pair of objects, where a is called the **first component**, and b is called the **second component**.

> A **Cartesian product** of two sets, A and B, creates a set of pairs of elements (a,b) where the first component of the ordered pair comes from set A, $(a \in A)$ and the second component of the ordered pair comes from set B, $(b \in B)$. It is written as $A \times B$.

Example: A = {Alice, Jane, Cynthia}, B = {John, Richard} Then let C be he set of all (women, men) combinations, i.e.

$A \times B = C =$ {(Alice, John), (Alice, Richard), (Jane, John),(Jane, Richard), (Cynthia, John), (Cynthia, Richard)}

Example: The set of all (x, y) coordinates in the x–y plane.

12.3.3 Visualizing Sets and their Operations and Relationships

The English Mathematician John Venn (1834–1923) introduced the idea of depicting sets using diagrams to help visualize their properties and relationships. They were originally introduced to help test the validity of a problem from logic called a syllogism, which we cover later on in Chapter 15.

A **Venn diagram** is a diagrammatic way of showing sets and their relationships. A Venn diagram is drawn with a rectangular region representing the universal set U. Contained within that rectangle are one or more circles representing sets. The individual elements of the sets can be shown or not shown in the diagram.

Examples:

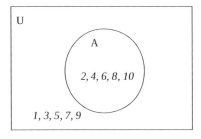
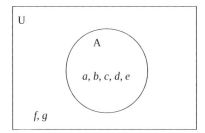

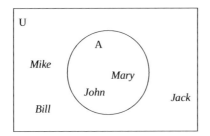 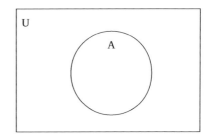

Venn diagrams are useful in visually illustrating the relationships between sets. For example consider each of the following.

Union $A \cup B$. The shaded region represents the union.

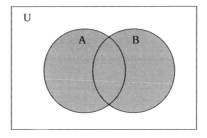

Intersection $A \cap B$, The shaded region represents the intersection.

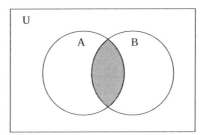

Difference $A - B$, The shaded region represents the difference.

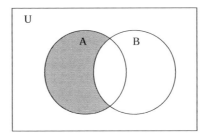

238

Complement A', The shaded region represents the complement.

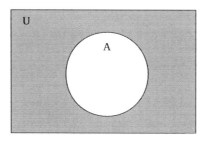

Subset A⊆B , (Note: this diagram also shows that A is a **proper subset** of B, or A⊂B)

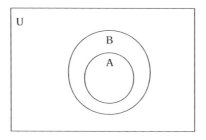

Examples of sets operations and Venn diagrams.

1. Ā∪B

a.   Ā

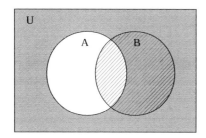

b.   B

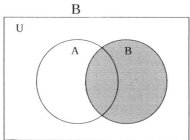

c.   Ā and B on same graph

d.   Ā∪B

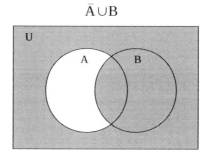

2. $A \cap \bar{B}$

a.   A

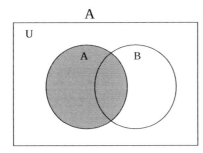

b. $\bar{B}$

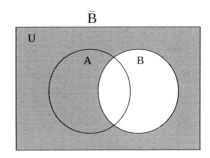

c.   A and $\bar{B}$ on same graph

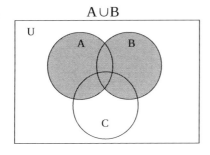

d. $A \cap \bar{B}$

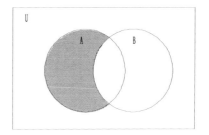

3. $(A \cup B) \cap C$

a. $A \cup B$

b.   C

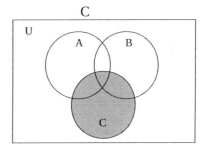

c. $A \cup B$ and C on same graph

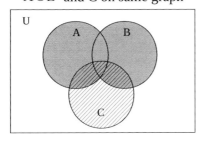

d. $(A \cup B) \cap C$

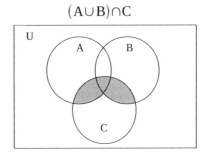

4. $A \cup (B \cap C)$

a.   A

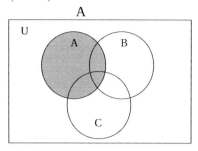

b. $B \cap C$

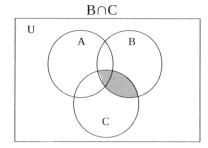

c.   A and $B \cap C$ on same graph

d. $A \cup (B \cap C)$

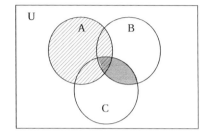

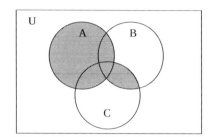

De Morgan's Law's

Starting with the basic concepts and definitions above, you can derive more elaborate relationships. For example, we can show two relationships between the union, intersection and the complement of sets. These are:

$$\overline{(A \cup B)} = \bar{A} \cap \bar{B}$$

> The complement of the union of two sets, is equal to the intersection of the complements of the two sets.

and

$$\overline{(A \cap B)} = \bar{A} \cup \bar{B}$$

> The complement of the intersection of two sets, is equal to the union of the complements of the two sets.

These two relationships are called **De Morgan's Laws**, named after the logician Augustus De Morgan (1806–1871).

Their proof can be illustrated visually using Venn diagrams.

$$\text{Proof that } \overline{(A \cup B)} = \bar{A} \cap \bar{B}$$

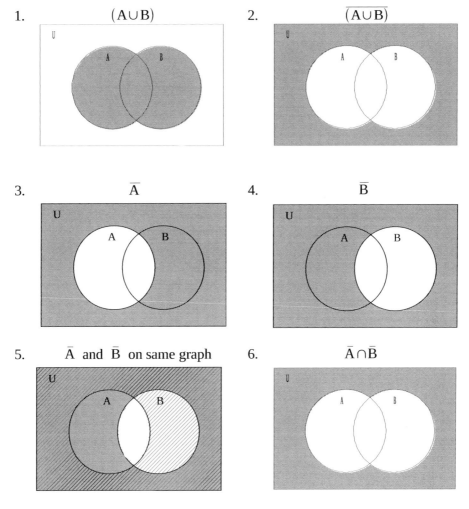

1. $(A \cup B)$
2. $\overline{(A \cup B)}$
3. $\overline{A}$
4. $\overline{B}$
5. $\overline{A}$ and $\overline{B}$ on same graph
6. $\overline{A} \cap \overline{B}$

Note step 2, the left hand side of the equation, and step 6, the right hand side of the equation, yield the same Venn diagram, so they are the same, and the result has been proven.

Proof that $\overline{(A \cap B)} = \overline{A} \cup \overline{B}$

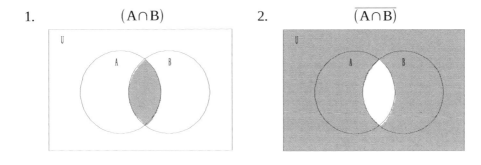

1. $(A \cap B)$
2. $\overline{(A \cap B)}$

3. $\overline{A}$

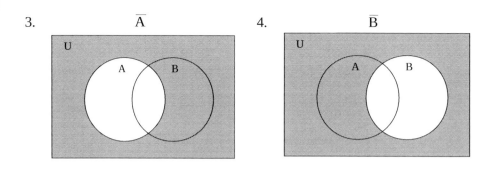

4. $\overline{B}$

5. $\overline{A}$ and $\overline{B}$ on same graph

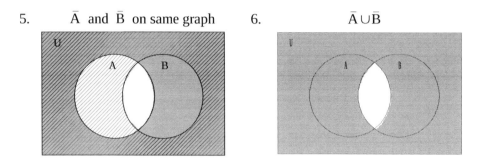

6. $\overline{A} \cup \overline{B}$

Again, note step 2, the left hand side of the equation, and step 6, the right hand side of the equation, yield the same Venn diagram, so they are the same, and the result has been proven.

These laws are helpful in making valid inferences in proofs, as well as in establishing logically equivalent statements that can be used to create the most efficient electronic logic gates in electronic devices.

12.3.4 Application of Venn Diagrams

Analyzing surveys

Venn diagrams have been used to visualize the results of a survey. We should point out that they do not provide any additional information, or help us solve any real problem, since we already have all the data we need from the survey, but they can be used more like a game to "solve" made-up problems. We will illustrate the process of analyzing surveys through Venn Diagrams using some examples. This will also help us to become more familiar with working with these very visual diagrams.

Example 1: A survey of 100 people was performed to see which wines they liked with the following results:

> 75 people liked Pinot Grigio
> 55 people like Cabernet Sauvignon
> 82 people liked Chardonnay

45 people like Cabernet Sauvignon and Chardonnay
68 people liked Pinot Grigio and Chardonnay
40 people liked Pinot Grigio and Cabernet Sauvignon
33 like all three

How many people like:
a) none of the three?
b) Pinot Grigio, but not Chardonnay?
c) anything but cabernet Sauvignon?
d) only Pinot Grigio?
To answer these questions we need to construct a Venn Diagram of the three wines and peoples preferences.

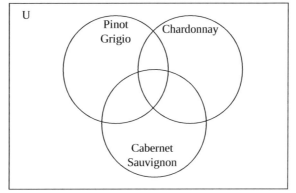

We need to fill in the numbers in each of the sets and their overlaps above. We start with the case where all three wines were liked (always start here if possible), which is 33, and fill in that part of the graph.

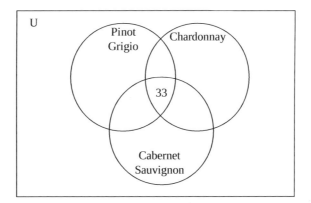

Next fill in the groups where two wines are liked at the same time. Just subtract the 33 from each of these numbers since this number is already accounted for, as follows:

244

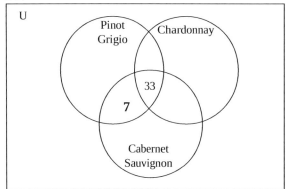 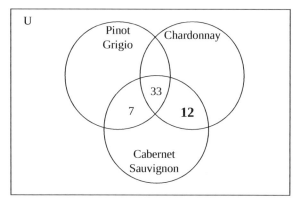

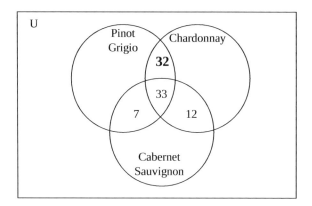

We have made the numbers that like Pinot Grigio and Cabernet Sauvignon add up to 40 (by adding 7 to the 33), then the numbers that like Chardonnay and Cabernet Sauvignon add up to 45 (by adding 12 to the 33), and finally the number of people that like Pinot Grigio and Chardonnay add up to 65 (by adding 32 to the 33).

Next, we focus on making the total numbers for each of the wines add up to the total that liked that particular wine.

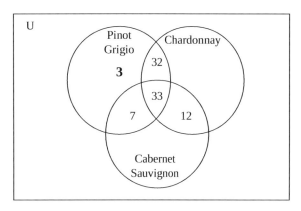 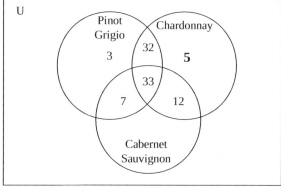

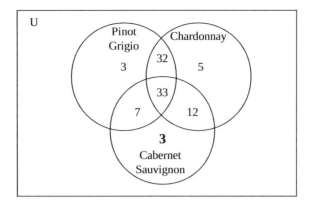

We add 3 to the Pinot Grigio set, so that the total adds up to 75. We add 5 to the Chardonnay set, so that the total here is 82, and 3 to the Cabernet Sauvignon, so that the total for this set is 65.

The last step is to find the number of people that don't like any of the wines. To find this number we simply add up all the numbers in the three sets of people we have filled in, which is $(3+32+33+7+12+5+3)=95$, and subtract it from the total number of people surveyed, which is 100, to obtain 5. We then write this in the region outside the three sets.

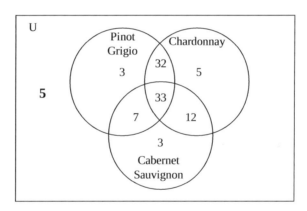

Now that we have our Venn Diagram, we can answer the questions:

How many people like:

b) none of the three?

From the Venn Diagram we can read off, 5 people

c) Pinot Grigio, but not Chardonnay?

From the Venn Diagram we can read off, (3 +7) or 10 people.

d) anything but cabernet Sauvignon?

From the Venn Diagram we can read off, (3+32+5+5) or 45 people.

246

e) only Pinot Grigio?

From the Venn Diagram we can read off, 3 people.

Example 2: A survey of 65 patients that were admitted to a cardiac unit of a hospital found that 44 had high blood pressure, 58 had high cholesterol, and 55 smoked cigarettes. Also, 38 patients had both high blood pressure, and high cholesterol, 48 had high cholesterol and smoked cigarettes, and 36 had high blood pressure and smoked. Finally 30 patients had high blood pressure, cholesterol and they smoked.

Answer the following questions:

a) How many patients only had high cholesterol and no other symptoms?

b) How many patients had none of the symptoms?

c) How many patients that did not smoke have high blood pressure?

We start by constructing the Venn diagram for the data.

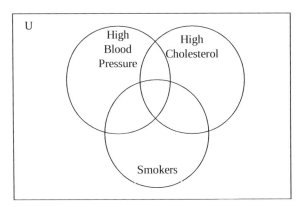

We need to fill in the numbers in each of the sets and their overlaps above. We start with the case where all three wines were liked (always start here if possible), which is 33, and fill in that part of the graph.

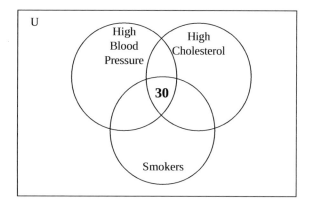

Next fill in the groups where two wines are liked at the same time. Just subtract the 30 from each of these numbers since this number is already accounted for, as follows:

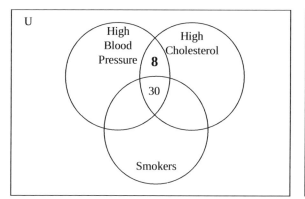

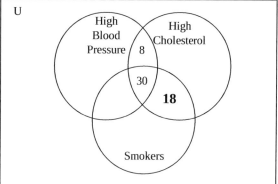

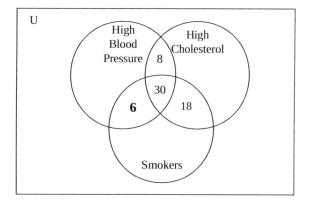

We have made the numbers that had two conditions add up to their respective values by subtracting 30 from the number, so high blood pressure and high cholesterol add up to 38 (by adding 8 to the 30), then the numbers that have high cholesterol and are smokers add up to 48 (by adding 18 to the 30), and finally the number of patients that have high blood pressure and are smokers add up to 36 (by adding 6 to the 30).

248

Next, we focus on making the total numbers for each of the conditions add to the total number in that group.

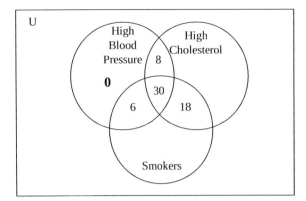
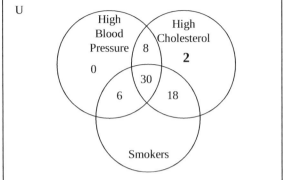

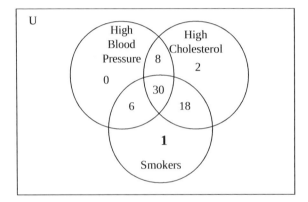

We add 0 to the high blood pressure set, so that the total adds up to 44. We add 2 to the high cholesterol set, so that the total here is 58, and 1 to the smokers, so that the total for this set is 55.

The last step is to find the number of people that don't like any of the wines. To find this number we simply add up all the numbers in the three sets of people we have filled in, which is $(0+6+30+8+2+18+1)=65$, and subtract it from the total number of people surveyed, which is 65, to obtain 0. We then write this in the region outside the three sets.

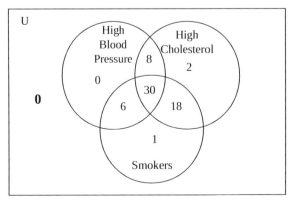

Now that we have our Venn Diagram, we can answer the questions:

a) How many patients only had high cholesterol and no other symptoms?

From the Venn Diagram we can read off, 2 patients

b) How many patients had none of the symptoms?

From the Venn Diagram we can read off, 0 patients.

c) How many patients that did not smoke had high blood pressure?

From the Venn Diagram we can read off, 8 patients.

d) How many patients has high cholesterol, but were not smoker?

From the Venn Diagram we can read off, 10 people.

As you can see from above, set theory can be used in different ways. Cantor created this powerful new tool to be able to better understand the concept of the infinite, and in the next section we show how it became indispensable for that purpose.

12.4 Making the Incomprehensible Comprehensible – Or Can We?

With the concepts developed above on set theory, we can now take a fresh look at infinity through the eyes of Cantor. We begin by looking at finite sets and then extend the ideas to infinite ones. The real strength of set theory is in its ability to provide meaningful comparisons between sets.

For example, if we wished to compare the size of two finite sets, without finding their cardinalities, we could simply set up a correspondence also called a mapping from each member in one set to a distinct member in another set.

$$A = \{1, 2, 3, 4, 5, 6, 7, 8, 9\}$$
$$\text{↑ ↑ ↑ ↑ ↑ ↑ ↑ ↑}$$
$$\text{↓ ↓ ↓ ↓ ↓ ↓ ↓ ↓}$$
$$B = \{a, b, c, d, e, f, g, h\}$$

In the two sets, A and B, above we can see that set A is larger than set B, because we can map every member in set B to set A, but set A has an extra member. This is different from simply finding the cardinality of each: $n(A) = 9$, and $n(B) = 8$, and seeing that 9 is greater than 8, and confirming that set A is larger. It is the process that is important, not a number.

From this simple exercise we can draw the conclusion that if we can set up a mapping from one set to another, where every point in one set has a corresponding member in the other set, with no members unaccounted for, then the two sets are equal in size. If, on the other hand, one set has elements that are unaccounted for, and the other set does not, then the set with unaccounted members is larger.

Examples of infinite set correspondences:

Let us now apply this concepts to infinite sets of numbers. The first thing we will notice is that we obviously cannot complete the process. Instead, we will be looking to see if we can see a repeated way to continue the pattern, without completion, that will still establish the relationship. For example, consider the two separate sets of even and odd numbers greater than zero.

$$O = \{1, 3, 5, 7, \ 9, \ 11, 13, 15, ...\}$$
$$\text{etc.}$$
$$E = \{2, 4, 6, 8, 10, 12, 14, 16, ...\}$$

We see from this comparison that for every odd number there is a corresponding even number. Furthermore, we can continue this process indefinitely, and never leave an odd or an even number out. Thus, we can reasonably conclude that there are the same number of odd numbers as there are even numbers. This is a very unsurprising result, since it agrees with our intuition.

Let's use this process to compare some other infinite sets. Consider the set of natural (counting) numbers and the set of even numbers. At first glance, it seems that surely the natural numbers must be larger than even numbers. After all there are twice as many natural numbers as even, aren't there?

Again we set up the same type of relationship and obtain:

$$N = \{1, 2, 3, 4, \ 5, \ \ 6, \ \ 7, \ \ 8, ...\}$$
$$\text{etc.}$$
$$E = \{2, 4, 6, 8, 10, 12, 14, 16, ...\}$$

The result is very interesting. It seems that we can find an even number for every natural number, without any numbers being left over! Obviously, we can continue this process forever, and not leave a number in either set out. Again, it is the process of comparing that is important, not trying to assign a specific number.

We have encountered our first strange fact about infinity. It seems that we can add two infinite sets together (namely the set of even numbers and the set of odd numbers) and still get an infinite set of the same size. This is quite contrary to our intuition from finite sets. We know that if we add two finite sets together that are the same size, we end up with a set that is twice as large.

Perhaps, this is telling us something about infinity. Maybe there is only one size to infinity, just as Galileo surmised. That way when we add two infinities, we get the same infinity back and avoid any contradictions. Let's test this thesis on other sets of numbers. Let's look at the set of integers versus the set of natural numbers.

$$N = \{1, 2, \ 3, \ 4, \ 5, 6, \ 7, \ 8, ...\}$$
$$\text{etc.}$$
$$Z = \{0, 1, -1, 2, -2, 3, -3, 4, ...\}$$

Again, we see the same result. The set of integers is the same size as the set of naturals. It seems that if we can "count" the elements in the set, then it is the same size as the counting numbers. It is a "countable" infinity.

What about the rational numbers. All those ratios of numbers, surely they must be larger than the natural numbers. They too can't be countable. This is where Cantor showed us his real genius and insight. He was able to come up with a way of representing the set of rational numbers (for us we'll just show the positive rational numbers, since it is easier to see and understand as a first introduction, but we can do the same for the negative rationals as well.)

Cantor created unending rows (some formulations of his proof use columns) of the positive integers all divided by the natural numbers whose value starts at one in the first row and increases by one in every subsequent row beneath it. Thus the first row has all the counting numbers divided by one, the next is all the counting numbers divided by two, etc.

We have shown this in the figure below. This arrangement will give us all the possible positive rational numbers. Look it over and see if you can convince yourself that it does. Come up with any rational number and see if it wouldn't be in the set given. You may notice that this representation does repeat numbers, but it doesn't leave any out. For example we have the number 2, but we also have 4/2, 6/3, etc., all of which are equal to 2. Thus, if anything this set would be larger than the set of rationals.

$$\frac{1}{1} \quad \frac{2}{1} \quad \frac{3}{1} \quad \frac{4}{1} \quad \frac{5}{1} \quad \frac{6}{1} \quad \cdots$$

$$\frac{1}{2} \quad \frac{2}{2} \quad \frac{3}{2} \quad \frac{4}{2} \quad \frac{5}{2} \quad \frac{6}{2} \quad \cdots$$

$$\frac{1}{3} \quad \frac{2}{3} \quad \frac{3}{3} \quad \frac{4}{3} \quad \frac{5}{3} \quad \frac{6}{3} \quad \cdots$$

$$\frac{1}{4} \quad \frac{2}{4} \quad \frac{3}{4} \quad \frac{4}{4} \quad \frac{5}{4} \quad \frac{6}{4} \quad \cdots$$

$$\vdots \qquad \vdots \qquad \vdots \qquad \vdots \qquad \vdots \qquad \vdots$$

The question then arises – can we count this set? Is there a way that we can count all the numbers in this set and not leave any numbers out? If so then we can say that the set of rational numbers, $\mathbb{Q}$, is countable. That means it is the same size as the natural numbers, and the integers. This would give more credence to the notion that there is only ONE infinity.

In the figure below we show how Cantor proposed counting them. It is quite ingenious. You start in the upper left corner, and move over one numbers, and then back done on the diagonal. When you reach the end on the left you go down one and then back up on the diagonal until you reach the top. You then move over one and repeat the process indefinitely. Just follow the arrows in the figure with the shaded number circles showing how to count the numbers.

As you can see you don't leave any numbers out, so the rational numbers are countable. This means the set of rational numbers is the same size as the set of natural numbers. This seems a bit odd to us finite beings, but at least it seems to send a consistent message about infinity. Infinity is infinity, it is just one concept and one size fits all!

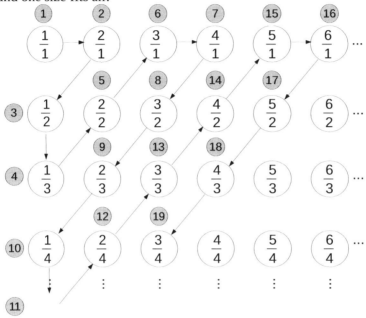

The problem, or the fascinating part, is that the story does not end here. What about all those irrational numbers from the previous chapter? Are they too countable?

Again, we rely on the genius of Cantor to guide us here. How can we represent the set of all irrational numbers in a way we can compare them with the set of natural numbers? The key lies in their decimal representation. We recall that a number is irrational if its decimal representation is non–terminating and non–repeating. So Cantor said, let us imagine that we could, by some unknown way, write down a list of all the irrational numbers. To make it even easier let's just say we write down all irrationals between zero and one, so the first number to the left of the decimal point is zero. Clearly this set could not be any larger than the set of all irrationals. Note that this set would exclude zero and one. Why?

Here is an example of such a list:

①   0 . 1 4 7 9 3 6 7 5 0 2 9 1 3 ...

②   0 . 3 0 6 1 9 4 4 1 1 7 8 0 8 ...

③   0 . 0 3 2 1 6 5 5 9 2 4 1 5 0 ...

④   0 . 9 9 4 0 5 8 4 2 7 5 4 8 9 ...

⑤   0 . 0 0 4 1 2 0 3 0 6 1 5 7 2 ...

⑥   0 . 4 5 6 8 2 3 7 3 3 0 0 4 1 ...

⑦   0 . 8 0 5 1 7 6 3 9 9 1 2 5 4 ...

⑧   0 . 2 2 0 5 9 3 6 1 1 7 2 5 5 ...

⋮   ⋮

This list goes on indefinitely. Now, we obviously couldn't construct the entire list, any more than we can write all the integers, but we can comprehend the idea of such a list.

We start by accepting the idea that such a list exists and that we have counted them all. Now, Cantor says let's consider the following. What if we were to start with the number at the top of the list and the first digit to the right of the decimal point. If we changed this number, then obviously the new number would be different from the current number. We then move down to the second number of the list and the second number to the right of the decimal and change this number. We now have a number different from the second. We continue this process indefinitely, moving down the list and over to the right.

This process is illustrated in the figure below, where each number we will be changing is circled:

①   0 . ⑴4 7 9 3 6 7 5 0 2 9 1 3 ...

②   0 . 3⓪6 1 9 4 4 1 1 7 8 0 8 ...

③   0 . 0 3②1 6 5 5 9 2 4 1 5 0 ...

④   0 . 9 9 4⓪5 8 4 2 7 5 4 8 9 ...

⑤   0 . 0 0 4 1②0 3 0 6 1 5 7 2 ...

⑥   0 . 4 5 6 8 2③7 3 3 0 0 4 1 ...

⑦   0 . 8 0 5 1 7 6③9 9 1 2 5 4 ...

⑧   0 . 2 2 0 5 9 3 6⑴1 7 2 5 5 ...

⋮   ⋮   ⋰

Now, here is a possible example of such a new number: 0 . 3 1 8 2 6 0 4 7 ...

We have changed the 1 to a 3, the 0 to a 1, the 2 to an 8, the 0 to a 2, the 2 to a 6, the 3 to a 0, the 3 to a 4, and the 1 to a 7. We would then continue the process indefinitely.

What have we accomplished? What Cantor has shown us is that we have found a new number that is not on our list. It is a number that is different from every other number on that list by at least one digit. It is a number not on the list, which means it has not been counted.

Think about this until you understand it. It is a little difficult to understand and absorb. Convince yourself of the validity of the argument. This is a strange and fascinating new result – The set of irrational numbers is not countable! Imagine this, a number so large that you can't even develop a way to begin counting it. At least with our natural numbers we can start one, two, three, etc., but with this set, we cannot systematically count its members. We can randomly count different members (irrational numbers) of the set, but there is no way to do so that will capture ALL the elements of the set.

This obviously meant that the size of the set of irrational numbers is greater than the size of the set of natural numbers. WHAT! Infinity comes in different sizes! How can that be?

With this simple demonstration Cantor had completely changed the way we looked at infinity. As far as infinity is concerned, one size does not fit all. Infinity comes in different sizes, and the smallest infinity is the size of the set of natural numbers, which Cantor called aleph, after the first letter in the Hebrew alphabet, and null, meaning the beginning value, or $\aleph_0$. This was the same size infinity as the number of integers and rational numbers.

The set of irrationals it turns out had a new infinite cardinality, which for lack of a better symbol we'll call C, for the continuum of real numbers. A new infinity, infinitely larger than the other infinity. Cantor did not end here, though. Cantor contemplated an infinite set of infinities, $\aleph_0, \aleph_1, \aleph_2, \aleph_3, \ldots$, and so was born the concept of **transfinite numbers**, numbers beyond the finite numbers, and even beyond the concept of a single infinity.

The first question that Cantor set out to answer was whether or not the cardinality, C, of the set of real numbers, which contained both the rationals and the irrationals, was the next largest infinity, aleph one, $\aleph_1$? This is called the **Continuum Hypothesis**. The hypothesis states that the next largest infinity above $\aleph_0$ (the number of natural numbers) is C, the cardinality of the set of reals. That is, does $C = \aleph_1$?

Cantor spent the rest of his tortured life trying to prove or disprove this result. There were times he thought he had proven it true, and others when he thought he had proven it false. This, amongst other things going on in his life at the time, drove him insane. It wasn't until forty five years after his death that a mathematician Paul Cohen in 1963 showed that the Continuum Hypotheses was undecidable. That meant it couldn't be proven either true or false. There will always be mysteries in mathematics.

Other even larger infinities have been put forth, such as the number of all possible two–dimensional curved shapes that can be drawn, or raising the number 2 to the power $\aleph_0$, C, or any other transfinite number. It has also been proven that the cardinality of the real numbers, $C = 2^{\aleph_0}$.

The following are all examples of infinities that are larger than $\aleph_0$, and larger than each other: $2^{\aleph_0}, 2^{\aleph_1}, 2^{\aleph_2}, \ldots$ or $2^C, 2^{2^C}, 2^{2^{2^C}}, \ldots$. We could go on forever constructing them.

Now while we can, for many of these types of infinities, tell which is larger and smaller than the other, we may never be able to put them in order, such as: $\aleph_0, \aleph_1, \aleph_2, \aleph_3, \dots$.

Transfinite numbers tell us that the concept of infinity is beyond anything we or the early Greeks could ever fathom. Amazingly Cantor showed that you could even do arithmetic with transfinite numbers, which has some strange properties, which we will not go into detail here.

The concept of the infinite takes us to the very edge of the wonder of mathematics and even beyond. It opens us a Pandora's box of strange and confusing results that have their roots way back to Ancient Greece and Zeno of Elea.

12.5 Paradox and the Infinite

Zeno of Elea (490–430 BCE) was a student of Parmenides (~540 BCE). Parmenides and Zeno both believed that the world consisted of one unchanging form. This form could not be divided, nor could it move or change in time. Division and motion were all just an illusion. There was no infinite, only the one. To show this, Zeno came up with a series of arguments showing why belief in infinity was not rational. These arguments are called paradoxes. Paradox comes from the Greek words "*para doxa*" meaning something contrary to opinion, or more precisely:

> A **paradox** is an illogical conclusion drawn from a sequence of a logical series of events.

We illustrate the idea of paradox through Zeno's original examples. Zeno wrote a book in which he had 40 such paradoxes. The book, however, is lost and only 8 paradoxes still remain, essentially due to the writings of Plato and Aristotle. Three of the more significant paradoxes of Zeno, related to what we are looking at, are: the Dichotomy paradox, Achilles and the tortoise paradox, and the Arrow in flight paradox.

We should point out that these paradoxes are subtle, and the average reader may not truly understand them and may dismiss them as crazy or nonsensical. This is not the case, they need to be looked at and understood in a very careful and thoughtful way.

Dichotomy Paradox

> *That which is in locomotion must arrive at the half–way stage before it arrives at the goal.*
> —Aristotle, Physics

This paradox can be viewed as making a trip across a room to a wall. Now everyone knows it is a task that can be completed without difficulty. The real question here is related to the underlying assumptions about the real world that allow us to do this. It we argue that space and time can be divided indefinitely it can cause some intuition problems, which is what Zeno was trying to highlight. Here is the paradox.

Let us say you are trying to walk to a wall. To accomplish this, you must first reach half the distance to the wall. Then you must reach half of the remaining distance to the wall which is 1/4 of the way. At this point you are 3/4 of the way to the wall. You must now move half of the remaining distance which is 1/8 of the way. You are now 7/8 of the way there. Again, you repeat the process and move

1/16, and are now located 15/16 the distance to the wall. You keep on moving half the distance to the wall, and since there are an infinite number of halfway points, you can never get to the wall in a finite amount of time, since you must have completed an infinite number of events before you got there! We have illustrated this with a picture below.

We should point out that from a mathematical perspective, this and the following paradoxes have been explained with satisfaction. However, this is not true from the real–world philosophical perspective. The reason why this can be explained mathematically is related to a part of mathematics that shows how the sum of an infinite set of numbers can actually equal a finite value and it can even occur in a finite time, provided we take the time needed to execute each step to be proportional to the size of the step. This strange new fact goes against what Aristotle and others have argued related to potential and actual infinities. This is a case of an actual infinity. Actual infinities do exist in mathematics. The question is – Do they exist in the real world?

More specifically it can be shown mathematically that:

$$\frac{1}{2} + \frac{1}{4} + \frac{1}{8} + \frac{1}{16} + \frac{1}{32} + \ldots = 1$$

The proof of this is beyond the scope of this book (it requires some calculus), so we will simply have to take it as a given fact. If we look carefully at this infinite sum of numbers we see that they are actually the same distances the walker must make as he or she approaches the wall. Thus, the walker moves a total of a unit distance, and furthermore, if the time it takes is proportional to one second times the proportion of the distance traveled to the wall at that particular step, then it will take the person only one second to make the trip.

Mathematically this is fine, but philosophically we have completed an infinite series of events. And according to the definition of infinity, a process that is never completed, this is not possible. This seems to contradict the very definition. We have completed that which by definition cannot be completed?

Aristotle's explanation was that there were logical problems with Zeno's reasoning in his paradoxes, but does not show where. Instead he argues on a more intuitive level that the paradoxes cannot be true, and dismisses them. Aristotle claims that space is a continuum and that you cannot divide it into discrete pieces as Zeno was suggesting (remember Aristotle and his view of the void). In Aristotle's eyes, it was the difference between a potential and an actual infinity. It was potentially possible to divide the distance forever, but to actually do it, as Zeno suggests, is not possible. Aristotle misses the point entirely, and fails to see the true objections raised by Zeno. It took over

2,000 years to find a satisfactory mathematical answer, and we still do not have a good philosophical one.

Achilles and the Tortoise Paradox

In a race, the quickest runner can never overtake the slowest, since the pursuer must first reach the point whence the pursued started, so that the slower must always hold a lead.
—Aristotle, Physics

In this paradox, there is a race between Achilles (one of fastest runners in Greece) and a tortoise (one of the slowest moving creatures). The tortoise is given a head start of 100 feet. After a finite amount of time Achilles gets to where the tortoise was when Achilles started running. However, during the time it took Achilles to get to that spot, the tortoise has moved another foot. Now Achilles must run to where the tortoise last was, and again the tortoise has moved an additional hundreth of a foot by the time Achilles reaches that spot. As you can probably tell, this paradox is really just a rephrasing of the Dichotomy paradox.

Every time Achilles runs, he must first reach where the tortoise was, but the tortoise has moved away during the time it took Achilles to get there. Therefore, Achilles cannot catch the tortoise, because he must always reach where the tortoise was last, but the tortoise has since moved, and this will have to go on indefinitely. There will be an infinite amount of steps before Achilles can catch the tortoise. Thus, an infinite amount of time must pass for Achilles to catch the tortoise, so he can never do so.

We illustrate this paradox with the figure below.

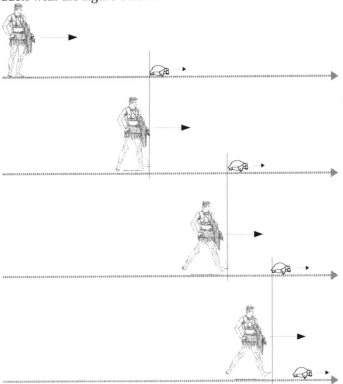

We, and Zeno, know that Achilles can actually catch the tortoise, but it raises questions about reality and the infinite that indicate we clearly do not understand how it all fits together. This is what a paradox is, something we know that can't be true, but logically we cannot state why it is true. What Zeno was trying to illustrate in this and the previous paradox is that you cannot divide things continually. In fact he was trying to prove that the world is just a single indivisible entity and that motion is merely an illusion. Achilles and the tortoise are actually part of the same thing, so of course he can catch the tortoise, because he is already a part of it and there already. According to Zeno, if we believe we can divide the distance into parts, then we will be able to do this indefinitely, so we can never end this process. It will take an infinite number of divisions, and we can't complete an infinity. The real problem according to Zeno is how we perceive reality.

This paradox is similar to the dichotomy paradox, only this deals with relative motion. The same basic ideas apply, though. Mathematically we can answer this the same way we did the dichotomy paradox. Aristotle's objections would also be the same, and philosophically we might digest and accept an unsatisfactory answer provided by Aristotle and others, but there are still some unanswered questions that need to be resolved.

The Arrow Paradox

If everything when it occupies an equal space is at rest, and if that which is in locomotion is always occupying such a space at any moment, the flying arrow is therefore motionless
—Aristotle, Physics

If you were to shoot an arrow into the air and watch its flight, this is what you would observe. At any instant of time (zero time passing) you would see the arrow suspended and motionless. At any point during its flight you would observe the same thing, a motionless arrow. If the arrow is motionless at every instant of time you observe it, then how can it move? Thus, the arrow never moves! Motion is nothing more than an illusion.

Zeno Today

These paradoxes are not tricks, and the answers given to try and solve them are not simple. They go to the very heart of reality, and the potential infinite divisibility of space and time. Even to this day, there are still open questions from a philosophical perspective.

As we have said above, Aristotle dismisses Zeno's paradoxes by pointing out that to be a paradox, they would have to accept the concept of an actual infinity. Aristotle claimed there were only potential and not actual infinities. Thus, Zeno's arguments must be wrong because he speaks of something that cannot be. In actuality Aristotle missed the entire point. However, at the time it was an argument that most accepted (because it was Aristotle) and society moved forward with Aristotle's view of infinity for the next two thousand years.

In 1967 Adolph Grünbaum published *Modern Science and Zeno's Paradoxes* in which he offered yet another possible philosophical solution to Zeno's paradoxes. Many others both before and after have been proposed. We have already stated that mathematically there are rational explanations of Zeno's paradoxes, but they still leave open some interesting philosophical questions. So contrary to what many have said, Zeno's paradoxes have not been explained to everyone's satisfaction. They have survived for nearly 2,500 years! Yes, we can describe a mathematical solution due to the calculus, but philosophically there is still a problem.

Some argue that regardless of what mathematics might tell us, you cannot break down any finite task, such as walking to the wall, into an infinite sequence of sub–tasks, since you by definition cannot complete all the sub–tasks. This is the nature of the paradox. It might also raise the question of whether or not infinity actually exists in the real world, or just in the world of mathematics.

The story about Zeno's paradoxes is far from over. This quote from Bertrand Russell sums up the controversy nicely.

> *In this capricious world nothing is more capricious than posthumous fame. One of the most notable victims of posterity's lack of judgment is the Eleatic Zeno. Having invented four arguments all immeasurably subtle and profound, the grossness of subsequent philosophers pronounced him to be a mere ingenious juggler, and his arguments to be one and all sophisms. After two thousand years of continual refutation, these sophisms were reinstated, and made the foundation of a mathematical renaissance ....*

> *Although they have often been dismissed as logical nonsense, many attempts have also been made to dispose of them by means of mathematical theorems, such as the theory of convergent series or the theory of sets. In the end, however, the difficulties inherent in his arguments have always come back with a vengeance, for the human mind is so constructed that it can look at a continuum in two ways that are not quite reconcilable.* Bertrand Russel, Principles of Mathematics, 1903.

Zeno was trying to say that if can divide something in half, then you can continue to divide it indefinitely, and dividing something indefinitely leads to a contradiction. Thus, the only answer is that we cannot divide anything. It is all one indivisible thing in Zeno's view.

Zeno's paradoxes essentially focus on the relation of the discrete to the continuous, an issue that is at the very heart of mathematics.

Paradoxes Beyond Zeno

Perhaps you are still not convinced about Zeno's paradoxes. If so then look at these two and see if you can gain any additional insight into the problem of paradoxes and the infinite.

Thompson's Lamp Paradox

Consider a lamp controlled by a switch. Turn the switch once, it turns the lamp on. Turn it again, it turns it off. Let us imagine there is a person with superhuman powers, and that they can turn the lamp on and off as fast as is needed. They in fact can complete an infinite number of "switchings" in a finite time.

First, she turns it on, and then at the end of one hour, she turns it off. At the end of an additional half hour she turns it on again. At the end of a quarter of an hour more she turns it off. In one eighth of an hour she turns it on again. And so on, turning the switch in the other direction each time after waiting exactly one–half the time she waited before switching it the last time. Applying the earlier mathematical discussion, we can show that all these infinitely many time intervals add up to exactly two hours (the sum of all half of half times is equal to a whole time).

The question is – At the end of two hours is the lamp on or off? Also, if the lamp was initially on instead of off, and the first time she turned the switch the lamp went off, would that make a difference on whether the lamp was on or off in the end?

Does this make you think? It should.

Finally we end we another paradox about the infinite. This one comes to us from the mathematician David Hilbert. It concerns a very special hotel.

Hilbert's Hotel Paradox

Consider a new hotel. Hilbert's Super– ∞ , paradoxical hotel shown below. The hotel has an infinite number of rooms, and it currently has a person occupying every room in the hotel. A young women comes by and asks if she can get a room, even though the no vacancy sign is on. The innkeeper says sure not a problem. Is he lying?

No he is not. He simply has to ask all those in the rooms to come out and move to the room next door, so the first room opens up for the new tenant. Problem solved. Since there are infinitely many rooms everyone currently in a room will get a new room, and the women can move in as well.

Now this is a very strange result. But wait, there is even more!

Now, an infinitely long bus pulls up with an infinite number of passengers on board. The driver goes inside and asks if there is enough room for all his passengers (remember, infinite in number). Again, the innkeeper says no problem. He must surely be lying now!

No he is not. He simply calls all the tenants and asks them to come out of their rooms and move down, leaving one empty room between each of them as they move down. Then all the passengers get off the bus and line up in front of every other open room, and they all are accommodated without any problem.

So how is it that a completely full hotel can accommodate an infinite number of new tenants? This is the amazing power of infinity. The power to create paradoxes, that is.

Aristotle's (also known as Galileo's) Paradox

Here is our final example of a paradox about the infinite. What we've shown below is an example of what has been called Aristotle's paradox. The author is listed as anonymous, but many believe the author was Aristotle, and so the name. It is also a paradox that Galileo wrestled with as well.

Consider two concentric circles of different radii, that are fixed together so that they cannot rotate independently. The circles then roll together to the right, and as they do so, they unroll the dark coating on their circumferences on to the two horizontal lines shown in the drawing. When the circles have turned one full rotation, we see that we would have two parallel horizontal lines of the same length. This means that the circumference of the inner circle, the length of the upper horizontal line, is equal to the circumference of the outer circle, the length of the lower horizontal line! Can this be?

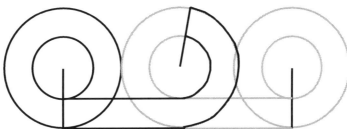

Two concentric circles rolling to the right

Think about it, can you explain what is happening? It is not so obvious, and it is very subtle. It is actually related to the concept of infinity and irrational numbers.

What we have tried to show in this section is just how strange and confusing the idea of the infinite can be. Just allowing it to exist in the real world opens up an array of potential paradoxes. These are things which cannot be explained. They follow from a logical series of events leading us to something that appears totally bizarre and nonsensical.

Today these paradoxes about infinity are curious and interesting artifacts of the "infinite" landscape of ideas. However, years ago, right before the Renaissance in Europe it was much more, for some, such as Bruno and Galileo, it was life threatening.

Banach-Tarski Paradox

Another interesting paradox is related to something called the axiom of choice.

The Axiom of Choice is a basic rule that was added to set theory after Cantor in 1904 by Ernst Zermelo to help prove a theorem he was working on. It is a very controversial axiom, with some mathematicians rejecting it and others wholeheartedly embracing it. The axiom basically states

> **Axiom of Choice:** Given a collection of nonempty sets, then there <u>exists</u> a way to <u>choose</u> a member from each set in that collection.

At first glance the axiom sounds harmless enough, and fairly intuitive, and one would wonder why anyone would question it, but there are some deep consequences that arise from it. The words "exists" and "choose" are critically important in this axiom. It simply says that a way to choose is possible, but does not show how to obtain that choosing process. No matter what the sets we are choosing from are, there is always a way to define the choice process, even if we don't actually find the way. For many mathematicians this is too much to ask for. Many constructivist mathematicians would argue that for the choosing process to exists, you have to be able to state very specifically what it is, and not simply claim that it exists.

Here is an elementary example to try and explain this further. This was originally provided by Bertrand Russell. Let's say that you have infinitely many pairs of shoes and you are asked to provide a rule (set of instructions) for picking one shoe from each pair, that can be repeated. The rule could be that you will always pick the right shoe. Thus, it is possible to define a specific rule for choosing. The rule exists, because you can state it without any ambiguity. In this case we do not need the Axiom of Choice in our set theory.

Alternatively, let us say that you have infinitely many pairs of identical socks and you are asked to provide a method for choosing one sock from each pair. What specific instructions can you give to pick a particular sock from each pair? You need to provide a unique rule, other than reach in and pick out one sock, since the process needs to be something that can be repeated by someone else, where they end up always choosing the same sock that you had picked. This requires the use of the Axiom of Choice, because you cannot provide a way to make the choice that can be exactly repeated by someone else. You are simply assuming that an approach does exist. The constructivist would say that I cannot simply say that it is possible, I instead need to find the exact rule, while the Axiom of Choice proponent relaxes this requirement, and says that although I cannot state what the rule is, I'll assume that there is such a rule (it exists) and proceed from there.

The constructivist says that the Axiom of Choice should not be introduced into set theory, since the actual choosing process may not be possible, while the mathematician in favor of the Axiom of Choice, likes the axiom because it can be useful in proving some theorems that the mathematician cannot find a way to prove otherwise.

If we admit the Axiom of Choice, we can prove some theorems about mathematics, but some strange results also come along. In 1924 two mathematicians, Stefan Banach and Alfred Tarski, proved a paradoxical result by assuming that the Axiom of Choice was true. Their original hope was to show that this axiom should not be added to set theory, since it gave "ridiculous" results. Some mathematicians agreed, but others became more entrenched and believed that Banach and Tarski had only shown that real numbers were richer than we originally thought.

The paradox is called the Banach-Tarski paradox which essentially states that:

> **Banach-Tarski Paradox**: It is possible to break up a sphere into distinct parts (finite in number), and then reassemble these parts into two different spheres that are each exactly the same size as the original sphere, without having any points missing in either of the two new spheres. The proof of this paradox requires the Axiom of Choice.

This result of creating something out of nothing seems crazy and impossible, which is why Banach and Tarski thought it would end the controversy about the Axiom of Choice, putting it to rest so to speak, but it did nothing of the sort. It instead fueled a debate that continues to this day.

Again, this paradox is related to the strange properties of irrational numbers and the fact that they are infinitely more numerous than the rationals.

An informal outline of the proof (borrowed from David Morgan – Mar at http://www.irregularwebcomic.net/2339.html) is as follows:

Consider a sphere, and two operations to move our sphere. You can rotate it in a counter-clockwise (CCW) direction about a line passing through the "equator" (see A below) or rotate CCW about a line through the "poles" (see B below). In addition to rotations A and B, consider rotations in the opposite clockwise (CW) direction that would reverse the effects of A or B, given as A^{-1}, or B^{-1} (read as A inverse and B inverse repsectively). All four of these rotations are shown below.

The motion of the sphere can be described by the sequence of rotations: B^{-1} A A B A^{-1} B A B … We shall choose to write these rotations right to left, with the right–most rotation performed first, and the left–most rotation performed last.

In addition, if we ever have pairs of rotations A A^{-1}, BB^{-1}, $A^{-1}A$, or $B^{-1}B$ together, then these pairs will cancel each other out, since they are simply rotations CW and then CCW, or CCW and CW, back to where the sphere was before the two rotations were performed. Thus, pairs of rotations like this will cancel each other out, and will not be in our sequence of rotations.

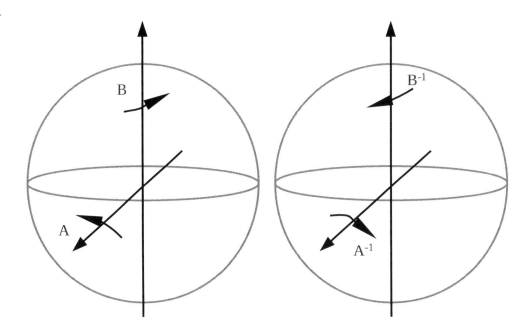

Furthermore, the arc–lengths of these rotations must also be irrational in size. Otherwise, if they were rational, you could possibly rotate the sphere back onto itself by continuously rotating it, until it returned to its starting value. For irrational lengths this can not happen.

These are the basic concepts and rules for moving a point around a sphere. We and now begin our argument ("proof") we suggested above.

Given any sphere there are 4 initial motions that can take place – either A, B, A^{-1}, or B^{-1}.

We let S equal the set of all possible rotations of the sphere by the 4 types of initial rotations from above.

Thus, if we let,

(set 1) = the set of all rotations starting with A
(set 2) = the set of all rotations starting with B
(set 3) = the set of all rotations starting with A^{-1}
(set 4) = the set of all rotations starting with B^{-1}

Then,

$$S = (set\ 1) + (set\ 2) + (set\ 3) + (set\ 4)$$

This means that we can break all the possible rotations of a sphere into four distinct and separate sets. But there is a subtle catch.

What if before we rotated (set 3) by A^{-1} we first rotated our sphere by A? Then the $A^{-1}A$ sequence would cancel out, and we would then have a set starting with either a B, A^{-1}, or B^{-1}. We would not have A since that would cancel out with the A^{-1} that was originally there, so it would not be at the start of (set 3). That means that (set 3) rotated by A first, actually contains all the rotations in (set 2), (set 3), and (set 4) combined!

So the set of all rotations,

$$S = (\text{set 1}) + (\text{set 3})\,A$$

In a similar way, we could take (set 4) and rotate by B first and show that,

$$S = (\text{set 2}) + (\text{set 4})\,B$$

This is a startling paradoxical discovery. We originally broke our rotations into four distinct sets that would cover all the rotations, but we were then able to separate these four sets into two sets that also contain all the rotations. We started with four, and created two identical copies that accomplish the same result!

Now consider a point of the sphere and see how it is moved by the rotations (see figure below).

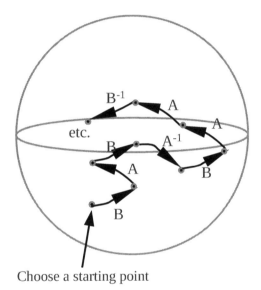

Choose a starting point

Imagine all the possible points you can reach on the surface of the sphere, with all the possible sequences of rotations. The set of points would be infinite, and they would be all over the surface of the sphere. However, not every point on the sphere would be reached in this way. There would still be an infinite number of points missing or not crossed on the surface (little infinitesimal holes throughout the surface). Thus, now choose a point that is not crossed and perform a similar set of rotations on that point. Again, some points will not be reached, so pick a point and continue the process. Each of these starting points creates an infinite set of points on the surface that we can reach. You will need to pick an infinite number of starting points to make sure we reach every point on the surface of the sphere.

Now consider a new set, call it P. The set P is formed by picking one point from each of the sets with different starting values. This is where the Axiom of Choice enters the picture. We are not stating how we choose the points in P, just that it is possible to do so.

Since the set of points in P contain a point that was reached by the set of rotations, then every point in P can be used to reproduce all the points on the surface by going through the rotations. Therefore, we can reach every point on the surface of the sphere using set P and our rotation sets 1, 2, 3, and 4 from above.

We now break the surface of the sphere into four pieces.

Piece 1: Contains all the points you can reach by starting with a point in P using (set 1) rotations.
Piece 2: Contains all the points you can reach by starting with a point in P using (set 2) rotations.
Piece 3: Contains all the points you can reach by starting with a point in P using (set 3) rotations.
Piece 4: Contains all the points you can reach by starting with a point in P using (set 4) rotations.

Thus, the

surface of the sphere = (piece 1) + (piece 2) + (piece 3) + (piece 4)

Now if we were to first rotate the sphere by A, then from above (set 3) is equivalent to (set 2), (set 3), and (set 4) combined, so (piece 3) is equivalent to (piece 2) + (piece 3) + (piece 4). This means that the

surface of the sphere = (piece 1) + (piece 3 after rotating by A first)

So we have created a complete copy of the surface by only using two pieces. We still have (piece 2) and (piece 4) left over.

Now (piece 4) can be turned into (piece 1) + (piece 3) + (piece 4), just as we did with (piece 3) above so, the

surface of the sphere = (piece 2) + (piece 4 after rotating by B first)

We created another copy of the sphere from the left over pieces 2 and 4!

We can extend this to the volume of the sphere, by taking the volume beneath each piece of the surface by taking:

Volume 1 = Volume under surface (piece 1)
Volume 2 = Volume under surface (piece 2)
Volume 3 = Volume under surface (piece 3)
Volume 4 = Volume under surface (piece 4)

and,

Volume of the sphere = (volume 1) + (volume 2) + (volume 3) + (volume 4)

Finally

Volume of the sphere = (volume 1) + (volume 3 after rotating it by A first)

and:

Volume of the sphere = (volume 2) + (volume 4 after rotating it by B first)

We have created two identical spheres from four separate volumes of the sphere!

Obviously, this is NOT a formal mathematical proof. The reasoning presented above has some flaws, but with some additional technical details this can be corrected. However, if we did so at the start it would make the visual argument above even more difficult to follow. For example, our argument does not cover the point at the center of the sphere nor the set of points beneath either of the rotation axes. For this we need to consider a fifth piece to break the sphere into. We will not do so, but just let you know that mathematically it is possible to prove this result without any such flaws in our reasoning.

We also need to point out that this is ONLY a mathematical result! It will not work for real physical objects such as bowling balls. All real physical objects are made up of a finite number of atoms. The volume of a mathematical sphere on the other hand is made up of an infinite number of points. This too is a strange result to consider, since how can an infinite number of points with zero volume each, add up to a finite volume? Again, the bizarre concept of infinity and its different sizes emerges. Cantor really did turn our mathematical world upside down.

Finally, we should point out that the pieces and volumes we used are not simply connected pieces and volumes, but are actually something called fractals (we visit these strange beasts in Volume II of this work). So we can't think of them as simple slices of an apple or an orange, or even oddly cut pieces like in a jigsaw puzzle. Fractals have really strange shapes.

The real issue with the paradox above is with the Axiom of Choice providing us with a way we can choose a specific real number form an infinite set of real numbers without ever having to specify exactly how we would do it. If we have to specify the process precisely, the proof would not work, and we'd only have a single sphere! This why many mathematicians reject the Axiom of Choice. However, those that do not reject it, feel that this is really a gateway into a new understanding of the concept of the infinite, and it also allows for some rather interesting and unique mathematical theorems to be proven.

Axiom of Choice, friend or foe, we may never know!

EXERCISES 12.1–12.5

ESSAYS AND DISCUSSION QUESTIONS:

1. Why was and is infinity such a difficult topic to understand? What are potential and actual infinities?

2. What are Zeno's paradoxes, and how are they related to infinity?

3. Who was Georg Cantor, and what role did he play in our understanding of infinity?

4. Are there different sizes to infinity? Explain your answer.

5. What is set theory? Who created it and why?

6. Can you name an infinite set that is not either a set of numbers or a set of geometric figures? Meaning, do infinite sets exist in the physical world outside of mathematics?

7. Explain Cantor's proofs on the countable infinity of rational numbers and the uncountable infinity of irrational numbers.

8. Does an actual infinity exist in the physical world?

9. What are transfinite numbers?

10. What is the Continuum Hypothesis? Explain what it is and who impacted its history.

11. What is the Axiom of Choice and how did it turn mathematics on its head?

12. How is the concept of infinity related to the Banach-Tarski paradox?

PROBLEMS:

12.3.2 Basics of Set Theory
Roster, Description, and Set Builder Notation
Write the other two missing forms of the set representations below.

1. {1, 3, 5, 7, … }

2. {0, 1, 2, 3, 4, 5}

3. The set of primary colors.

4. The set of all US states whose name begins with a T.

5. {n| n in a counting number between 1.5 and 7.5}

6. {p| p in the set of all countries with a population greater than one billion people}

7. {a, e, i, o, u}

8. {a, b, c, d}

9. The set of all US car companies.

10. The set of all the planets in the Solar System.

11. {x| x is a mountain at least 28,000 feet above sea level}

12. {c| c in the set of all basic natural hair colors}

Finite and Infinite Sets
Determine which sets are finite and which are infinite.

13. The set of all real numbers between zero and one.

14. The set of all integers between five and twenty.

15. The set of all the letters in the alphabet.

16. The set of all the planets in the Milky Way Galaxy.

17. The set of all counties in New York State.

18. The set of all integers less than zero.

<u>Cardinality</u>

Determine the cardinality of each set. If infinite, just state infinite.

19. The set of all real numbers between zero and one.

20. The set of all integers between five and twenty.

21. The set of all the letters in the alphabet.

22. The set of all the planets in the Solar System.

23. The set of primary colors.

24. The set of all integers less than zero.

25. $\{1, 3, 5, 7, \dots \}$

26. $\{0, 1, 2, 3, 4, 5\}$

27. $\{a, e, i, o, u\}$

28. $\{a, b, c, d\}$

29. $\{\dots -3, -2, -1, 0\}$

30. $\{$Red, Yellow, Blue$\}$

<u>Universal Sets</u>

31. Describe what a universal set is, and give three examples.

Each of the following could be a universal set. Explain how.

32. The set of all the people in the US.

33. The set of planets in the Milky Way Galaxy.

34. The set of all Hollywood actors and actresses.

35. The set of all major league baseball players.

36. The set of all Caucasian people.

37. The set of high cholesterol food.

38. The set of all smokers.

<u>Empty Sets</u>

Which of the following are empty sets?

39. The set of all integers between zero and ten.

40. The set of all four sided triangles.

41. The set of all integers between 1/10 and 999/1000.

42. The set of all men greater than 18 feet tall

43. The set of people in the US that are not citizens.

270

44. {2, 3 ,4 ,5}

45. { Ø }

46. {x| x is a counting (natural) number less than zero}

47. {y| y is an integer between 1 and 2}

48. The set of all circles with an integer radius between 1 and 10.

49. The set of all women presidents or leaders of countries.

50. The set of all circles whose area is equal to one.

Membership

Determine if the following statements are true, given:

      A = {1, 2, 3, 4, 5, 6, 7, 8, 9},   B = {2, 4, 6, 8}
      C = {John, Joe Jake, James, Jeremy},   D = {Sue, Sally, Sarah, Samantha}

51. $2 \in A$	54. $John \in C$	57. $Jason \notin C$	60. $Sissy \in D$
52. $5 \in B$	55. $Sally \in B$	58. $Sally \notin B$	61. $James \notin C$
53. $-2 \notin A$	56. $3 \in B$	59. $0 \in A$	62. $8 \in A$

Union, Intersection, Complement, Difference, Disjoint, Subset, and Proper Subset

Given:

      A = {1, 2, 3, 4, 5, 6, 7, 8, 9},   B = {2, 4, 6, 8}
      C = {John, Joe Jake, James, Jeremy},   D = {Sue, Sally, Sarah, Samantha}
      E = {John, Sue, Joe, Sarah},   F = {1, 3, 5, 7, 9}

Determine the following:

63. $A \cup B$	69. $C \cup D$	75. Is $B \subset B$?
64. $A \cap B$	70. $C - E$	76. Is $F \subseteq A$?
65. $B \cup F$	71. $B - F$	77. Are B and F disjoint?
66. $A - B$	72. Is $B \subset A$?	78. Are C and D disjoint?
67. $C \cap E$	73. Is $B \subseteq A$?	79. Are C and E disjoint?
68. $C \cap D$	74. Is $A \subset F$?	80. Are A and B disjoint?

Given:

$U = \{a, b, c, d, e, f, g, h, i\},$ $A = \{a, c, e\},$ $B = \{b, d, f, h\},$ $C = \{f, g, h, i\}$

Determine the following:

81. $A \cup B$
82. $A \cap B$
83. $\bar{A}$
84. $U - A$

85. $U - \bar{A}$
86. $\bar{A} \cap B$
87. $\overline{A \cup B}$
88. $\bar{C}$

89. $\bar{U}$
90. $B \cap C$
91. $(A \cup B) \cup C$
92. $\bar{C} \cap B$

93. Which sets are disjoint?

94. Which sets are subsets of each other?

95. Which sets are proper subsets of each other?

<u>Cartesian Product</u>
Given:

$A = \{1, 2, 3\},$ $B = \{2, 4, 6\}$
$C = \{John, Joe, James, Jeremy\},$ $D = \{Sue, Sally\}$

Determine the following:

96. $A \times B$
97. $B \times A$

98. $C \times D$
99. $D \times C$

100. $A \times D$
101. $C \times B$

12.3.3 Visualizing Sets and Their Operations and Relationships

<u>Set Operations and Venn Diagrams</u>
Illustrate the following set operations using Venn diagrams.

1. $A \cup B$
2. $A \cap B$
3. $\bar{A} \cap B$
4. $\bar{B} \cup B$

5. $A - B$
6. $\overline{A - B}$
7. $(A \cap B) \cap C$
8. $(A \cup B) \cap C$

9. $(\bar{A} \cap B) \cup C$
10. $(A \cap B) \cap C$
11. $(A - B) \cap C$
12. $\bar{C} \cup (A \cap B)$

Identify the possible set operations indicated by the Venn Diagrams below. (Note: answers can vary)

13.

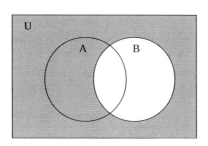

14.

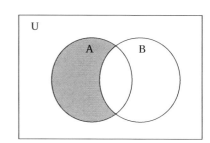

15.
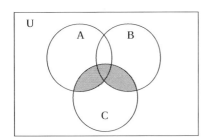

17.
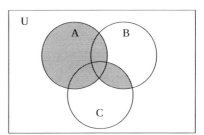

16.
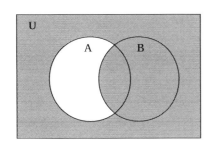

18.
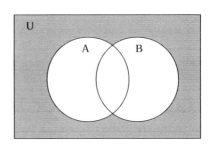

De Morgan's Laws

19. State the two forms of De Morgan's Laws.

20. What are De Morgan's Laws useful for?

12.3.4 Application of Venn Diagrams

Surveys

Use Venn diagrams to answer the following.

21. A psychology major at a university took a survey of a group of 55 students regarding their preference to female singers with the following results:

17 like Pink
17 like Shakira
23 like Taylor Swift
6 like Pink and Shakira
8 like Pink and Taylor Swift
10 like Shakira and Taylor Swift
2 like all three singers

How many students like
a) exactly two singers?
b) exactly one singer?
c) None of these singers?
d) Pink, but neither of the others?

22. A survey of a group of female students at a college regarding their preference for three male bands/singers with the following results:

12 like Justin Bieber, My Chemical Romance and 30 Seconds to Mars
13 like 30 Seconds to Mars and Justin Bieber
28 like 30 Seconds to Mars
3 like 30 Seconds to Mars, but not My Chemical Romance
2 like Justin Bieber, but not 30 Seconds to Mars and My Chemical Romance
8 Don't like any of the artists

16 like 30 Seconds to Mars, but not Justin Bieber

a) How many female students were surveyed?

Find the number of female students that
b) like 30 Seconds to Mars.
c) like Justin Bieber.
d) like Justin Bieber and 30 Seconds to Mars.
e) don't like Justin Bieber.
f) don't like My Chemical Romance.

23. A survey of a group of 105 adults regarding their use of texting, emails, or cell phone calls found the following results:

54 text
69 use cell phones
51 send emails
30 use texting and cell phones
20 use texting and emails
28 use cell phones and emails
10 use all three

How many adults communicate using
a) exactly two technologies?
b) exactly one technology?
c) None of these technologies?
d) texting, but none of the others?

24. A survey of a group of 88 people in the mall about their candy preferences fond that:

55 like chocolate
26 like hard candy
48 like gum
15 like chocolate and hard candy
29 like gum and chocolate
16 like gum and hard candy
12 like all three

How many people like
a) exactly two types?
b) exactly one type?
c) None of these candies?
d) chocolate, but neither of the others?

25. A survey of 120 heart patients at a clinic found that:

66 smoke cigarettes
79 had high cholesterol
90 had high blood pressure
55 smoked and had high blood pressure

57 had high blood pressure and cholesterol
50 smoked and had high cholesterol
43 had all three conditions

How many patients had
a) exactly two symptoms?
b) exactly one symptoms?
c) None of these conditions?
d) high cholesterol, but neither of the others?

References

1. Aczel, A.D., (2000). The mystery of the Aleph: Mathematics, the Kabbala, and the Human Mind. New York: Four Walls Eight Windows Publishing.

2. Benardete, J., (1964). Infinity, Oxford.

3. Burova, I.N., The development of the problem of infinity in the history of science (Russian), Nauka (Moscow, 1987).

4. Cantor, G., (1955, 1915). Contributions to the Founding of the Theory of Transfinite Numbers. New York: Dover. ISBN 978-0486600451

5. Dauben, J.W., (1977). Georg Cantor and Pope Leo XIII: Mathematics, Theology, and the Infinite. Journal of the History of Ideas 38.1.

6. Dauben, J.W., (1979). Georg Cantor: his mathematics and philosophy of the infinite. Boston: Harvard University Press.

7. Dauben, J.W., (1983). Georg Cantor and the Origins of Transfinite Set Theory. Scientific American 248.6:122-131.

8. Dauben, J.W., (1993, 2004). Georg Cantor and the Battle for Transfinite Set Theory. in Proceedings of the 9th ACMS Conference (Westmont College, Santa Barbara, CA) (pp. 1–22).

9. Davenport, A.A., (1997). The Catholics, the Cathars, and the Concept of Infinity in the Thirteenth Century. Isis 88.2:263–295.

10. Ewald, W.B. (ed.), (1996). From Immanuel Kant to David Hilbert: A Source Book in the Foundations of Mathematics.

11. Grattan-Guinness, I., (1971). Towards a Biography of Georg Cantor. Annals of Science 27:345–391.

12. Grattan-Guinness, I., (2000). The Search for Mathematical Roots: 1870–1940. Princeton University Press.

13. Hallett, M., (1986). Cantorian Set Theory and Limitation of Size. New York: Oxford University Press.

14. Halmos, P., (1998, 1960). Naive Set Theory. New York & Berlin: Springer.

15. Hill, C. O. & Rosado Haddock, G. E., (2000). Husserl or Frege? Meaning, Objectivity, and Mathematics. Chicago: Open Court.

16. Johnson, P.E., (1972). The Genesis and Development of Set Theory. The Two-Year College Mathematics Journal 3.1:55–62.

17. Maor, E., (1987). To infinity and beyond : A cultural history of the infinite, Boston, MA.

18. Meschkowski, H., (1983). Georg Cantor, Leben, Werk und Wirkung (George Cantor, Life, Work and Influence, in German). Wieveg, Braunschweig.

19. Moore, A.W., (1995, April). A brief history of infinity. Scientific American.4:112–116.

20. Penrose, R., (2004). The Road to Reality, Alfred A. Knopf.

21. Purkert, W. & Ilgauds, H.J., (1985). Georg Cantor: 1845–1918. Birkhäuser.

22. Reid, C., (1996). Hilbert. New York: Springer-Verlag.

23. Rodych, V., (2007). "Wittgenstein's Philosophy of Mathematics" in Edward N. Zalta (Ed.) The Stanford Encyclopedia of Philosophy.

24. Rucker, R., (2005, 1982). Infinity and the Mind. Princeton University Press.

25. Russell, B., (1996) [1903], The Principles of Mathematics. New York, NY: Norton

26. Russell, B., (1967). Modern Science and Zeno's Paradoxes of Motion, in: R.M. Gale (ed.), The Philosophy of Time. A Collection of Essays. New York: Anchor Doubleday Books, pp. 422–494.

27. Snapper, E., (1979). The Three Crises in Mathematics: Logicism, Intuitionism and Formalism. Mathematics Magazine 524:207–216.

28. Suppes, P., (1972, 1960). Axiomatic Set Theory. New York: Dover.

29. Vilenkin, N.Y., (1995). In search of infinity, Boston, MA.

30. Wallace, D.F., (2003). Everything and More: A Compact History of Infinity. New York: W.W. Norton and Company.

31. Weir, A., (1998). Naive Set Theory is Innocent!. Mind 107.428:763–798.

CHAPTER 13

Imaginary Numbers

Adam and Eve are like imaginary number, like the square root of minus one... If you include it in your equation, you can calculate all manners of things, which cannot be imagined without it.

Philip Pullman

13.1 Introduction

We have seen the concept of number grow and grow, so what more could surprise us. Our definition of number was stretched by applying numbers beyond simply counting objects. When applied to keeping track of portions of a whole, we were led to fractions, or rational numbers. If we wanted to keep track of debts, we needed the concept of negative numbers. Applying the numbers to geometry we discovered the need for irrational numbers.

Rene Descartes, a French mathematician, showed us how to link geometry and algebra, so that applying geometry to numbers is equivalent to applying algebra to numbers. Thus solving the geometric problem of Pythagoras' Theorem for a right triangle with both sides equal to one, is equivalent to solving the algebraic equation:

$$x^2 = 2$$

which we have come to solve and write as $x = \sqrt{2}$, an irrational number.

The next amazing thing is that all these numbers, that were arrived at for different reasons, can all be placed on a single entity called a number line. They are related by an order, although when we consider the irrational numbers this order concept becomes very abstract, involving the concept of infinity. After the concept of irrational numbers and the infinite, you might consider the question of any other new type of number closed. I mean we have filled in the number line completely. We even gave this complete set of numbers a name, calling them the Real numbers. However, there was still another problem that bothered mathematicians, and it was as simple as the equation above. They would from time–to–time arrive at problems such as:

$$x^2 = -1 \quad \text{or equivalently} \quad x^2 + 1 = 0$$

That is, find the square of a number that is equal to a negative one. This is counter to anything we have developed earlier. We all know that to square a number is just a shorthand way of saying we multiply a number by itself, and furthermore, if that number is positive, the resulting product is positive, and if that number is negative the resulting product is still positive, since a negative times a negative, we have learned, although with some reservations, is positive! Thus, this problem doesn't make any sense, and can't be solved.

Mathematicians, however, do not like having problems without solutions. So they said, let's imagine that the problem $x^2 = -1$ does have a solution. If it did then the new number that made the equation true would have a very special property. It seems that whenever we multiplied this new number by itself we would always get a negative number. Since we are just imagining this as a possibility, let's call this number an imaginary number, and let us write it as, i, for imaginary.

When we do this, all sorts of new and interesting things emerge. We'll discuss this further below, but we'll first view the historical beginnings of this new and strange idea.

13.2 Brief History of Imaginary/Complex Numbers

In 1545 the Italian mathematician Girolamo Cardan (1501–1576) first proposed a new type of number written as $\sqrt{-15}$. He combined this new entity with real numbers and would write expressions like $5 + \sqrt{-15}$ and $5 - \sqrt{-15}$ when solving algebraic equations. Cardan noticed that when he multiplied these two quantities together he got the real number 40, which happened to solve the problem he was working on! However, he only identified this as an interesting curiosity that was "as subtle as it was useless."

A little less than 30 years later another Italian mathematician, Rafael Bombelli (1526–1572) provided justification for Cardan's approach and introduced a new number, which he wrote as:

$$a + \sqrt{-b}$$

The approach gave him correct answers to algebraic problems, convincing him of the validity of these new numbers. Others, however, were not so convinced. Rene Descartes was the first to call the square root of the negative number, "imaginary." He did so when he was talking about roots to polynomial equations, which has both real and these "imagined" solutions.

Gottfried Leibniz (1646–1716) worked at applying the laws of arithmetic and algebra to these new numbers. Leonard Euler (1707-1783) was the first to use i to represent $\sqrt{-1}$. The combination of the real numbers, a, with these new imaginary numbers, $b\,i$, became known as complex numbers, or

$$a + b\,i$$

The concept of imaginary and complex numbers continued to find use and rapidly developed. Work on them led to the fundamental theorem of algebra. They were used extensively in various branches of geometry, in solving certain equations in number theory, as well as in quantum mechanics, fluid dynamics, and electricity and magnetism. What many people once believed to be impossible, ridiculous, and even fictitious, became a reality. Their properties helped to explain many real physical phenomena, even though they were not thought of as "real."

13.3 Mathematics of Imaginary/Complex Numbers

Imaginary numbers are different from real numbers. You cannot add an imaginary number to a real number and get either an imaginary or a real number. Instead you get a new number called a complex number, $\mathbb{C}$, with both a real and an imaginary part. To illustrate the relationship between the real the imaginary and the complex numbers we use a two–dimensional graph. The graph has both a real axis and an imaginary axis that are perpendicular to each other. Together, they form the space of complex numbers, also called the complex plane, as illustrated in the graph below.

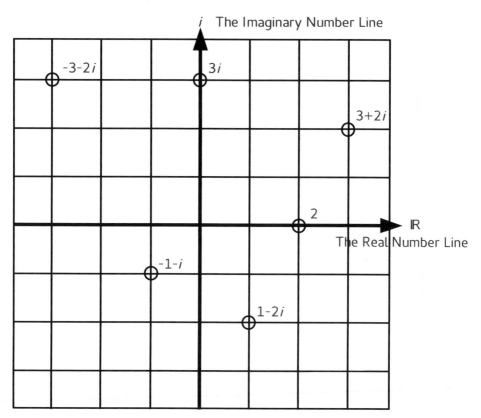

Notice that we have identified several complex numbers on this graph. This is similar to plotting points on a Cartesian grid. Only now instead of writing an (x, y) pair, we write a complex number a+bi. The "a" value is the distance along the horizontal real axis, and the "b" value is the distance along the vertical imaginary axis. The set of complex numbers contains both the set or pure real numbers, and pure imaginary numbers as subsets.

Just as numbers in a grid are identified by an x and y value with the ordered pair (x, y), so too can a complex number be identified with a pair of numbers (a, b), where a is the real part, and b is the imaginary part.

The Arithmetic of Complex Numbers

Adding and Subtracting Complex Numbers

To add or subtract two complex numbers you simply add or subtract their real parts from each other, and their imaginary parts from each other.

For example: To add the complex number $4+2i$ to the complex number $2+3i$. We simply add the 4 and the 2 (real parts), and the 2i and the 3i (imaginary parts), as shown below:

$$(2+3i)+(4+2i)=(2+4)+(3+2)i=6+5i$$

On the other hand, to subtract the complex number $4+2i$ from the complex number $2+3i$. We subtract the 4 from the 2 (real parts), and the 2i from the 3i (imaginary parts), as shown below:

$$(2+3i)-(4+2i)=(2-4)+(3-2)i=-2+i$$

Adding and subtracting complex numbers is a lot like adding and subtracting linear binomial terms, as we did in basic algebra.

Multiplying Complex Numbers

Multiplying i times a complex number

If we multiply a complex number by i, we see that the real part becomes the imaginary part and the negative of the imaginary part becomes the real part.

For example:
$$i(2+3i)=2i+3i^2=2i+3(-1)=2i-3=-3+2i$$

or

$$i(-4-7i)=-4i+(-7)i^2=-4i+(-7)(-1)=-4i+7=7-4i$$

Multiplying two complex numbers

To multiply two complex numbers, we can also mimic what was done for the binomials back in algebra. We can use the FOIL method, or the "distribute term–by–term" method to arrive at our answer.

For example: To multiply the two complex numbers $2+3i$, and $4+2i$ we carry out the following procedure:

$$(2+3i)\cdot(4+2i)=(2)(4)+(2)(2i)+(3i)(4)+(3i)(2i)$$
$$=8+4i+12i+6i^2$$
$$=12+16i+(6)(-1)$$
$$=6+16i$$

As you can see we multiplied the two First terms in each complex number (2) and (4), then the two Outer terms (2) and (2i), the two Inner terms (3i) and (4), and finally the two Last terms (3i) and (2i) (notice the FOIL approach). The last multiplication leads to an i^2 term which we replace with (–1) and after multiplying by (6), subtract it from the 12 up front. This gives a real part of 6 and an imaginary part of 16i.

Division is a bit more complicated and beyond the scope of what we want to do here. However, for those that are interested we include it, but others can skip this part, and it won't impact the flow of the book.

Dividing Complex Numbers

To divide a complex number by another complex number, we look at two types of examples. The first shows how to divide by a pure imaginary number, and the second shows how to divide by a complex number with both a real and imaginary part.

Dividing a Complex Number by a Pure Imaginary Number

To divide by a pure imaginary number, you multiply the numerator and denominator by i.

EXAMPLE 1:

Divide $7-6i$ by i.

$$\frac{7-6i}{i}$$

You first multiply the numerator and denominator of the quotient by i, and then carry out the arithmetical operations as shown below:

$$\frac{7-6i}{i}=\frac{7-6i}{i}\cdot\frac{i}{i}=\frac{(7-6i)i}{i^2}, \qquad \text{Multiply by } \frac{i}{i}$$

$$=\frac{7i-6i^2}{-1}=\frac{7i+6}{-1}, \qquad \text{Distribute } i$$

$$=-7i-6=-6-7i, \qquad \text{Reorder}$$

EXAMPLE 2:

Divide $-4+8i$ by $2i$.

$$\frac{-4+8i}{2i}$$

You first multiply the numerator and denominator of the quotient by i, and then carry out the arithmetic as shown below:

$$\frac{-4+8i}{2i} = \frac{-4+8i}{2i} \cdot \frac{i}{i} = \frac{(-4+6i)i}{2i^2}, \quad \text{Multiply by } \frac{i}{i}$$

$$= \frac{-4i+8i^2}{-2} = 2i-4, \quad \text{Distribute } i$$

$$= -4+2i, \quad \text{Reorder}$$

Complex Conjugate

To divide a complex number by another complex number, you again use the process of multiplication, but this time you multiply by something called the **complex conjugate**.

The **complex conjugate** of a complex number is obtained by simply changing the sign of the imaginary part of the complex number.

For example, the complex conjugate of: $4+2i$ is $4-2i$,

and the complex conjugate of: $12-7i$ is $12+7i$.

We can now use this approach to divide two complex numbers, which we illustrate with an example.

Dividing Two Complex Numbers

EXAMPLE 3:

Divide $2+i$ by $1+i$.

$$\frac{2+i}{1+i}$$

You start by multiplying both the numerator and denominator of the quotient above, by the complex conjugate of the denominator term, $1-i$, and then carry out the arithmetical operations as shown below:

$$\frac{2+i}{1+i} = \frac{2+i}{1+i} \cdot \frac{1-i}{1-i} = \frac{(2+i)(1-i)}{(1+i)(1-i)}, \qquad \text{Multiply by } \frac{1-i}{1-i}$$

$$= \frac{2-2i+i-i^2}{1-i+i-i^2} = \frac{2-i-(-1)}{1-(-1)}, \qquad \text{Distribute } 1-i$$

$$= \frac{3-i}{2} = \frac{3}{2} - \frac{i}{2}, \qquad \text{Rewrite}$$

What we wanted to accomplish in this chapter is to show you that the concept of number can be broadened even further. It came about in a natural investigative way, trying to find solutions to problems in algebra, which had their origins in geometry. We learned how to add, subtract, and multiply these numbers, similar to what we've done with other types of number. The question that comes to mind – are these new complex numbers, with their imaginary part, just a nice invention of mathematicians for helping them solve problems, or are they more than that? Are complex numbers necessary for us to understand the physical world, or do they just provide a convenient, human invented, mechanism to do so? Are we using notation to describe something that actually exists, or are we inventing something to simply describe what we think we see?

13.4 The Physical Significance of Imaginary Numbers

We have now extended our numbers to a new level of abstraction beyond our real numbers. A number that seems to have no connection to physical reality, and contradicts what we have learned previously, an imaginary number! Little did the original discoverers or inventors know, just how "real" this imaginary number would become. In fact, without it, we wouldn't have all these cell phones and portable electronic devices we have become addicted to (maybe they aren't so good after all).

Complex numbers are intimately tied to geometry and space. They help unravel the mysteries of everything from AC electrical current, to quantum physics, relativity, wormholes, tachyons, non–Euclidean geometry, imaginary matter, spin and reality in general. All this from a number that nobody even believed in.

Again, we come to a question we have asked many times before – Do imaginary numbers exist independent of humans, or are they just our creation? How is it that something used to solve a simple quadratic equation, can apply to a multitude of physical problems?

EXERCISES 13.1–13.4

ESSAYS AND DISCUSSION QUESTIONS:
1. What is an imaginary number?

2. Do you think imaginary numbers were discovered or created? Explain your response.

3. What real physical things have imaginary numbers been useful in investigating?

PROBLEMS:
Find the imaginary numbers.

1. $\sqrt{-4}$ 4. $\sqrt{-25}$ 7. $-\sqrt{-36}$

2. $\sqrt{-16}$ 5. $\sqrt{-49}$ 8. $-\sqrt{-100}$

3. $\sqrt{-9}$ 6. $\sqrt{-121}$

Plot the following complex numbers in the complex plane.

9. $2+i$ 12. $-3i$ 15. $-3-4i$

10. $-1-i$ 13. 4 16. $3+6i$

11. $4i$ 14. -5 17. $1-6i$

Add or subtract the following complex numbers as indicated.

18. $(2+i)+(5-2i)$ 24. $(2+i)-(5-2i)$

19. $(7+2i)+(-3+4i)$ 25. $(-3-2i)-(-2-6i)$

20. $(-6+5i)+(-4-5i)$ 26. $(-7-3i)-(-6i)$

21. $(6+3i)+(-2-3i)$ 27. $(2)-(-9i)$

22. $(4+i)+(-4-2i)$ 28. $(13-15i)-(15-21i)$

23. $(-7+3i)+(7-2i)$ 29. $(22+16i)-(25+17i)$

Multiply the following complex numbers as indicated.

30. $(2+i)(5-2i)$ 36. $(2+7i)(2+9i)$

31. $(3-2i)(4-6i)$ 37. $(-2i)(3-i)$

32. $(i)(5-2i)$ 38. $(-1-i)(-1+i)$

33. $(-4)(-3-2i)$ 39. $(14-11i)(1+2i)$

34. $(3+5i)(3-5i)$ 40. $(7+i)(2-i)$

35. $(-1-4i)(-2-6i)$ 41. $(8-3i)(2+2i)$

Find the complex conjugate of the following complex numbers.

42. $2+i$

43. $-3-12i$

44. $21-33i$

45. $-16-4i$

46. $3i$

47. $-6i$

48. -5

49. 13

50. $3-22i$

51. $64+8i$

52. $9+17i$

53. $17+21i$

Divide the following complex numbers as indicated.

54. $\dfrac{(2+i)}{i}$

55. $\dfrac{(3-i)}{-2i}$

56. $\dfrac{(2-i)}{(3-2i)}$

57. $\dfrac{(1+i)}{(1-i)}$

58. $\dfrac{(8+4i)}{(1+3i)}$

59. $\dfrac{i}{(1-i)}$

60. $\dfrac{3}{(4+7i)}$

61. $\dfrac{7}{(7+9i)}$

62. $\dfrac{(3+3i)}{(3-3i)}$

63. $\dfrac{(9-6i)}{(2-5i)}$

64. $\dfrac{(-1-5i)}{(-2-i)}$

65. $\dfrac{(-8-3i)}{(-1-4i)}$

References

1. Asimov, I., (1942). The Imaginary. Super Science Stories. Nov. 1942. Reprinted in The Early Asimov, Book One. Del Rey, 1986.

2. Brown, D., (2003). The Da Vinci Code. New York, Doubleday.

3. Burton, D.M., (1995). The History of Mathematics (3rd ed.), New York: McGraw-Hill.

4. Conway, J. H. and Guy, R. K., (1996). The Book of Numbers. New York, Springer-Verlag.

5. Ebbinghaus, H.D., (1991), Numbers, Springer.

6. Frucht, W. (Ed.), (2000). Imaginary Numbers: An Anthology of Marvelous Mathematical Stories, Diversions, Poems, and Musings, 2nd ed. New York, Wiley.

7. Katz, V.J., (2004). A History of Mathematics, Brief Version, Addison-Wesley.

8. Mazur, B., (2003). Imagining Numbers (Particularly the Square Root of Minus Fifteen), Farrar, Straus and Giroux,.

9. Nahin, P.J., (1998). An Imaginary Tale: The Story of $\sqrt{-1}$, Princeton University Press.

Part II Summary – Making the Connections

In summary, we have learned that the concept of number is very broad indeed. The names and symbols we give them are indisputably a human creation, however, their existence may be beyond us. They seem to have a life of their own, almost as if they are already there and if we look hard enough we can discover them and their properties. It is truly amazing that the simple concept of counting can lead to accurate descriptions of how the world works, enabling us to predict the future. That the relationships between numbers can lead to true and certain knowledge is an amazing fact.

We end with a very famous formula known to mathematicians. It is called Euler's Formula. It is appropriately named after the mathematician, Leonhard Euler (1707–1783), who first discovered/created it.

$$e^{i\pi} = -1$$

Here we have four very important numbers – Euler's number, e, associated with continuous growth: π, associated with the geometry of circles and cyclic patterns, -1, a number that allows us to reverse time and direction, and the imaginary number, i, associated with the deeper secrets of space, matter, and time. Why or how could these four totally different numbers be so simply connected?

All of these unique numbers and concepts that when first created/discovered had no apparent connection, but have all been shown to be intimately related. It makes you think – are they truly linked by nature and god? Is there a mathematical foundation to this world, or is this just a lucky creation of humanity? Is it just a coincidence, that not by design, but by chance, these numbers are all related?

It is truly amazing that numbers, the symbols for basic counting, have grown into helping us explain and understand the very fabric of our universe. Along the way, we have seen our definition of what we call a number grow by leaps and bounds. As a civilization we have come from marking our hunting kills on bones with simple tally marks some 30,000 years ago, to trying to use the expanded number concept to explain the strange wave–particle duality of matter.

From the beginning we have seen a progression to more and more abstract concepts. Starting from the development of natural numbers, zero, negative numbers, rational numbers, exponential numbers, irrational numbers, infinity, and even imaginary numbers. Mathematics has been used to better understand and describe the world in which we live. For some it is the language of the universe or for others it is the creation of a supreme being and to try and understand this, is a part of who we are. A truly educated person needs to be aware of what we have achieved in this discipline we call mathematics.

ESSAYS AND DISCUSSION QUESTIONS:

Write an essay on each of the following, or be prepared to discuss these questions:
1. Discuss the new abstract concept of number.

2. What is Euler's formula, and why it is so unique?

PART III: HISTORICAL DEVELOPMENT OF REASONING (LOGIC)

We have seen the philosophical side of mathematics. The deeper connection it seems to have with questions about what the universe is and why it is the way that it is. We have also seen how the concept of number broadened to include unbelievable ideas and hidden relationships that are a part of the very fabric of the world, and in some sense seemingly beyond it. In this part of the book we will embark upon a journey about a completely different side of mathematics. The side that is associated with how we even begin to reason, to question, and how we think. Amazingly, mathematics has a lot to say here as well. It is through mathematics that we can pursue reasoning and logic, and see what impact this vast subject has on questions we all try to ask and answer. Again we will see the remarkable story, that is mathematics, unfold in interesting and enlightening ways.

CHAPTER 14

Introduction to Types of Reasoning – What is Proof?

Why, you might just as well say that 'I see what I eat'
is the same thing as 'I eat what I see'!

Lewis Carroll

14.1 Introduction

Is the world logical, or is it humans that want to perceive it that way? Perhaps it is as nonsensical as Lewis Carroll's Wonderland, but we just don't realize it. In this chapter we look at the fascinating subject of logic. Why, because it is important in everything that we do. Without logic the world as we know it would crumble into chaos. Logic helps us makes sense of our world. The patterns we see fit together in an understandable way, and what's more we can prove it. Prove it! What does that mean anyway? What is proof and why do we even care?

14.2 What is Proof?

To prove something means you are able to provide irrefutable evidence that what you are claiming is true. This is a very strong assertion. How is it possible to give such evidence? How often have you heard someone claim something is true, only to find out later that it really wasn't? For the longest time, many people believed the earth was flat. I mean it is obvious when you look at it that it must be flat, isn't it? People also believed that the earth was the center of the universe. After all, we see all the stars, the sun, and the moon rotating around us every day, so the Earth must be located at the center, right? How do we prove or disprove assertions like these?

The problem is that the idea of proof can take on many forms. In science we create laws of nature that can be used to predict how the world behaves. The more accurate the prediction, the better the model. We can never actually prove the laws are irrefutably true, although some may try to argue otherwise. We instead, lean in their favor, based upon the consistency and accuracy of their predictions.

In mathematics we can prove relationships between things. For example, consider the figure below:

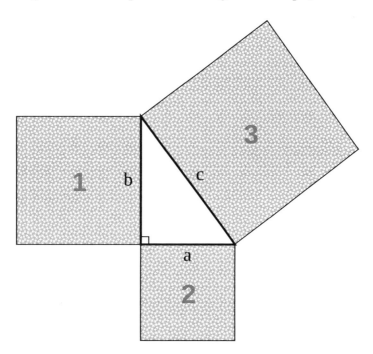

Consider the three squares shown above, let's assume that they are plates of gold. Given the choice, would you rather have both squares 1 and 2, or just square number 3 by itself? Which is worth more?

As you may well have discovered, this is a trick question. It turns out that they are the same size – neither is smaller or larger than the other.

Some may have recognized this for what it really is, the Pythagorean theorem:

$$a^2 + b^2 = c^2$$

The real question is – how can we prove this? The Egyptians were aware of this relationship, and had discovered triples of natural numbers that satisfied this, e.g. (3, 4, 5), or (5, 12, 13), or (20, 21, 29). They never asked the question of whether or not this was a provable fact. They just used the results and created lengths satisfying this relationship, and then they knew they could form a right angle in this way. The fact that you can prove this is truly amazing!

In our legal system we often hear the word proof mentioned. Civil law requires "proof by a preponderance of evidence," or "clear and convincing evidence." What does this mean? Ask any judge or lawyer and they will tell you that this is just slightly in favor of one position over the other or what they sometimes characterize as 51% versus 49%.

Our criminal legal system on the other hand requires "proof beyond a reasonable doubt." What does this mean? This is where the legal scholars change their tune and avoid quantification. In fact the majority believe that reasonable doubt cannot be quantified, believing it is more of a qualitative than a quantitative concept. "Proof beyond a reasonable doubt is a quantum without a number." (see Rembar) It is shocking that something so critical in our legal system can't even be quantified. However, some analysis has been performed and it seems that beyond a reasonable doubt is somewhere around 2/3 or approximately 66.7% sure of guilt. This is probably why the legal scholars avoid the quantification argument, since it means that there could be as many as 1/3 or approximately 33.3% people convicted wrongly. That's an awfully large number, especially if it is you that has been wrongly convicted!

The idea of proof plays a very important role in our lives, and every educated person needs to understand what it is and what it may or may not mean. It gives us a level of certainty and clarity. The goal is always absolute certainty and total clarity, but this is seldom achieved. With all the confounding examples given above, it would seem that absolute proof of anything is impossible. Wouldn't the truth or falsity of something always be an open question? Isn't it possible that our senses might deceive us? Can we ever truly know anything? Amazingly the answer to this last question is yes we can, and we can find such answers in mathematics. That is why mathematics is so unique, and why it has become a model (the gold standard, if you will) for all methods of inquiry. If done correctly a mathematical proof can remove the mystery and lead to complete certainty – a pure knowledge that harbors no doubts or questions.

We can trace the beginnings of mathematical proof back to the ancient Greek philosopher and mathematician Thales of Miletus, who was the first person to conceptualize the idea of proof. To Thales proof was the pinnacle of rational thought. Thales lived during the 6th century BCE and was believed to have traveled extensively throughout the world. In his travels to Egypt he was exposed to the many geometric formulas for computing areas and volumes that the Egyptians used to survey land or build impressive structures like the pyramids. To the Egyptians, however, geometry was just a tool. It provided them with useful, but seemingly unrelated formulas to calculate quantities that were important to them. They never seemed to question why the formulas were the way they were or how they were created. They seemed to work, and that was all that mattered. This is pretty much how mathematics is viewed by most people today.

Thales, on the other hand, began to question why. For him, it was not enough for someone to tell you something was true you had to see it for yourself. This perspective is called **skepticism**, and it is one of the key ideas that set ancient Greece apart from other societies. It is a major achievement for humankind to begin to ask why things were the way they were. It was a breath of fresh air to go beyond simple description and move to inquisitive inquiry. What is the world about and why is it the way that it is?

To answer the why question, there are two distinct rational/logical approaches one can take – they are called inductive and deductive reasoning.

14.2.1 Inductive Reasoning (Indirect Reasoning)

Inductive reasoning is a form of reasoning that comes from our observations. After numerous observations we may see a pattern forming. Thus, we reason that the pattern will continue. This is usually the first form of reasoning and is connected to our senses. It is how many species learn to survive. If every time you go to a certain place you get attacked, you stop going to that place. If every time you eat a certain food you become ill, you stop eating that food.

> **Inductive reasoning** starts with specific observations and leads to a general conclusion. In inductive reasoning you come to probable or approximate conclusions, but not definite or exact ones.

For the most part, science develops through the process of inductive reasoning. We observe nature and from the observations we form laws that can be used to predict what will happen.

Although mathematics, as we will show shortly is predominantly a deductive discipline, we sometimes first discover a mathematical pattern through inductive reasoning. For example, some early civilizations observed a relationship between the circumference and diameter of a circle. The circumference of a circle is approximately equal to three times its diameter (a little more than 3 actually).

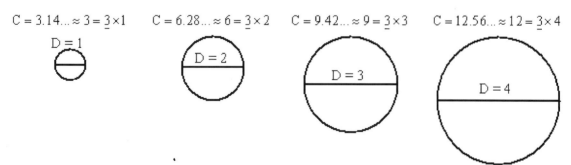

$$C = 3.14... \approx 3 = \underline{3} \times 1 \qquad C = 6.28... \approx 6 = \underline{3} \times 2 \qquad C = 9.42... \approx 9 = \underline{3} \times 3 \qquad C = 12.56... \approx 12 = \underline{3} \times 4$$

So some earlier civilizations used the following empirically determined law:

$$\text{Circumference} = 3 \times \text{Diameter}$$

or

$$C = 3 \times D$$

This law was improved upon by later civilizations:

$$C = \frac{25}{8} \times D \quad \text{Egyptians and Babylonians}$$

to

$$C = \frac{22}{7} \times D \quad \text{Greeks (Archimedes)}$$

However, it wasn't until the late 1700's that both Lambert of Switzerland and Legendre of France, both independently proved with absolute certainty that the number in front of the diameter, D, was an irrational number (and we now know just how different they are) we call π, and the exact law became:

$$C = \pi \times D$$

How did they achieve such certainty called a proof? They used something called deductive reasoning.

14.2.2 Deductive Reasoning (Direct Reasoning)

Deductive reasoning begins with some undefined terms, definitions, and irrefutable facts called axioms or postulates, and proceeds through a step–by–step process using rules of inference to a conclusion. This sounds rather complicated, but we'll try and simplify it a bit.

We will not go into detail here explaining rules of inference, since, we will explain these in full detail when we discuss logic later on in the book. For now we will think of them loosely as the rules of the "game." Deductive reasoning is sometimes called the axiomatic approach.

> **Deductive reasoning** starts with generally accepted facts and leads to specific conclusions. Unlike inductive reasoning, deductive reasoning always leads to certain, not probable, conclusions.

One of the most famous forms of deductive reasoning is called a **syllogism**. Here is a popular example:

> All men are mortal.
> Socrates is a man.
> Therefore, Socrates is mortal.

This syllogism is an example of deductive reasoning that we will describe in greater detail in Chapter 15.

Not all deductive reasoning comes in the form of a syllogism, though. Here is another form of deductive reasoning.

> We are absolutely certain that the murder took place sometime between 2:00 AM and 4:00 AM. The Defendant can prove beyond any doubt that between 2:00 AM and 4:00 AM he was over 400 miles away from the scene of the murder. Therefore, the defendant could not have committed the murder.

Since inductive reasoning only leads to probable conclusions and deductive reasoning leads to certain conclusions, the early Greek skeptics/philosophers/mathematicians were drawn more to deductive reasoning. It was a clear and direct path to certain knowledge and truth. To this day the sum of the squares of the sides of a right triangle equals the square of the hypotenuse. This will never change. We won't suddenly discover that it was only partly or approximately true. It has been proven irrefutably true forever! (provided we are talking about a Euclidean Geometry, which we'll talk about in Volume II) The power of this statement is unbelievable. Anything else we talk about always has degrees of certainty – levels of possible truth–value. However, in mathematics this is not the case. Once proven true it remains true forever!

This then is the elegance and beauty of mathematics and truth. Proof lies upon the throne of reason. For skeptics who question all that they see, it is the reasoned answer. To all the rest, it gives a level of comfort that what was said or presented is true beyond any doubt. How do we obtain such proof?

Deductive Proof (The Process)

We first illustrate the process of proof with a simple of a childhood game created by Lewis Carroll (Charles Dodgson, see section 15.5 below), called Word Ladder. The rules of the game are simple, you start with a word and change a single letter in that word to make a new word. The goal is to eventually transform the starting word into a different final word. For example, let's say we wanted to change the word "car" into "bus."

> We would start with "car" and then we could change this to "far",
> > then to "fur",
> > then to "fun",
> > to "bun",
> > to "bus."

> Thus, we have the sequence of: car – far – fur – fun – bun – bus.

> We accomplished this by changing a single letter at each step, always forming a word, until the final word "bus" was created.

This process (game) is in many ways similar to a deductive proof. In a proof we try and show that through a sequence of logical, follow–the–rules, steps we can go from one result to another. It is the game mathematicians play, and for many, this game is quite addicting.

You've heard the saying, "you can't get blood from a stone" – or can you. Can you transform the word "stone", step–by–step into the word "blood?"

Here is one example (proof):

stone – stole – stale – state – slate – plate – plane – plank – plunk – slunk – slink – blink – blind – blond – blood.

This "proof" took fourteen steps! Is this the only possible series of words that works? Can you find a different sequence? Is your solution smaller in the number of total steps or larger than 14? What is the smallest number of letter changes possible? Is it possible to prove that this minimum number is really possible or impossible? These are questions that can arise out of the process, which can lead to more and more results and proofs.

We should remark that just like in a mathematical proof, it is possible to go both ways to try and come up with a solution. That means if you are not making progress by starting with "stone" you could start with "blood" instead and try and work backwards, or try it from both directions to see if you can find a sequence connecting the two words. Then if you find one sequence that works, you look for another one that can be performed in fewer steps.

Understanding What is Being Asked

Another important aspect of proof is in understanding what you are trying to prove. Consider this example:

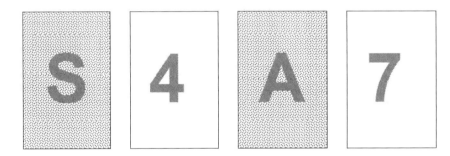

The four cards above have two sides. On one side is a letter and on the other side is a number. It is given that:

"If a card has an even number on one side, then it must have a vowel on the other."

What is the minimum number of cards you have to turn over to see if this assertion is true?

This example teaches us the importance of understanding what is being asserted, without reading more into it than there should be. The assertion does NOT say that if a card has a vowel on one side, then the other side must have an even number. So it is possible to have an A on one side and the number 13 on the other. We must read the assertion accurately. If we see an even numbered card, it will have a vowel on the other side! Not the other way around.

Thus, in the example above, we only need to turn over two cards, the S and the 4. We need to turn over the S to validate that the number on the other side is odd otherwise the assertion would not be true. We need to turn over the 4 to make sure there is a vowel on the other side. Turning over card A gives us no useful information, since the number is allowed to be even or odd. Turning over the 7 gives us no useful information, since the letter is allowed to be a vowel or a consonant.

<u>Example of a Deductive Proof in Mathematics:</u>
Consider the statement: "The sum of two even natural numbers is always even." Is it always true? We first must start with the question – What does it mean for a number to be even? If you recall, an even number is a number "evenly" divisible by two. That means the result of dividing the number by two gives a natural number (1, 2, 3, 4, 5, …) without a remainder.

Let's test the sum of two even natural numbers assertion by looking at a few examples of adding even numbers together, and then testing the sum to see if it is even:

$$2+2=4$$
$$\frac{4}{2} = 2, \text{ a natural number, therefore it is even}$$

$$6+14=20$$
$$\frac{20}{2} = 10, \text{ a natural number, therefore it is even}$$

$$102+274=376$$
$$\frac{376}{2} = 188, \text{ a natural number, therefore it is even}$$

$$2,214+32,012=34,236$$
$$\frac{34,236}{2} = 17,118, \text{ a natural number, therefore it is even}$$

It definitely seems true by our inductive exploration, but can we prove it? Are we sure that it is always true? We could spend our entire lives working out example after example, but this would only build our confidence level, it wouldn't actually prove anything. Instead we have to approach this systematically and this is where deductive reasoning comes into play.

First let's define an even number algebraically. As we said a number is even if it is divisible by two with no remainder, or stated in a similar way, two is a factor of the number. This means we can write any even number as the product of two times a natural number, or

Any even number $= 2n$, where n is a natural number, $1,2,3,...$

We have created a formula to represent any even number. Now if we want to consider two different even number possibilities in the same discussion, we need to use two different letters to represent the natural numbers, i.e. say $2n$ for one number and $2m$ for the other number. That way with different natural numbers n and m we can still consider all the possible combinations of even numbers in the same discussion. If we used only one, say $2n$, then the numbers would always be the same, for the same value of n.

We now have an algebraic way to represent two even numbers. Next we introduce the basic axioms of arithmetic. These are the rules–of–the–game we will play. They tell us what we are allowed to do to derive other results. Sort of like, changing one letter at a time to create a new word that eventually leads to a final target word, from our word game above.

Axioms of arithmetic:
There are three axioms (rules) of addition.

1. The commutative law, is illustrated by the equation: $a + b = b + a$
    This rule states that the order in which you add two numbers does not change the end result.
    For example, $6 + 8$ and $8 + 6$ both mean the same thing.

2. The associative law which is written $a + (b + c) = (a + b) + c$. This rule suggests that grouping numbers when adding them, also does not affect the sum.

3. The closure property which states that the sum of two natural numbers a and b, $a + b$, is a natural number.

There are also three axioms of multiplication.

1. The commutative law of multiplication illustrated by the equation $a \times b = b \times a$.
    This rule states that we can multiply numbers in any order, and not change the result.

2. The associative law of multiplication denoted by $a \times (b \times c) = (a \times b) \times c$.
    This rule allows us to multiply groups of terms in any order.

3. The closure property of multiplication, which states that $a \times b$ is a real number.

Another important rule related to both addition and multiplication is the distributive property, illustrated by the equation $(a + b) \times c = (a \times c) + (b \times c)$.

A Basic Proof

Now using these axioms/rules, let's show how to add two even numbers together, and see if we can prove that the sum is indeed an even number. We add the two even numbers $2n$ and $2m$.

$$2n + 2m = 2(n + m)$$

Notice we have used the distributive property in reverse order, also called factoring, to rewrite the sum above. It is also assumed that n and m can take on any natural number value independently of each other, so this expression really represents all the possible additions of all possible even natural numbers in one simple compact form.

This is a level of abstraction that was not as obvious in ancient times, as it probably is to some of you today. So instead of "testing" specific instances of the sum of even numbers, we can now "test" this expression, and this represents ALL possible even number sums!

Now divide the result by 2 to test for "eveness" to obtain:

$$\frac{2(n + m)}{2} = (n + m)$$

Notice that the common factor of 2 cancels out. Also, since n and m are natural numbers, we use the closure rule of addition, so that the sum of two natural numbers is always a natural number Thus, the sum of two even numbers has been proven to be even, because it is divisible by 2! Since, when we divide by two, the factor of 2 cancels and we get a natural number.

This is what simple direct deductive proofs are about – Plugging up all the possible loopholes in the argument and then demonstrating the desired result. It is like a game or a puzzle that needs to be solved. It can be quite fun and rewarding when you accomplish it, but when you have to reproduce these results for a test it is often very stressful, which is why many students are afraid of proofs and go to great lengths to avoid them.

This process has another benefit. It trains the mind to organize, and critically analyze. You become better problem solvers, since you are trained to look closely at what is being asked and then look for loopholes or exceptions that you might encounter, i.e. you are trained to act with accuracy and skepticism.

Another Basic Proof

What about the product of two odd natural numbers, is it even or is it odd?

$$3 \times 5 = 15$$
$$9 \times 7 = 63$$
$$25 \times 13 = 325$$

Again, from our inductive reasoning it looks like it is an odd number. Can we prove it? First, we will need a way to represent all possible odd numbers. We notice that to get an odd number we can simply add one to an even number,

$$2 + 1 = 3$$
$$26 + 1 = 27$$
$$98 + 1 = 99$$

Since this is true, let's use our algebraic representation of an even number from above and add one to it to make it an odd number, i.e.

Any odd number = 2n + 1, where n is a whole number; 0, 1, 2, 3, ...

Now let's multiply two potentially different odd numbers:

$$(2n+1)(2m+1) = (2n)(2m)+(2n)(1)+(1)(2m)+(1)(1)$$ Use the distributive property for binomials
$$= 4nm+2n+2m+1$$ Multiply the terms out
$$= 2 \cdot 2nm + 2n + 2m + 1$$ Factor 4, notice that 2 is a common factor
$$= 2(2nm+n+m)+1$$ Factor out the common factor of 2
$$= 2p+1$$ Rewrite using p=2nm+n+m

The end result is in the form of an odd number: $2p + 1$

Therefore, we can conclude that the product of ANY two odd numbers will always be odd.

EXERCISES 14.1–14.2

ESSAYS AND DISCUSSION QUESTIONS:

1. What is proof?

2. Where do we need proof?

3. Can proof ever be totally certain? Explain your answer

4. What are inductive and deductive reasoning?

5. Why is mathematical proof unique?

PROBLEMS:

Determine if the following are a form of either deductive or inductive reasoning.

1. You observe that the sun rises every morning, so you conclude that it will rise tomorrow morning.

2. Based upon the known motion of the earth and the sun derived from fundamental laws of planetary motion, we conclude that the sun will appear to rise on the horizon tomorrow morning.

3. It has rained every day I've been on vacation, therefore I expect it will rain tomorrow as well.

4. Every time I fly my flight is late, therefore I can say that all airline flights are late.

5. If you take ibuprofen for your headache, it will go away. You took ibuprofen. Therefore, your headache will go away.

6. Susan's first two children were girls. If she has another baby it will be a girl.

7. I've never heard of Facebook sharing information with outside companies. Thus, I am sure my private information concerning what I like will not be sold by Facebook to other companies.

8. If you smoke you increase your risk of lung cancer. Michele smokes, so she is more likely to get lung cancer.

9. Too much radiation can harm your body. Jason received twenty times the recommended limit for a dose of radiation. Jason is more likely to become ill from radiation poisoning.

10. The last several times I went out in the sun I got a sunburn. It is sunny and I'm going out today, so I expect to get burnt.

11. All plants need water, otherwise they will die. You have not watered the plant, so it will probably die.

12. The typical voter is uniformed. It is likely that John is voting for someone he knows little about.

13. Whenever I go to the grocery store I always run in to Rita. I'm on my way to the grocery store, so I should see Rita there.

If possible, perform each of the Word Proofs:

14. Prove that you can turn a *car* into a *van*.

15. Prove that you can turn the *sun* into a *top*.

16. Prove that you can turn a *bun* into a *cup*.

17. Prove that you can turn *red* into *one*.

18. Prove that you can turn *run* into *out*.

19. Prove that you can turn a *gun* into a *mop*.

20. Prove that you can turn the *moon* into a *star*..

21. Prove that you can turn *mind* into *body*.

22. Prove that you can turn *wine* into *soda*.

23. Prove that you can turn *life* into *poor*.

24. Prove that you can turn *debt* into *cash*.

  25–27 are challenge problems

25. Prove that you can turn *blood* into *water*.

26. Prove that you can turn a *horse* into a *tiger*.

27. Prove that you too can get *Blood* from a *Stone*. (Use your own words, not the same as those presented in the example above.)

Prove each of the following facts about numbers:

28. Prove that the sum of two odd natural numbers is always even.

29. Prove that the product of an odd and an even natural number is always even.

30. Prove that when you add one to an even natural number you always get an odd number.

31. Prove that when to add one to an odd natural number, you always get an even number.

32. Prove that when you divided an even natural number by 2 you can get any natural number.

33. Prove that when you divide any odd natural number by 2 you get a number that is neither even nor odd.

References

1. Cupillari, A., (2001). The Nuts and Bolts of Proofs. Academic Press.

2. Krantz, S.G., (2007). The History and Concept of Mathematical Proof.

3. Franklin, J. and Daoud, (1996). A. Proof in Mathematics: An Introduction, Quakers Hill Press.

4. Polya, G., (1954). Mathematics and Plausible Reasoning. Princeton University Press.

5. Rembar, (1980). The Law of the Land, Simon and Schuster.

6. Solow, D., (2004). How to Read and Do Proofs: An Introduction to Mathematical Thought Processes. Wiley.

7. Velleman, D., (1994). How to Prove It: A Structured Approach. Cambridge University Press.

8. Vickers. J., (2006). The Problem of Induction. The Stanford Encyclopedia of Philosophy, Winter Ed.

9. Zarefsky, D., (2002). Argumentation: The Study of Effective Reasoning Parts I and II, The Teaching Company.

CHAPTER 15

Aristotle – The Father of Propositional Logic

I know what you're thinking about, but it isn't so, no how. Contrarywise, if it was so, it might be, and if it were so, it would be. But as it isn't, it ain't. That's logic

Lewis Carroll

Nothing new had been done in Logic since Aristotle!

Kurt Godel

15.1 Introduction

Again, we take ourselves back to Ancient Greece. It was a time when people were questioning and trying to understand the world. In Athens the goal was to create model citizens, and a well educated ruling class; well versed in geometry, music, rhetoric (the art of persuasion), and logic. The concept of logic was created as a tool by which one could seek out certain knowledge and truth. Athenians would argue back and forth defending different positions, views, and beliefs.

It was Aristotle who created the foundation for propositional (word) logic by organizing the subject into well defined concepts and rules. For his work, Aristotle is known as the Father of Logic. Aristotle essentially created the subject of logic we know today through his writings called the Organon. The Organon was a series of books (essays) that help to make the art of arguing more precise.

Language can be very ambiguous and imprecise. Oftentimes a sentence can take on different meanings depending upon how you interpret it.

As examples, consider these humorous and imprecise newspaper headlines:

- Arson Suspect is Held in Massachusetts Fire
- Astronaut Takes Blame for Gas in Spacecraft
- Man Struck by Lightning Faces Battery Charge
- Two Soviet Ships Collide, One Dies

Here are some other ambiguous and imprecise statements (some rather entertaining).

- When removing the rods from the nuclear reactor, one can't remove them too quickly.
- Include your children when baking cookies.

- Kids make nutritious snacks.
- I saw her duck.
- He gave her cat food.
- The lady hit the man with the umbrella.
- Woman without her man is a savage.
- The chicken is ready to eat.
- French history teacher.
- Kicking baby considered to be healthy
- Please wait for the hostess to be seated.
- Did you see the girl with the telescope.
- Tuna are biting off the Washington coast.
- After the Governor watched the lion perform, he was taken to Main Street and fed twenty–five pounds of red meat in front of the Theater.
- In exchange for painting my house, I promise to pay him $2,500 and give him my car only if he finished the job by next Tuesday. He finished the day after, what does he get?

In addition to such obvious (or not so obvious) ambiguities, you can have contradictory statements, or statements that are obviously (or not so obviously) incorrect (fallacies), as well as arguments with correct reasoning that lead to absurd conclusions (paradoxes).

With all the potential pitfalls of language, how do we create an argument that no one could disagree with, that leads to a new understanding? Language needs to be precise to avoid things such as contradictions, paradoxes, and fallacies (which we introduce and discuss further in this chapter.) This is why logic is important. It is at the very heart of the reasoning we use to live and understand our lives. The ability to reason correctly is very important.

Propositional logic is important for another reason. It is related to what is called critical thinking. Critical thinking or reasoning is the process by which we rationally try to understand and analyze anything and everything. The process of critical thinking can be described as:

Critical Thinking, Harold Fawcett, 1938

1. Select significant words and phrases in any statement that are important, and ask that they be defined carefully.
2. Require evidence to support conclusions you are asked to accept
3. Analyze that evidence and distinguish facts from assumptions.
4. Recognize stated and unstated assumptions essential to the conclusion.
5. Evaluate these assumptions, accept some and reject others.
6. Evaluate the argument, accept or reject the conclusion.
7. Constantly re–examine the assumptions that are behind their beliefs and actions.

Propositional logic is a subset of critical thinking that focuses on creating a precise structure to analyze certain types of logical word arguments, whereas critical thinking can be applied to any thought or reasoning process. In this chapter we present the basic ideas behind Aristotle's approach to constructing valid word arguments. We focus on propositional reasoning, the subset of what we are more broadly concerned with, critical thinking. To be a better critical thinker, you need to understand the basic art of reasoning, i.e. propositional logic.

15.2 Historical Development of Propositional Logic

In around 500 BCE, Zeno and the Eleatics in Ancient Greece developed the art of argumentation, but did not formulate any laws or rules on how to proceed. They did start to categorize the process, though. They defined two basic aspects of arguing: rhetoric and dialectic:

> **Rhetoric**, the art of persuasion.

and

> **Dialectic**, an exchange of propositions and counter propositions resulting in a consensus.

Rhetoric and the dialectic were very useful in the Greek political state, where officials tried to persuade other officials in their Senate to come up with policies beneficial to the state. The approach is similar to our government today. This should not be a surprise, since our founding fathers looked to the Greek political system as a model for our new government.

Socrates refined the process beyond what Zeno and the Eleatics had developed. He devoted his entire life to critical reasoning and a search for truth. As we have stated before, he felt strongly that "the unexamined life was not worth living." He often questioned the truthfulness of popular opinions, and would analyze and contemplate the real meaning of words. What does the word "good" or the word "just" really mean? His constant questioning and encouraging others to question, is what got him into trouble.

Socrates perfected what is known as the Socratic Dialectic or Method, which as we have discussed earlier, became the beginnings of constructivist learning. It was Aristotle, though, that formalized the rules of logic to make argumentation more precise. Aristotle wrote a series of six treatises (essays) which together were called the Organon. Organon is the Greek word meaning tool, instrument, or organ. Thus, the Organon contained the tools one could use to argue effectively.

Organon
The Categories: A study of the ten kinds of basic concepts used to identify the subject and the predicate of a proposition. They include substance, quantity, quality, relation, place, time, position, state, action, passion or affection

On Interpretation: Establishes a relationship between language and logic

Prior Analytics: A formal analysis of a valid argument and syllogism

Posteriori Analytics: A discussion of correct reasoning in general and its relation to scientific knowledge
Topics: A discussion of issues involved in constructing valid arguments.

Sophistical Refutations: 13 fallacies identified by Aristotle: accent or emphasis, amphibology, equivocation, composition, division, figure of speech, accident, affirming the consequent, converse accident, irrelevant conclusion, begging the question, false cause, fallacy of many questions.

These works established the rules for creating a correct argument and for performing sound philosophical reasoning. They defined the categories of logic and distinguished between the two types of reasoning, deductive and inductive, as we discussed in Chapter 14. Since deductive reasoning leads to certain conclusions, the focus of propositional logic turned exclusively to the deductive approach.

15.3 Fundamentals of Propositional Logic

Propositional logic is concerned with deductive and not inductive reasoning. It begins by defining the basic building blocks of an argument, and then proceeds to study various forms of arguments to determine which are correct, and which are not. To understand an argument you first need to understand and analyze its basic parts, and then consider how they come together to form an argument.

Basic Definitions

The statement is the most basic part of a logic. It is what all the other concepts are built upon or derived from.

> **Statement:** A declarative sentence that is either true or false.

Examples: In each of the examples below we can assign a truth value, meaning we can declare the statement to be either true or false.

1. Today is Sunday
2. It is raining.
3. The sun is shining and it is cold outside.
4. She intends to buy either a pair of jeans or a dress at the store.

Examples of sentences that are not statements are **commands**, such as:

1. Stop at the light!
2. Turn left at the corner.
3. Pay your insurance,

or **questions**, such as:

1. What time is it?
2. Can I have the last piece of cake?
3. How often do you come here?

Statements are declarative sentences that are either true or false, so commands and questions do not qualify.

Statements can be simple or compound.

A **simple statement** expresses a single idea.

Examples of simple statements are:
1. Today is Sunday
2. It is raining.

A **compound statement** combines two or more ideas in a sentence using logical connectives such as AND, OR, as well as the modifier, NOT.

Examples of compound statements are:
1. The sun is shining and it is cold outside.
2. She intends to buy either a pair of jeans or a dress at the store.

Statements that make claims whose truth value needs to be supported are called conclusions.

Conclusion: A statement that is designed to be supported or defended.

Examples:
1. The economy is improving.
2. Women are better voters than men.
3. Mathematics is discovered and not created.
4. You have cancer.

Sometimes these statements are preceded by certain words that indicate the statement is a conclusion.

Some typical **Conclusion Indicators** are: therefore, thus, hence, consequently, etc.

Examples:
1. Therefore you should buy the Ford.
2. Hence, this candidate is better than the other.
3. Consequently, the storm will cause over $1 billion worth of damage.

Statements that provide supporting evidence for a certain belief or position are called premises.

Premises: The statements that support a conclusion.

Examples:
1. Sales have increased over last quarter.

2. Unemployment has declined.
3. Your white blood count is low.

Oftentimes premises will begin with certain key words to indicate that they should be thought of as the premises of an argument.

Some possible **Premise Indicators** are: since, because, for, etc.

Examples:
1. Since the plane landed late.
2. Because you live near a toxic waste site.

The argument is what you are trying to analyze. It is a position you are asked to support or agree with. It is made up of premises that are designed to lead you to a specific conclusion.

> **Argument:** A group of statements (premises) leading to a conclusion.

Examples:

1. Because failure to stop at a stop sign is against the law, and   PREMISE
   you did not stop at the stop sign,   PREMISE
   you deserve the ticket.   CONCLUSION

2. All men are mortal.   PREMISE
   Socrates is a man.   PREMISE
   Therefore, Socrates is mortal.   CONCLUSION

3. The road was wet.   PREMISE
   You were driving too fast.   PREMISE
   You should have anticipated that you would lose control of your car.   CONCLUSION

4. Since the world has a beginning in time,   PREMISE
   it is also limited (not infinite) in regards to space.   CONCLUSION

5. Since every thing we observe has a primary cause, and   PREMISE
   since the universe is an observable object, then   PREMISE
   there must be a primary cause to the universe called God.   CONCLUSION

6. If I study more than 8 hours for a test when I'm not tired,
   I always get an A.   PREMISE
   I studied for 9 hours for this test.   PREMISE
   I was not tired.   PREMISE
   Therefore, I should get an A on this test.   CONCLUSION

7. We are on the road to economic recovery,   CONCLUSION
   since, sales have improved over last quarter, and   PREMISE

the unemployment rate has declined.	PREMISE

8. All men are pigs.   PREMISE
   John is a man.   PREMISE
   Therefore, John is a pig.   CONCLUSION

9. Oswald did not act alone in killing president Kennedy, since   CONCLUSION
   Kennedy was shot multiple times, and   PREMISE
   one person shooting from the same location could not have done this.   PREMISE

Notice that the conclusion does not have to occur at the end of the argument, and also that you can have one or more premises in an argument. Sometimes key indicator words for premises and conclusions are used, but now always.

Some of the above arguments appear to be reasonable, and you would quickly agree with them. Others seem less certain. While some you might reject altogether. How do we sort through all different arguments to find those that are correct and those that are not correct?

The first thing we will find is that the reasoning of an argument can be fine, but the conclusion we arrive at may be nonsensical. This is because there are two things needed for an argument to be believable. The first aspect addresses whether or not the argument proceeds in a logically connected way. The other aspect deals with the truthfulness of the premises used in the argument. Thus, we introduce two new concepts – the valid and sound argument.

> A **Valid Argument**: An argument in which conclusions follow from premises, otherwise it is **Invalid**.

Aristotle and others analyzed arguments and found certain forms that would always lead to a valid conclusion. These arguments were based upon using expressions such as: IF statement P is true, THEN the statement Q must follow.

Here are some examples of arguments:

If it rains then the we cannot work.
It rained.
Therefore, we cannot work.

If you purchased a new car, then your credit was okay.
Your credit was not okay.
Therefore, you did not purchase a new car.

If you go to the store, then you will buy a ring.
If you buy a ring, then you will ask her to marry you.
Therefore, if you go to the store, then you will ask her to marry you.

Either you will wait five years, or retire now.
You don't retire now.

Therefore, you wait five years to retire.

If sales improve over last quarter and the unemployment rate goes down, then we are on the road to economic recovery.
Sales have improved since last quarter, and the unemployment rate has gone down.
Therefore, we are on the road to economic recovery.

Aristotle noticed that many arguments fit distinct patterns and that there are forms of arguments that always lead to valid conclusions. Four common types of such valid arguments are shown below.

1. Modus ponens

   If P, then Q
   P
   Therefore Q

2. Modus tollens

   If P, then Q
   not Q
   Therefore not P

3. Hypothetical syllogism

   If P, then Q
   If Q, then R
   Therefore If P then R

4. Disjunctive syllogism

   P or Q
   not Q
   Therefore, P

Even though an argument is valid it does not mean that it is true. For example the following argument is valid.

> If you are a cat, then you have wings.
> Zack is a cat.
> Therefore, Zack has wings.

The conclusion of the argument is absurd, but the conclusion does follow from the premises. This is an example of a valid argument, but one that is not sound. To be a sound argument, the argument must first be valid, meaning the conclusion follows from the premises; however, it must also have true premises to be a sound argument.

> A **Sound Argument**: A valid argument with true premises.

Here are examples of valid arguments that are both sound and unsound.

Sound Arguments

> If you are a dog, then you have a tail.
> Spot is a dog.
> Therefore, Spot has a tail.

> You can get a DWI if and only if you drink and drive.
> You do not drink and drive.
> Therefore, you cannot get a DWI.

> Either you live inside or outside the US.
> You don't live outside the US.
> Therefore, you live in the US.

Unsound Arguments

> If you are an elephant, you can fly.
> Babar is an elephant.
> Therefore, Babar can fly.

> If you are a man, you are over 15 feet tall.
> Scott is a man.
> Therefore, Scott is over 15 feet tall.

> You are either an extraterrestrial or a human.
> You are not a human.
> Therefore, you are an extraterrestrial.

When we get to the next chapter, Chapter 16, we will come back to the question of valid and invalid arguments with a more powerful tool. The question of the soundness of an argument, however, will always require good reasoning skills and abilities. This is why we have to learn how to be aware of and to avoid problems in how we reason. Fallacies, contradictions and paradoxes, all create issues with how we reason.

15.4 Problems with Reasoning – Fallacies and Paradoxes

The first real issue with how we reason, concerns arguments that have strong persuasive power, but are not based upon valid reasoning. These arguments are called fallacies.

> A **Fallacy**: An argument that is psychologically persuasive, but completely invalid.

Some Common Types of Fallacies:

1. <u>Ad Hominum Attack</u>: Attacking the person to reject their position, rather than the position itself.

   The person that supports this position has other unrelated problems,
   so the position must be wrong.

   Example:
   > You can't believe Alex,
   > after all, he is convicted drug dealer.

2. <u>Appeal to Force</u>: The use or the implied use of force – either physical, psychological, legal, or other type – to make you accept a conclusion.

   Either you agree with the position,
   or some harm will come to you.

   Example:
   > If you don't agree with the CEO regarding these changes,
   > you will not be working here tomorrow!

3. <u>Appeal to Inappropriate Authority</u>: Using the testimony or opinion of someone who is an expert in a field other than the one in question.

   This position must be true,
   since other popular, non–experts, believe it is true.

   Example:
   > Wheel of Fortune host, Pat Sajak, is a registered Republican,
   > so from now on we should only vote the Republican ticket.

4. <u>Argument from Ignorance</u>: Arguing a position is true, because it hasn't been proven false, or false because it hasn't been proven true.

   This position must be true (or false),
   because it has not been proven false (or true).

   Example:
   > There must be a god,
   > otherwise someone would have proven their wasn't one by now.

5. <u>Begging the Question</u>: Assumes the argument you are trying to prove. The conclusion is just a restatement of the premises.

   This position is valid,
   because this position is valid.

Example:

> There is a god, since the Bible states so, and the Bible is the word of God.

6. Gambler's: Assumes that a previous pattern will impact the short term.

Something has happened
The something is not expected to occur over the long term
Therefore, that something will change.

Example:

> I have not won the lottery in the last five years,
> so I'm due to win it shortly.

7. Hasty Generalization: Using unusual or rare cases to support a general point covering all cases.

This general position must be always be true,
because I was able to find a specific case that was true.

Example:

> Since Cynthia lied to me,
> it is obvious that you can never believe what any woman says.

8. Red Herring: An irrelevant topic is presented to divert attention from the actual position.

This position can't be true,
because this unrelated event took place.

Example:

> Grading on a curve is the most logical thing to do. Since, classes are more productive when the students and the professor like each other.

9. Straw Man: Substituting a simplistic argument (straw man) in place of the actual argument, and defeating the straw man argument and not the actual argument.

This position can't be true,
since this overly simplified case is not true.

Example:

> A US Senator says that we should cut funding for the Department of Homeland Security. I disagree entirely. I can't understand why he wants to leave us dopen to an attack from ISIS.

This list is not exhaustive, as there are many more types of fallacies. This list, however, is simply meant to give you an idea of what fallacies are, and how to identity them.

Examples:

Identify the type of Fallacy in each of the following.

1. People have been trying for centuries to prove that God exists. But no one has yet been able to prove it. Therefore, God does not exist.

   Argument from Ignorance

2. I know for a fact that gambling is evil. My parents gambled all the time and this caused us to lose our house and security in life.

   Hasty Generalization

3. My barber says that genetically modified food is bad. Therefore all such products should be pulled off the market.

   Appeal to Inappropriate Authority

4. Of course the equal rights amendment must be defeated. Do you want men and women sharing the same bathroom.

   Straw Man

5. You shouldn't listen to Osama's argument, after all he's not really one of us.

   Ad Hominum Attack

6. If we don't agree with the boss on this, none of us will be working here tomorrow.

   Appeal to Force

7. A priest never lies, so it must be true when my priest says capital punishment is wrong.

   Begging the Questions

Sometimes problems with reasoning are not due to any fallacious or contradictory statements, but are due to a fundamental flaw in the structure of reasoning itself. This is what are called logical paradoxes.

<u>Logical Paradoxes</u>

In Chapter 12 we presented the paradoxes of Zeno. In those cases, we found them to be related to the strangeness of infinity. Paradoxes are not limited to confronting the infinite, though. They are also found as a part of language. Consider the truth value of the following statement:

> This sentence is false.

This is an example of what is called the Liar's Paradox. If it is false, then it is surely true. However, if it is true, then it is most surely false. Which is it? This is another form of paradox which, like Zeno's, has no clear answer. It is a paradox due to self–referencing. Self–reference is when a statement says something about itself. If this happens you need to be aware of potential inconsistencies.

Most paradoxes stem from either considering the infinite, or by self–reference. The paradox of self–reference, as we shall see in Chapter 17 has some significant implications regarding the foundations of mathematics. It seems that even the path of logic can lead to some confusing and inconsistent results and ideas. For some the idea of inconsistency and confusion can become a marvelous playground for the mind.

15.5   Having Fun with Meaning – Alice's Adventures in Wonderland

Did you know that the books *Alice's Adventures in Wonderland,* and *Through the Looking Glass,* were actually written by a mathematician? Charles Dodgson, whose pen name was Lewis Carroll,

was a mathematician very much interested in logic. The books he authored intertwined nonsense and logic, and on some level seemed to question the very meaning of meaning by indirectly asking – What do we mean by the idea of meaning? Nothing seems to make sense down the rabbit hole.

Although the books are essentially for children; written as fascinating and fanciful adventures into a world of seeming nonsense and contradictions, they also take a deeper satirical look at the search for meaning, and its connection to logic.

Charles Dodgson (see the 1864 photograph of him to the left), A.K.A. Lewis Carroll was a close friend of the Liddell family, who's parents had three young girls named Edith, Lorina, and Alice. Charles, on outings with the girls, used to tell the sisters stories and one day while rowing on a lake he told the story of the precocious Alice and her adventures.

After his telling several more parts to the story, young Alice Liddell (see the 1858 picture of young Alice below to the right) begged him to write it down, and these imaginative classics were born.

Dodgson preferred the company of young children. He once stated that once he had to tip his hat to a passerby, that they had graduated to adulthood, and that he could no longer be close to them. He never married, he also suffered from a stuttering problem, and these occurrences along with others, have caused a great deal of unconfirmed speculation about his personal life, which we will leave for others to discuss. We are more concerned here with his vision of logic.

Although he never achieved the rank of a renowned mathematician, he did obtain a level of immortality through his non–mathematical writings. He showed a fun and entertaining side to mathematics. He helped us look at the world in a different way, and for that we are grateful. He also helped to bring the discipline of logic to life, even for young children.

It is no accident that Dodgson's work is often cited in books on the fundamentals of logic. Dodgson takes us and Alice down a rabbit hole where logic is turned on its head. It is in the sequel to *Alice in Wonderland, Through the Looking Glass,* however, that some have argued Dodgson creates an imaginative work with logic as its central theme (see Saksteder reference below). In *Through the Looking Glass,* Dodgson pushes the bounds of logic to create a world understood only by the children he related so well to. It is a world in which he takes us through a fanciful journey that could only be understood through the mind of a child. In many ways it is the antithesis of rationalism and logic.

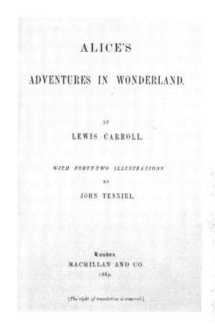

EXERCISES 15.1–15.5

ESSAYS AND DISCUSSION QUESTIONS:

1. What is propositional logic?

2. Who is considered the father of propositional logic, and what did he do to deserve the title?

3. What is an argument?

4. What is a valid argument?

5. What is the difference between a valid and a sound argument? Provide an example of each.

6. What are logical fallacies? Provide a couple of examples.

7. What is a logical paradox?

8. Who was Charles Dodgson, and how is he related to the discipline of logic?

PROBLEMS:

Premises and Conclusions
Fill in the blank spaces with the appropriate premises or conclusions.

1.

  Premise 1: If technologically advanced aliens existed, they would have already made contact with us.

  Premise 2:

  Conclusion: Therefore there are not technologically advanced aliens.

2.

  Premise 1: If the President had any integrity he or she would not pander to the demands of lobbyists.

  Premise 2: The President continues to deliver on the requests of lobbyists.

  Conclusion:

3.

  Premise 1: We cannot know what we do not experience.

  Premise 2: We do not have any experience of life after death.

  Conclusion:

4.

  Premise 1: Anyone who kills indiscriminately is evil.

  Premise 2:

  Conclusion: Therefore, Timothy McVeigh was evil.

5.

  Premise 1: We should not break the law.

  Premise 2:

  Conclusion: We should not commit acts of civil disobedience.

6.

  Premise 1: The food at fast food restaurants has been linked to cancer.

  Premise 2:

  Conclusion: Therefore, Johnathan will probably get cancer.

7.

Premise 1:

Premise 2: There is evil in the world.

Conclusion: Therefore, God does not exist.

8.

Premise 1:

Premise 2: Hiring a prostitute is for your own selfish needs.

Conclusion: Therefore, it is wrong to hire a prostitute.

Identify the premises and the conclusions in the following arguments.

9. All dogs have four legs. All four-legged creatures have wings. It follows, then, that all dogs have wings.

10. If Hamlet was not written by Shakespeare, then it was written by another great writer. Francis Bacon was a great writer who lived during the time of Shakespeare. Therefore, Bacon must have written Hamlet.

11. The funding of medical care by a government is a form a socialism. Socialism is not good. Therefore, government funding of medicare should stop.

12. The US Constitution guarantees us the right to life, liberty, and the pursuit of happiness. But lack of appropriate health care makes it impossible for some to be able to pursue happiness. Therefore, health care is a right, and not a privilege, guaranteed by the Constitution.

13. You should not judge others unfairly, since all humans are imperfect.

14. All rational beings are responsible for their actions, and since all human beings are rational, it follows that all human beings are responsible for their actions.

15. Many of the ingredients found in fast-foods have been shown to cause cancer in animals. Therefore, it is reasonable to expect that eating at fast-food restaurants can lead to cancer.

16. Women are better voters than men. My mother told me so, and mothers don't lie

17. Laura claims that he saw a flying saucer land in her back yard. But Laura is still in elementary school. She is completely unaware of what others have written about flying saucers and aliens, so she couldn't possibly be telling the truth.

18. All alcoholics started with a first drink. If you have that first drink, then you will become an alcoholic.

19. It is immoral to kill animals for food, because killing is immoral, and to make the food available you must kill the animal.

20. Twitter is a good thing for a democracy. It provides a forum for people to openly present and debate ideas which is the basis of a democratic government.

21. President Obama acted as a king, because he used too many executive actions.

Logical Fallacies
Identify the logical fallacies in the following passages.

22. I've never heard anyone prove that God does not exist. He must exist, otherwise someone would have proven it by now.

23. Mathematics must be created, since it requires a human to produce it.

24. We should not let illegal immigrants gain a legal path to citizenship. After all, do you want to be flooded with immigrants from all these other countries.

25. Giving illegal immigrants a path to citizenship is the right thing to do. Because if we make them legal, then they wouldn't be violating the law.

26. We've never heard anything bad about this candidate. Therefore he must be truthful and good.

27. Michael says I should invest in this property, but how can I trust the word of a Democrat.

28. There is no direct evidence that Julian Assange "leaked" the secret files to the press, so it couldn't have been Mr. Assange who did it.

29. All the women in my family always cry during a romantic movie. My father, brother and I never do. Thus, we can safely say that all women are too sentimental!

30. If you don't want to get beaten up, you will agree with what I say.

31. Police should be allowed to shoot and kill whoever they want. After all do you want to live in anarchy.

32. Police should not be able to shoot anyone. Since once they start they may not be able to stop and everyone is at risk.

33. My views on not having any gun control have been endorsed by some of Hollywood's most notable actors: Kirk Douglas, Jimmy Stewart, and Charlton Heston, among others. How could you not agree with me?

34. Opposition to the North American Free Trade Agreement amounts to nothing but opposition to free trade.

35. I've flipped this coin three times in a row and it landed on heads. The next time it has to land on tails.

36. The leader of North Korea must always be right. Since anyone who disagrees with him is executed for treason.

37. Feminists want to ban all pornography and punish everyone who reads it! But such harsh measures are surely inappropriate, so the feminists are wrong: porn and its readers should be left in peace.

38. We must liberate Iraq from a dictatorial leader. If we do, then sooner or later Iranians will also gain their freedom and eventually the entire Arab world will become democratic.

39. President Bush's spokesman Marlin Fitzwater, when asked what he thought about an investigation of charges that the 1980 Reagan campaign made a deal with Iran to delay the

release of American hostages until after the Presidential election gave this response: "There's no reason to investigate something that never happened."

40. The candidate supports tax breaks for the rich. This doesn't surprise me. He has always been against the poor.

41. If we allow for these tax breaks to go forward, the poor will become rebellious and democracy in this country could be threatened.

42. There is nothing wrong with depriving people of their liberties, since it is necessary to lock up dangerous criminals, and the mentally unstable.

43. If you continue to argue against everything I say, I will not argue with you any further because it is clear that you will argue against anything, and will never give in to any counter argument no matter how strong it is.

44. Democracy is the best form of government. If it wasn't then we wouldn't be living in one.

45. Pain medications are habit forming. Therefore, if you allow your doctor to treat your pain with pain killers you will become a drug addict.

46. You can't come into this restaurant dressed like that. I don't care that there isn't a sign posted. If you don't leave now, I'll call the police and have you forcibly removed.

47. This horse has lost his last three races, he is due for a win.

48. All foreign made automobiles are gas guzzlers. Because the Bugatti and the Bentley are foreign made automobiles with lousy gas mileage.

49. There are many poor people that are Republicans. My approach is not to talk politics with them, since I assume a minimum level of intelligence in my friends.

50. My best friend told me her mathematics class was hard, and the one I'm in is hard, too. All mathematics classes must be hard!

51. We should abolish the death penalty. Many respected people, such as the actor Jeremy Irons, have publicly stated their opposition to it.

52. A female author has written several books arguing that pornography harms women. However, the author is an ugly, bitter person, so you shouldn't listen to her.

53. Active euthanasia is morally acceptable. It is a decent and ethical thing to help another human being escape suffering through death.

54. Murder is morally wrong. So active euthanasia is morally wrong.

55. Our last four children have been boys. The next will surely be a girl.

56. Government by the people is ideal because democracy is the least inadequate form of government.

57. You are so stupid your argument couldn't possibly be true.

58. I figured that you couldn't possibly get it right, so I ignored your comment.

59. How can you take Rush Limbaugh's views seriously. The man, after all, is a convicted drug user.

60. "Freedom of speech is when you talk." - former Dodgers manager Tommy Lasorda arguing before a congressional committee why the First Amendment should not protect flag burning, Newsweek, July 20, 1998, p. 21.

61. As Yogi Berra used to say "it ain't over till it's over," so it must not be done yet.

62. I am a good worker, because Frank says so. How can we believe Frank? Simple: I will vouch for him.

63. In the 4 years that I have been practicing my own, innovative, cognitive-humanistic-dynamic-behavioral-deconstructive-metaregressive-deontological psychotherapy, CHDB DMDP, there has not been a single study showing that it fails to work, or that it has ever harmed a patient. Therefore, it is clearly one of the best treatment techniques available!

64. Since he supports the separation of church and state, he also must support godless atheistic communism? See how well that worked out in Russia, China, and Cuba?

65. Those who believe in behavior modification obviously want to try to control everyone by subjecting them to rewards and punishments.

66. She believes that the Ground Zero Mosque should be allowed to be built. This is because she is an America-hating liberal.

67. Since no evidence has been collected on UFOs, then they must not exist.

68. Scientists don't know exactly what happened in the Big Bang, so it must not be true.

69. Ladies and gentlemen of the jury, I urge you to convict John Jones of this crime of murder. We need to put him where he can never commit any crimes. If you don't convict him, you may be his next victim.

70. Mr. superintendent, you should cut the school budget by $150,000. I need not remind you that past school boards have fired superintendents who did not keep down costs.

71. I have not won a hand all night, therefore I'm due to win big so I must keep on playing.

72. "Please your Majesty,' said the Knave, 'I didn't write it, and they can't prove I did: There's no name signed at the end.' 'If you didn't sign it,' said the King, 'that only makes the matter worse. You must have meant some mischief, or else you'd have signed your name like an honest man.'" - Lewis Carroll, Alice's Adventures in Wonderland

73. "Anyone who believes for a moment that the state would honor a pledge to replace the revenue lost to local government would believe pledges of fidelity made in a brothel. Eliminating the vehicle license fee is an election-year buzz-word scam touted by an Assembly member who salivates over the thought of no government to impede his motivations, and a governor who flat out hates counties." - letter to editor, Fresno Bee, May 12, 1998

References

1. Morris, K., (1972). Mathematical Thought From Ancient to Modern Times. Oxford University Press

2. Barwise, J. (ed), (1977). Handbook of Mathematical Logic. North-Holland Publ. Co., Amsterdam.

3. Boolos, G., (1993). The Logic of Provability, Cambridge University Press, Cambridge.

4. Carroll, L. (1865). Alice's Adventures in Wonderland.

5. Carroll, L. (1871). Through the Looking Glass.

6. Chang, C.C., Keisler, J.J., (1990). Model Theory. North-Holland Publ. Co., Amsterdam.

7. van Dalen, D., (2002). Intuitionistic Logic. In: Gabbay, D. and F. Guenthner (eds.) Handbook of Philosophical Logic. 5.(Second ed.) Kluwer , Dordrecht.

8. Davis, M., (1958). Computability and Unsolvability. McGraw Hill, New York.

9. Devlin, K (2010). The Hidden Math Behind Alice in Wonderland. Devlin's Angle. Mathematical Association of America, March 2010.

10. Dummett, M., (1977). Elements of Intuitionism. Oxford University Press, Oxford.

11. Girard, J.Y., (1987). Proof Theory and Logical Complexity. I. Bibliopolis, Napoli.

12. Girard, J.Y., Lafont, Y., Taylor, P., (1989). Proofs and Types. Cambridge University Press, Cambridge.

13. Gray, B.G., (1995). Alice in Wittgenstein: Inside the great mirror. The Journal of Value Inquiry 29:77-88, Kluwer Academic Publishers.

14. Negri, S., von Plato, J., (2001). Structural Proof Theory. Cambridge University Press.

15. Sacksteder, W., (1967). Looking Glass: A Treatise on Logic. Philosophy and Phenomenological Research, 27, 338–55.

16. Troelstra, A.S., Schwichtenberg, H., (1996). Basic Proof theory. Cambridge University Press.

CHAPTER 16

Beyond Aristotle – Symbolic Logic and the Insight of Boole

We ought no longer to associate Logic and Metaphysics, but Logic and Mathematics.

George Boole

16.1 Introduction

The logic of Aristotle developed very little over the 2,000 plus years since it was originally founded, until 1854 when George Boole brought it back to life. Boole created an algebra out of Aristotle's propositional logic, called Symbolic Logic; A major new breakthrough for the field of logic. He essentially "mathematized" logic.

Boole's new system of logic created rules of logic that could now be approached algebraically, and the analysis of arguments could proceed in a precise analytical way. In creating this new branch of mathematics, Boole had become the Father of Symbolic Logic. In what follows below, we show how Boole transformed the logic of Aristotle using symbols instead of words.

16.2 Transforming Propositional Logic Into Symbolic Logic

As we've already discovered, a logical argument has several basic parts. We have the statements, the logical connectives, the premises and the conclusion. How to symbolize these objects and ideas is the subject of this section. In what follows we develop the Boolean algebra of word logic and open up new possibilities and uses for logic.

Again, we start with the statement. Recall that a statement, also called a proposition, is a complete declarative sentence that expresses a single idea that has a distinct truth value. Statements can be represented symbolically using letters. We typically start with p and continue down the alphabet with, q, r , s, t, …

> Boole initially used the letters x and y for statements in his original formulation of the rules of logic into their symbolic form. Then in 1897 Giuseppe Peano (1858–1932) used the letter p to represent a proposition (perhaps p for proposition.) In 1903, in his *Principles of Mathematics,* Bertrand Russell, who was corresponding with Peano at the time, used the letters p, q, and r, to represent propositions, and this designation was adopted universally.

For example: The following are statements, and can be associated/identified with specific logical variables.

1. "Today is Monday" = p
2. "The sun is shining" = q
3. "John has heart disease" = r
4. "Mathematics is created" = s
5. "Women are more likely to smoke then men" = t

Simple and Compound Statements

As we've learned from propositional logic in Chapter 15, statements can be **simple** statements, if they express a single idea, or they can be **compound** if they express two or more ideas in the same sentence. A compound statement needs connectives, to link the different ideas.

In language, we have specific words that "connect" two or more statements together or even modify a specific statement. Some of the more common connectives are AND, and OR, along with NOT which is used to negate a simple statement.

For example, you might say, *I went to the store AND I brought my friend*. The AND in this sentence connects the simple statement, *I went to the store,* with the simple statement, *I brought my friend*.

Or we may wish to negate a statement and say, *I did NOT go to the store*. Thus, if we want our new symbolized form of logic to accurately represent word logic, we need to be able to represent some fairly complex sentences symbolically. Logical connectives help us do this.

16.2.1 Logical Connectives, Operators and Their Symbols

Conjunction

The AND logical connective is called a **conjunction**, and is represented symbolically as $\wedge$.

Examples of conjunctions

1. The example above, *I went to the store and I brought my friend,* can be expressed symbolically as follows:

   First, we identify the simple statements in this compound statement, and assign logical variables to the word statements.

   Let,   p = *I went to the store*

   q = *I brought my friend*

   Next, we join the them together with the logical AND connective as: $p \wedge q$

So the compound statement, *I went to the store and I brought my friend*, is symbolically equivalent to, $p \wedge q$.

2. *It is raining and I brought an umbrella*, can be expressed symbolically as follows:

Let, p = *It is raining*

        q = *I brought an umbrella*

Join the them together with the logical AND connective as: $p \wedge q$

Disjunction

The OR logical connective is called a **disjunction**, and is represented symbolically as $\vee$.

Examples of disjunctions

1. *I will go to the store or I will go to a movie*, can be expressed symbolically as follows:

First, we identify the simple statements in this compound statement, and assign logical variables to the word statements.

Let, p = *I will go to the store*

        q = *I will go to a movie*

Next, we join the them together with the logical OR connective as: $p \vee q$

So the compound statement, *I will go to the store or I will go to a movie*, is symbolically equivalent to, $p \vee q$.

2. *I will buy a shirt or a pair of pants*, can be expressed symbolically as follows:

Let, p = *I will buy a shirt*

        q = *I will buy a pair of pants*

Join the them together with the logical OR connective as: $p \vee q$

Negation

The NOT logical operator is called a **negation**, and is represented symbolically as $\sim$.

Examples of negations

1. *I will not go to the store*, can be expressed symbolically as follows:

First, we identify the simple statements in these statements, and assign logical variables to the word statements.

Let,   p = *I will go to the store*

Negate this as   ~ p

So the statement, *I will not go to the store*, is symbolically equivalent to, ~ p .

2.   *I do not have cancer*, can be expressed symbolically as follows:

Let,   p = *I have cancer*

Negate this as   ~ p

So the statement, *I do not have cancer*, is symbolically equivalent to, ~ p .

In addition to the AND, OR, and NOT connectives, we also have connectives that are based upon a condition happening.

For example, you might want to say – If I go to the store, then I will take a taxi.

We are now connecting the two statements, "I go to the store," and " I take a taxi," with the IF … THEN conditional connective. Meaning, a condition must first be satisfied, before the next part will be executed. The IF part of the conditional is called the **antecedent**, and the THEN part is called the **consequent**.

Conditional

The IF...THEN logical connective is called a conditional, and is represented symbolically as →, and arrow pointing from left to right.

Examples of conditionals

1.   *If I go to the store then I will take a taxi*, can be expressed symbolically as follows:

First, we identify the simple statements in this compound statement, and assign logical variables to the word statements.

Let,   p = *I go to the store*

    q = *I will take a taxi*
Next, we join them together with the logical IF … THEN conditional as: $p \rightarrow q$

So the compound statement, *If I go to the store then I will take a taxi*, is symbolically equivalent to, $p \rightarrow q$.

2. *If he owns a car it's a Ford*, can be expressed symbolically as follows:

Let,   p = *He owns a car*

q = *It's a Ford*

Notice that the word *then* is not explicitly stated. It is, however, implied, so we include it, and write this as $p \rightarrow q$

So the compound statement, *If he owns a car it's a Ford*, is symbolically equivalent to, $p \rightarrow q$.

Bi–conditional

In addition to the IF … THEN conditional statement, we also have a **bi–conditional** connective, which represents the idea of, IF AND ONLY IF (also written as IFF).

The IF AND ONLY IF logical connective is called a bi–conditional, and is represented symbolically as $\leftrightarrow$, an arrow pointing in both directions.

For example, consider the following sentence:

I'll buy you lunch if and only if you pass the exam.

This means that if you pass the exam, then I'll buy you lunch, but it also means if I buy you lunch then you passed the exam. The IF AND ONLY IF conditional goes both ways.

Examples:
1. *You are permitted to vote if and only if you are registered,* can be expressed symbolically as follows:

First, we identify the simple statements in this compound statement, and assign logical variables to the word statements.

Let,   p = *You are registered*

q = *Your are permitted to vote*

Next, we join the them together with the logical IF AND ONLY IF conditional as: $p \leftrightarrow q$
So the bi–conditional statement, *You are permitted to vote if and only if you are registered*, is symbolically equivalent to, $p \leftrightarrow q$. This means that, *If you are registered then you can vote*, as well as, *If you have voted then you are registered*.

2. *Sally will go to work if and only if she gets a raise*, can be expressed symbolically as follows:

Let,   p = *Sally gets a raise*

q = *Sally goes to work*

Next, we join the them together with the logical IF AND ONLY IF conditional as: $p \leftrightarrow q$

So the bi–conditional statement, *Sally will go to work if and only if she gets a raise*, is symbolically equivalent to, $p \leftrightarrow q$. This means that, *If Sally goes to work then she will get araise*, as well as, *If Sally got a raise then she went to work*.

Quantifiers

Words such as *all, each, every, no, and none* are called **universal quantifiers**.

Words and phrases like *some, there exists,* and *for at least one* are called **existential quantifiers**.

Negation of quantified statements

Negating quantified statements can be a bit tricky. To negate the *all* quantifier we must choose a quantifier that will make the appropriate statement false. Certainly the quantifier *none* works, but this is the strongest case, and we need the weakest possible quantifier that will make it false. In this case, it is the *some* quantifier. So the logical negation of *all* is *some*.

Thus, the negation of *All do* is *Some do not*.

To negate the *some* quantifier we use the *none* quantifier.

Thus, the negation of *Some do* is *None do*.

Quantifier	Negation
All	Some, Not all
Some	None, All do not

Examples: Write the negation of each statement.

1. *Some dogs have fleas.*

    "Some" means "at least one," therefore, the above statement is equivalent to "At least one dog has fleas." The negation of this statement is "No dog has fleas."

2. *All US manufactures are efficient.*

    This statement claims that every manufacturer in the US is efficient, so the negation is that there is at least one manufacturer that is not efficient, or "Some US manufacturers are not efficient."

16.2.2 Translating Word Statements into Symbolic Form

The first part of symbolic logic involves the translation process. You need to be able to translate a compound word statement into symbolic form. Consider the examples below:

Translate the following word statements into symbolic form.

Example 1:

If it is a rainy day and I brought the umbrella, then I'll stay dry

Let   p = It is a rainy day, and
        q = I brought an umbrella
        r = I'll stay dry

Then the proposition becomes: $(p \land q) \rightarrow r$

Example 2:

He is neither a democrat or a republican.

Let   p = He is a democrat
        q = He is a republican

Then the proposition becomes: $\sim p \lor \sim q$

Example 3:

I will either vote for the resolution or stage a protest.

Let   p = I will vote for the resolution
        q = I will stage a protest

Then the proposition becomes: $p \lor q$

Example 4:

Math is a mystery and will always lead to great discoveries.

Let   p = Math is a mystery
        q = Math will lead to great discoveries

Then the proposition becomes: $p \land q$

Compound statements lead the way to arguments with premises and conclusions. Coni=sider the following examples of translating arguments.

Example 5:

If math is discovered then it is not created.
Math is discovered.
Therefore, it is not created.

Let   p = Math is discovered
          q = Math is created

Then the word argument becomes:

$$p \rightarrow \sim q$$
$$q$$
Therefore, $\sim q$

Example 6:

Princess Diana was assassinated or was killed in an accident.
Princess Diana was not assassinated.
Therefore, Princess Diana was killed in an accident .

Let   p = Diana was assassinated
          q = Diana was killed in an accident

Then the word argument becomes:

$$p \vee q$$
$$\sim p$$
Therefore, q

Example 7:

All insects have wings.
Wood lice are insects.
Therefore, wood lice have wings.

Here we have to rewrite the "all" statement to "If you are an insect then you have wings." This is how we will handle the "all" statement.

Let   p = You (or any creature is) are an insect
          q = You (or any creature) have wings

Then the word argument becomes:

$$p \rightarrow q$$
$$p$$
Therefore, q

We now consider the truth value of the compound expressions for various.

16.3 Truth Values of Compound Statements

Depending upon the logical truth value of the simple statements in a logical expression, the compound statement also has a truth value.

Consider the compound statement:

It is raining and I brought an umbrella.

We have the two simple statements,
$$p = It\ is\ raining$$
$$q = I\ brought\ an\ umbrella$$
joined by the logical AND connective.

If it is true that it is raining, and true that I brought an umbrella, then you would say that I was telling the truth when I made the compound statement. Thus, the truth value of this compound statement would be true, or
$$T \wedge T = T$$

If on the other hand, It is raining, but I did not bring and umbrella, you would say I lied, since I did not bring the umbrella, although I committed to bringing it, or

$$T \wedge F = F$$

Thus, the AND connective is only true when both of the simple statements are also true. It is false otherwise.

The OR connective, however, can be viewed in two different ways when we wish to determine the logical value of a compound statement. The OR, as we'll now show, can be thought of "inclusively" or "exclusively."

Inclusive and Exclusive OR Statement

In language, the OR connective can cause a bit of confusion. For example, the statement,

You can have either tea or coffee,

from a waiter at a restaurant usually means you can have one or the other, but not both, as does the statement,

I will drive the Ford or the Chevy to work today.

You would never expect both to be true at the same time.

On the other hand, the statement,

I will have coffee or tea with dinner,

could be interpreted that you either had coffee, tea, or even both, without being thought of as a false statement or a lie, as does the statement,

The problem with your car could be the brakes or the shock absorbers.

You would not be surprised if the mechanic said both were bad (in fact you probably would be surprised if he didn't say this).

These statements highlight an ambiguity in language that needs to be clarified. In symbolic logic, this is done by creating two types of OR statements. The **inclusive** OR and the **exclusive** OR statement. The first two statements above are called exclusive, since we exclude the possibility of both instances happening at the same time. The next two statements are inclusive OR statements, since we allow the possibility of both happening simultaneously.

Since this is only a basic introduction to symbolic logic, we will only consider the inclusive OR in all that follows. In more advanced treatments of logic the exclusive OR is also considered. Thus, whenever we see an OR statement in this book, we will consider it to be true if either the simple statements are true, or if both are true. It is false only when both are false. Let's now examine statements with the inclusive OR.

Consider the compound statement:

Some day I plan to buy a car or a house.

This is two simple statements,

$$p = \textit{I plan to buy a car}$$
$$q = \textit{I plan to buy a house}$$

joined by the logical connective OR.

If it is true that you bought a house, but false that you bought a car, the statement would not be considered a lie, i.e.

$$T \vee F = T$$

On the other hand, if it is false that you bought a house, but true that you bought a car, the compound statement would also be considered truthful.

$$F \vee T = T$$

Now, if you bought both the house and the car, we interpret the OR inclusively, thus, we would consider:

$$T \vee T = T$$

We will not use the exclusive OR interpretation of,

$$T \vee T = F,$$

although it is something that needs to be addressed in a more thorough introduction to symbolic logic.

In this introduction to symbolic logic we will interpret the OR connective inclusively. The exclusive OR will have to be considered in a more complete discussion on logic than we plan to introduce here. However, you should be aware of the difference and also be able to determine which compound OR statements can be interpreted inclusively and which can be interpreted exclusively, and which are hard to say.

More Complicated Compound Statements

It is possible to have more than one connective in a compound statement. As this occurs, however, it becomes more difficult to determine the overall truth value of the statement. Consider the following example.

It is warm and sunny, or I did not go to the beach.

We have three simple statements,
$$p = \textit{It is warm}$$
$$q = \textit{It is sunny}$$
$$r = \textit{I went to the beach} \text{ (notice we are taking the positive of}$$
the statement here and not its negative)
joined by the logical AND and OR connectives, as well as modified by the negation NOT.

If it is true that it is sunny, true that is warm, and true that I went to the beach, then you would say that I did not lie when I made the compound statement. The resulting truth value of the compound statement would be true, or

$$(T \wedge T) \vee (\sim T) =$$
$$T \quad \vee \quad F \ =$$
$$T$$

If, on the other hand, it was warm, but not sunny and I went to the beach I would be lying, since

$$(T \wedge F) \vee (\sim T) =$$
$$F \quad \vee \quad F \ =$$
$$F$$

By replacing the simple statements by their individual truth values we can see that it is a bit easier to keep track of the truth value of the overall compound statement. Without these individual truth values it is harder to judge the truth value of the overall compound statement. This is especially true when the negation operator is present.

If we approached determining the truth value of all compound statements in this way, it would be very time consuming. There is, however, another approach that makes the determination more systematic, and easier to accomplish. This is through something called a truth table.

16.4 Truth Tables

One way to identify all the logical outputs of a logical expression is through something called a Truth Table. A truth table shows all the possible logical outputs of a logical expression given all the possible logical inputs into that expression. As we will also demonstrate, truth tables provide us with a direct and easily understandable method of testing the validity of an argument.

A truth table has at least two parts. The set of logical inputs, and the resulting logical outputs of the logical expression we are trying to evaluate, based upon the specific values of the logical inputs. Consider the symbolic expression:

Logical Inputs

If the logical expression we are evaluating has one logical proposition, p, associated with it, then we have two possible logical inputs – true and false.

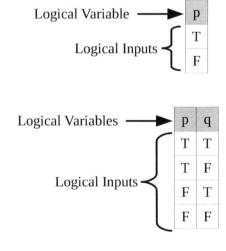

334

If the logical expression has two logical inputs, p and q, then we have four possible logical input combinations as shown below.

If the logical expression has three logical inputs, p, q, and r, then we have eight possible logical input combinations.

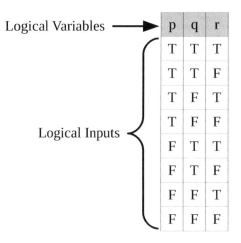

Logical Variables ⟶

Logical Inputs

Each additional logical input variable doubles the preceding amount of possible logical input combinations. Thus, if we had 5 logical variables we would need $2^5 = 32$ logical input combinations.

> In general, for n input variables we have 2^n logical input combinations to evaluate.

Furthermore,

> To obtain the proper T/F input combinations we:
>
> 1. Start in the p column and fill in the first half (4 if you have three variables) with T's and the second half with F's.
>
> 2. We move to the q column and label the first two (half of the half) with T's and the next two with F's and repeat until the column is completely filled in.
>
> 3. In the last r column, we just alternate between T's and F's until the column is filled in.
>
> This approach works for any number of logical variables. Using this approach guarantees that you will have all the required logical T/F combinations, and that you will not duplicate or leave any out.

Logical Outputs

Logical outputs will be shown in the last column. They are often the culmination of a series of logical evaluations. If the logical expression we are evaluating is fairly complex, although initially we'll look at simple expressions, and they do not require any intermediate columns, so we just have logical inputs and outputs to consider, as shown in the chart below.

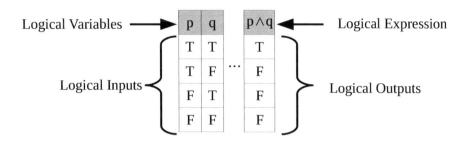

We illustrate the use of truth tables by analyzing the truth values of the logical connectives.

The Negation Operator: ~p

The truth table for this operator has only two possible inputs, true of false, since there is only one logical variable associated with this operator.

p	~p
T	F
F	T

The first column contains the logical input values for the variable p and the second column contains the logical outputs for ~p that are associated with the logical inputs. The negation operator changes True inputs in False outputs, and False inputs into True outputs. It negates the logical input value.

The AND Connective: p∧q

The truth table for this operator has two logical variables, p and q, and from above we have four possible combinations of logical input values.

p	q	p∧q
T	T	T
T	F	F
F	T	F
F	F	F

The last column provides the logical output of this connective associated with the logical input values of p and q.

> The AND statement is true if both the input values are true, otherwise it is false.

We illustrate how we obtained this with the following example:

Consider the following statement. "It is raining, and I brought an umbrella."

The two simple logical statements are:

p = *It is raining*
q = *I brought an umbrella*

We need to assess the truth value of the compound statement: "It is raining, and I brought an umbrella," for each of the following combinations of truth values of p and q. We need to determine if the person making the statement is lying given the actual circumstances of whether it is or is not raining, and whether I did or did not bring an umbrella.

1. If both p and q are true, meaning it is raining and I did bring an umbrella, then I did not lie and the above compound statement is true.

2. If p is true and it is raining, but q is false, meaning I did not bring an umbrella, then I would be lying if I said, "It is raining, and I brought an umbrella," so the output for this combination is false.

3. If p is false and it is not raining, but q is true, meaning I did bring an umbrella, then I would be lying if I said, "It is raining, and I brought an umbrella," so the output for this combination is false.

4. If p is false and it is not raining, and q is also false, meaning I did not bring and umbrella, then I would be lying if I said, "It is raining, and I brought an umbrella," so the output for this combination is false.

The OR Connective: $p \vee q$

The truth table for this operator has two logical variables, p and q, and from above we have four possible combinations of logical input values. In this case we consider the inclusive or.

p	q	$p \vee q$
T	T	T
T	F	T
F	T	T
F	F	F

The last column provides the logical output of this connective associated with the logical input values of p and q.

The OR statement is false if both the input values are false, it is true otherwise.

We illustrate how we obtained this with the following example:

Consider the following statement. "I'll go to the store, or see a movie."

The two simple logical statements are:

$$p = I'll \ go \ to \ a \ store$$
$$q = I'll \ see \ a \ movie$$

As we did earlier, we need to assess the truth value of the compound statement: "I'll go to the store, or see a movie," for each of the following possible combinations of truth values of p and q.

1. If both p and q are true, meaning I went to the store and I saw a movie, then I did not lie and the above compound statement is true.

2. If p is true and I went to the store, but q is false, meaning I did not go to the movies, then I would not be lying if I said, "I'll go to the store, or see a movie," since I at least did one of them. Thus, the output for this combination is true.

3. If p is false and I did not go to the store, but q is true, meaning I did go to the movies, then I would not be lying if I said, "I'll go to the store, or see a movie," since I at least did one of them. Thus, the output for this combination is true.

4. If p is false and I did not go to the store, and q is also false, meaning I did not go to the movies, then I would be lying if I said, "I'll go to the store, or see a movie," since I didn't do either thing I said I would. Thus, the output for this combination is false.

Conditional IF ... THEN Connective: $p \rightarrow q$

The truth table for this operator has two logical variables, p and q, and from above we have four possible combinations of logical input values.

p	q	$p \rightarrow q$
T	T	T
T	F	F
F	T	T
F	F	T

The last column provides the logical output of this connective associated with the logical input values of p and q.

> The IF ... THEN statement is false if the first input is true and the second is false, it is true otherwise.

We illustrate how we obtained this chart with the following example:

Consider the following statement. "If it is raining, then I'll bring an umbrella."
The two simple logical statements are:

p = *It is raining*
q = *I'll bring an umbrella*

We need to assess the truth value of the compound statement: "If it is raining, then I'll bring an umbrella," for each of the following possible combinations of truth values of p and q. Given the truth or falsity of the antecedent and the consequent, you must be able to assess if the statement is a lie or not. If it is a lie then it is false, if it is not a lie then we will call it true.

1. If both p and q are true, meaning it is raining and I brought an umbrella, then I did not lie and the above compound statement is true. I did not lie because I made the commitment and kept it.

2. If p is true and it is raining, but q is false, meaning I did not bring an umbrella, then I would be lying if I said, "If it is raining, then I'll bring an umbrella," since I made a commitment and I did not keep it. Thus, the output for this combination is false.

3. If p is false and it is not raining, but q is true, meaning I did bring an umbrella, then I would not be lying if I said, "If it is raining, then I'll bring an umbrella," since I never made a commitment. My commitment of bringing an umbrella is contingent upon it raining, and it is not. Thus, I did not lie. The output for this combination is not false (a lie), so it is true by default.

4. If p is false and it is not raining, and q is also false, meaning I did not bring an umbrella, then I would not be lying if I said, "If it is raining, then I'll bring an umbrella," since I never made a commitment. My commitment of bringing an umbrella is contingent upon it raining. Since it is not raining it doesn't whether I bring an umbrella or not. Thus, the output for this combination is not false so it is true by default.

The truth table for this conditional is quite often confusing, due to the truth values of the last two rows. It seems counter–intuitive that the last two truth values are true.

One way to think of this is as follows. The p–part (the antecedent) is the commitment, and the q–part (the consequent) is whether or not I kept my commitment. Thus, for row one in the truth table, I made the commitment and kept it, so I told the truth. In row two I made the commitment, but I did not keep it, so I lied. In rows three and four I did not make any commitment, so I cannot be found in a lie, so it must be true. If no commitment is made I am not on the hook for keeping it, so it is not a lie. We consider not lying as telling the truth, and this is the confusing part. In logic we only have one of two choices, either false if we lie, or true if we don't lie.

Bi–conditional, IF AND ONLY IF, Connective: p ↔ q

The truth table for this operator has two logical variables, p and q, and from above we have four possible combinations of logical input values.

p	q	p ↔ q
T	T	T
T	F	F
F	T	F
F	F	T

The last column provides the logical output of this connective associated with the logical input values of p and q.

> The IF AND ONLY IF statement is true if both the inputs are true or if both are false, it is false otherwise.

We illustrate how we obtained this chart with the following example:

Consider the following statement. "I'll bring an umbrella if and only if it is raining."

The two simple logical statements are:

$$p = It\ is\ raining$$
$$q = I'll\ bring\ an\ umbrella$$

We need to assess the truth value of the compound statement: "I'll bring an umbrella if and only if it is raining," for each of the following possible combinations of truth values of p and q. Given the truth or falsity of p and q, you must be able to assess if the statement is a lie or not. If it is a lie then it is false, if it is not a lie then we will call it true.

1. If both p and q are true, meaning it is raining and I brought an umbrella, then I did not lie and the above compound statement is true. I did not lie because I made the commitment and kept it.

2. If p is true and it is raining, but q is false, meaning I did not bring an umbrella, then I would be lying if I said, "I'll bring an umbrella if and only if it is raining,"

since I made a commitment and I did not keep it. Thus, the output for this combination is false.

3. If p is false and it is not raining, but q is true, meaning I did bring an umbrella, then I would be lying if I said, "I'll bring an umbrella if and only if it is raining," since I said I would only bring an umbrella if it is raining. Since it is not raining and I have an umbrella, I lied. The output for this combination is false (a lie).

4. If p is false and it is not raining, and q is also false, meaning I did not bring an umbrella, then I would not be lying if I said, "I'll bring an umbrella if and only if it is raining," since I carried through on my commitment of not bringing an umbrella if it was not raining. Thus, the output for this combination is true since I did what I said I'd do.

You may have noticed that the bi–conditional can be obtained by using the conditionals: $p \rightarrow q$ and $q \rightarrow p$, and requiring that both conditionals be true at the same time. (We'll talk more about this later when we discuss something called equivalence.) Look at the truth table above and try this, by reading the conditionals both ways, and if one way is not true the truth value is false.

16.5 More Complex Logical Expressions and Their Truth Tables

Up to this point the logical expressions we have considered involved only one logical connective. However, as we know language is more complicated than that. For example what if I made the statement: "I will take my car and buy a shirt or I will go to the movies." Notice that we now have two logical connectives, AND and OR, in the same compound statement.

At the same time we have also added a new level of uncertainty (or ambiguity) to the problem. Symbolically we can write this expression in two different ways. To show what we mean let us write:

$$p = I \ take \ my \ car$$
$$q = I \ buy \ a \ shirt$$
$$r = I \ go \ to \ the \ movies$$

Then using symbolism we can write either $\quad (p \wedge q) \vee r \quad$ or $\quad p \wedge (q \vee r)$

That is, we could mean that I

(take my car and buy a shirt) or go to the movies,

or instead, we may want to say that I

take my car and (buy a shirt or go to the movies).

The later interpretation seems more likely, but how can we be sure? Perhaps we are taking the car to buy a shirt, and then come home and somebody may be picking us up for a movie, or we are taking the car to buy a shirt, or take the care to the movies. To help with this ambiguity we can use commas to group the statements. Thus, we could write

I will take my car and buy a shirt, or I will go to the movies. $(p \wedge q) \vee r$

Or we could write

I will take my car and, buy a shirt or I will go to the movies. $p \wedge (q \vee r)$

As you can see getting the meaning correct in a compound logical language expression can be tricky, and we have to be careful. What we also see is that this ambiguity goes away when we write it symbolically. You may ask – does it really make a difference? They sound almost the same. Is there any real difference, logically, between the two?

We'll answer this question shortly, but for now we want to start out with an easier introduction so we'll keep the logical variables involved in the expression to only two (p and q). We'll come back to this question at the end of this section, when we are more skilled at evaluating logical expressions, and can add additional logical statements.

Let's learn how to determine the logical outputs of some more complicated compound logical expressions involving only two variables. We will not be concerned with their word meaning just yet, only their logical true or false value.

> In all that follows we will have to know and apply the truth tables for the AND $(\wedge)$, OR $(\vee)$, NOT (~), IF...THEN $(\rightarrow)$, and IF AND ONLY IF $(\leftrightarrow)$, operators.
>
> We should commit these to memory, or at least understand the rules for finding their truth tables, so we can do this quickly and easily.

Example 1: Find all the possible truth values of the logical expression $p \wedge (q \vee p)$

To find the logical outputs from all possible logical inputs, means we need to construct a truth table. However, we now have two logical connectives instead of one. The way we will approach this is similar to how we approach problems in algebra, by using order of operations. We can think of the AND, $\wedge$, and the OR, $\vee$, operators as arithmetic operations, such as multiplication and addition, and treat the parentheses, brackets and braces the same way we do in arithmetic.

We have two logical inputs, p and q, so we start with the following set of logical inputs:

p	q
T	T
T	F
F	T
F	F

In the above expression we will first evaluate the parentheses term $q \lor p$ and then evaluate $p \land (q \lor p)$. The first expression is just the OR truth table, with the truth variables reversed which can be easily found as:

p	q	$q \lor p$
T	T	**T**
T	F	**T**
F	T	**T**
F	F	**F**

The last column combined with the first column, then becomes the input into the final expression $p \land (q \lor p)$. In this expression we are evaluating the AND operator. We already evaluated the OR in the previous column.

p	q	$q \lor p$	$p \land (q \lor p)$
T	T	T	$T \land T = \mathbf{T}$
T	F	T	$T \land T = \mathbf{T}$
F	T	T	$F \land T = \mathbf{F}$
F	F	F	$F \land F = \mathbf{F}$

The last column reads T, T, F, F, and this is the final answer. We know that the full range of logical inputs will result in outputs of the form:

p	q	$q \lor p$	$p \land (q \lor p)$
T	T	T	T
T	F	T	T
F	T	T	F
F	F	F	F

Notice that the logical output is the same as the p input column, meaning q could be anything.

Notice that the logical expression had two logical operators, $\land$ and $\lor$. Also notice that we needed two columns beyond the input column to evaluate this. You will see that for every logical operator we will use a column to evaluate that particular logical operator.

Example 2: Find the truth table for the logical expression $(p \land q) \lor \sim p$

In this example we have three logical operators: $\land$, $\lor$, and $\sim$. We follow the same process we did in the previous example, but now we have three columns after the input p and q columns, one column for each operator.

We will evaluate $p \wedge q$ first, and then evaluate $\sim p$, and finally $(p \wedge q) \vee \sim p$. We will do this in four steps, to show how we fill in the truth table

1. Set up the logical inputs p and q
   (the p column is two T's and two F's, the q column is alternating T, and F values)

p	q	p∧q	~p	(p∧q)∨~p
T	T			
T	F			
F	T			
F	F			

2. Evaluate $p \wedge q$ using p and q values

p	q	p∧q	~p	(p∧q)∨~p
T	T	T		
T	F	F		
F	T	F		
F	F	F		

3. Evaluate $\sim p$ using p values

p	q	p∧q	~p	(p∧q)∨~p
T	T	T	F	
T	F	F	F	
F	T	F	T	
F	F	F	T	

4. Evaluate the $\vee$ in $(p \wedge q) \vee \sim p$ using $p \wedge q$ and $\sim p$ values

p	q	p∧q	~p	(p∧q)∨~p
T	T	T	F	T
T	F	F	F	F

F	T	F	T	T
F	F	F	T	T

Example 3: Find the truth table for the logical expression $[(p \lor q) \land \sim q] \to p$

In this example we have four logical operators: $\land$, $\lor$, $\sim$, and $\to$. We follow the same process we did in the previous examples, but now we have four columns after the input p and q columns, one column for each operator.

Using the order of operations we will evaluate $p \lor q$ first, and then evaluate $\sim q$, and then the $\land$ in $(p \lor q) \land \sim q$, and finally the $\to$ in $[(p \lor q) \land \sim q] \to p$. We will do this in five steps, to show how we fill in the truth table

1. Set up the logical inputs p and q, and all the logical operations

p	q	p∨q	~q	(p∨q)∧~q	[(p∨q)∧~q]→p
T	T				
T	F				
F	T				
F	F				

2. Evaluate $p \lor q$ using p and q values

p	q	p∨q	~q	(p∨q)∧~q	[(p∨q)∧~q]→p
T	T	T			
T	F	T			
F	T	T			
F	F	F			

3. Evaluate $\sim q$ using Q values

p	q	p∨q	~q	(p∨q)∧~q	[(p∨q)∧~q]→p
T	T	T	F		
T	F	T	T		

p	q	p∨q	~q		
F	T	T	**F**		
F	**F**	F	**T**		

4. Evaluate the ∧ in $(p \lor q) \land \sim q$ using $p \lor q$ and $\sim q$ values

p	q	p∨q	~q	(p∨q)∧~q	[(p∨q)∧~q]→p
T	T	T	F	**F**	
T	F	T	T	**T**	
F	T	T	F	**F**	
F	F	F	T	**F**	

5. Evaluate the → in $[(p \lor q) \land \sim q] \to p$ using $(p \lor q) \land \sim q$ and p values
Remember when doing the evaluations for the conditional you must work them from right to left and not left to right, as the columns are in reverse order here from what we have seen in the past.

p	q	p∨q	~q	(p∨q)∧~q	[(p∨q)∧~q]→p
T	T	T	F	F	**T**
T	F	T	T	T	**T**
F	T	T	F	F	**T**
F	F	F	T	F	**T**

Notice that the last column has all True values. This is a special occurrence that we'll discuss in greater detail in the next section. It is called a **tautology**.

We close this section by working through the original problem we started with containing three logical variables, p, q, and r. This means that we have $2^3 = 8$ rows to fill in.

We need to fill in the logical inputs for p, q, and r, as well as the logical expressions we will evaluate in our truth table.

The final expression $(p \land q) \lor r$ contains only two logical operators, ∧ and ∨, so we will have only two columns after the input variables

2. Set up the logical inputs p, q, and r, and all the logical operations

p	q	r	p∧q	(p∧q)∨r
T	T	T		

p	q	r	p∧q	(p∧q)∨r
T	T	F		
T	F	T		
T	F	F		
F	T	T		
F	T	F		
F	F	T		
F	F	F		

3. Evaluate p∧q using p and q values

p	q	r	p∧q	(p∧q)∨r
T	T	T	**T**	
T	T	F	**T**	
T	F	T	**F**	
T	F	F	**F**	
F	T	T	**F**	
F	T	F	**F**	
F	F	T	**F**	
F	F	F	**F**	

4. Evaluate (p∧q)∨r using p∧q and R values

p	q	r	p∧q	(p∧q)∨r
T	T	T	T	**T**
T	T	F	T	**T**
T	F	T	F	**T**
T	F	F	F	**F**
F	T	T	F	**T**
F	T	F	F	**F**
F	F	T	F	**T**
F	F	F	F	**F**

We follow the same process, but this time we use the other logical expression with the parentheses around the OR rather than the AND operator, $p \wedge (q \vee r)$.

1. Set up the logical inputs p, q, and r, and all the logical operations

p	q	r	$q \vee r$	$p \wedge (q \vee r)$
T	T	T		
T	T	F		
T	F	T		
T	F	F		
F	T	T		
F	T	F		
F	F	T		
F	F	F		

2. Evaluate $q \vee r$ using q and r values

p	q	r	$q \vee r$	$p \wedge (q \vee r)$
T	T	T	T	
T	T	F	T	
T	F	T	T	
T	F	F	F	
F	T	T	T	
F	T	F	T	
F	F	T	T	
F	F	F	F	

3. Evaluate the $\wedge$ in $p \wedge (q \vee r)$ using $q \vee r$ and p values

p	q	r	$q \vee r$	$p \wedge (q \vee r)$
T	T	T	T	T
T	T	F	T	T
T	F	T	T	T
T	F	F	F	F

F	T	T	T	F
F	T	F	T	F
F	F	T	T	F
F	F	F	F	F

Putting them both together we have:

p	q	r	$(p \wedge q) \vee r$	$p \wedge (q \vee r)$
T	T	T	T	T
T	T	F	T	T
T	F	T	T	T
T	F	F	F	F
F	T	T	**T**	**F**
F	T	F	F	F
F	F	T	**T**	**F**
F	F	F	F	F

We see that logically they are not the same. They are almost the same, but there are two instances in which they are different. Both of which are shown in bold in the table above. This is typical of propositional logic. We need to be very careful with how we set up an argument symbolically, but once we set it up correctly, the symbolism helps us to avoid the mistakes that the word logic can often lead to. Using *negations* along with *and's, or's*, and *if...then's* in a word statement can be very confusing. Especially when you need to determine the final truth value.

As we have shown above Aristotle, and others after him, were able to come up with valid arguments for propositional forms. It would be great if symbolic logic could be used here, as it might help to simplify the process. It just so happens that it can, and this is its real strength.

16.6 Valid Arguments – Tautologies

A logical argument is an argument that follows from the premises. If the conjunction (combining them all together with AND connectives) of all the premises is true and they are equivalent to the conclusion, then the argument is valid.

This means that the result of analyzing an argument should lead us to all true statements. The premises make the conclusion true. If the values of a logical expression are always true, regardless of the truth values of the inputs, we call this a tautology.

> A **tautology** is a logical statement that is always true, no matter what the truth value of its components are.

The final column in the truth table of a tautology contains all true values. Thus, to test to see if a logical expression is a tautology, we simply construct a truth table and observe the last column.

Example:

Is the expression, $(p \vee q) \vee \sim q$ a tautology?

An argument is said to be a valid argument if it leads to a tautology. Thus to test to see if an argument is valid or not, we can simply construct its truth table to see if it is a tautology.

How then do you take a verbal argument with premises and a conclusion and construct its symbolic equivalent form? The answer is quite simple. For the argument to be valid you are assuming that the premises must lead to the conclusion.

> An argument is **valid** if all the premises logically lead to the conclusion.
> Otherwise, the argument is **invalid**, or a **fallacy**.

All the premises leading to the conclusion, means that the premises must be connected by conjunctions (AND connectives), and they all must point to the conclusion (IF … THEN), i.e.

IF (premise 1 AND premise 2 AND premise 3, etc.) THEN (conclusion)

We'll illustrate the process through examples.

Example 1:

> *If I go to the park then it is Saturday.*
> *I went to the park.*
> *Therefore, it is Saturday.*

We begin by defining the simple statements. Let
$$p = \textit{I go to the park}$$
$$q = \textit{It is Saturday}$$

The argument becomes,

$$
\begin{array}{ll}
p \rightarrow q & \text{premise 1} \\
p & \text{premise 2} \\
\hline
q & \text{conclusion}
\end{array}
$$

We then construct the compound expression for our argument,

IF (premise 1 AND premise 2) THEN (conclusion)

$$[(p \rightarrow q) \wedge p] \rightarrow q$$

Finally, set up and evaluate the truth table for this expression.

p	q	p→q	(p→q)∧p	[(p→q)∧p]→q
T	T	T	T	T
T	F	F	F	T
F	T	T	F	T
F	F	T	F	T

The last column contains all true values, so this is a tautology. Since it is a tautology the argument is valid. Notice that this is the Modus Ponens argument from Chapter 15 on word logic.

Example 2:

> *If I owned a Lexus I would be rich.*
> *I do not own a Lexus*
> *Therefore, I am not rich.*

We begin by defining the simple statements. Let

> p = *I own a Lexus*
> q = *I am rich*

The argument becomes,

> p→q   premise 1
> ~ p   premise 2
> ──────
> ~ q   conclusion

We then construct the compound expression for our argument,

IF (premise 1 AND premise 2) THEN (conclusion)

$$[(p \rightarrow q) \wedge \sim p] \rightarrow \sim q$$

Finally, set up and evaluate the truth table for this expression.

p	q	p→q	~p	(p→q)∧~p	~q	[(p→q)∧~p]→~q
T	T	T	F	F	F	T
T	F	F	F	F	T	T
F	T	T	T	T	F	F
F	F	T	T	T	T	T

The last column does not contain all true values, so this is not a tautology. Since it is not a tautology the argument is invalid. Notice that this is the invalid form of the Modus Tollens argument from Chapter 15 on word logic. The reason that this is not a valid argument can be seen from the truth table above. If p is false, meaning you don't own a Lexus, it does not follow that you are not rich. This is why the final truth value is false. We would say that owning a Lexus is sufficient, but not a necessary condition for being wealthy.

16.7 Logical Equivalence (Optional Section)

The bi–conditional expresses a form of equivalence called **material equivalence**, and is represented symbolically as $p \leftrightarrow q$. You may also see the symbol $\Leftrightarrow$.

Material equivalence means that one is true provided the other is true and one is false provided the other is false. It means that p and q just happen to have the same truth values.

For example,

Amanda will buy a car, if and only if, it is a Ford

Notice Amanda and a Ford are only linked through this example, and not at a more fundamental level.

Also consider,

Mary will go to the concert, if and only if, her favorite band is playing

Again, Mary and the concert and band are only related through this example, and not more fundamentally.

Material equivalence is different from another type of equivalence called **logical equivalence**. In logical equivalence the two statements are fundamentally related, and one is just a restatement or alternative definition of the other. Logical equivalence is represented symbolically as $p \equiv q$. It means that both p and q must always have the same truth value.

For example,

A triangle is equilateral, if and only if, it has three equal angles

In this example, an equilateral triangle is also defined to be a triangle with three equal angles. The two are fundamentally linked and are always true, independent of this example.

Also consider,

A bachelor is an unmarried male

This example is a definition, and definitions are always logically equivalent. The two always mean the same thing, regardless of this example.

In both cases, material and logical equivalence, have the same truth table. The reason they have the same truth values, however, are different. One is based upon material facts which can change from example to example, and the other is based upon a fundamental logic that never changes, making them always equivalent.

This is a subtle distinction, so in this first introduction to logic we will consider them both to be the same. However, in a more advanced treatment on logic, you may see this corrected, and a greater distinction made.

16.8 Related Conditional Statements (Optional Section)

In addition to the conditional IF … THEN statement introduced in 16.2, there are several related statements that are very useful in proving theorems in mathematics. They are the **converse**, the **inverse**, and the **contrapositive**. These statements are related to interchanging and negating the parts of the conditional or **direct statement**.

For example the direct statement:

If you go to the movies, then I will go out to eat

Could be rewritten as:

If I go out to eat, then you will go to the movies

We have switched the antecedent (you go to the movies) and the consequent (I go out to eat). This new conditional is called the **converse** of the original statement.

We could also rewrite the original conditional as:

If you do not go to the movies, then I will not go out to eat

We have negated both the antecedent and the consequent of the original direct statement. This is called the **inverse**.

Finally, we could switch the antecedent and the consequent, as well as negate them both, or write,

If I do not go out to eat, then you will not go to the movies
This new form of the conditional is called the **contrapositive**.

Related Conditional Statements		
Direct Statement	$p \rightarrow q$	If p, then q
Converse	$q \rightarrow p$	If q, then p
Inverse	$\sim p \rightarrow \sim q$	If not p, then not q
Contrapositive	$\sim q \rightarrow \sim p$	If not q, then not p

Example 1.

Given the direct statement,

If Aaron buys a sports car, then it will be a Ford,

Write the **(a)** Converse, **(b)** Inverse, and **(c)** Contrapositive.

(a) Converse: Reverse the antecedent and the consequent (p and q values).

If Aaron buys a Ford, then it will be a sports car.

(b) Inverse: Negate the antecedent and the consequent.

If Aaron does not buy a sports car, then it will not be a Ford.

(c) Contrapositive: Reverse and negate the antecedent and the consequent.

If Aaron does not buy a Ford, then it will not be a sports car.

Let's investigate how these statements are related to each other logically by examining their truth tables.

		Direct	Converse	Inverse	Contrapositive
p	q	$p \rightarrow q$	$q \rightarrow p$	$\sim p \rightarrow \sim q$	$\sim q \rightarrow \sim p$
T	T	T	T	T	T
T	F	F	T	T	F
F	T	T	F	F	T
F	F	T	T	T	T

If you look at the table you will see that the converse and the inverse give the same truth values, or they are both logically equivalent (see the optional section 16.8 if you have not done so already.)

$$q \rightarrow p \equiv \sim p \rightarrow \sim q$$

In the same way, the direct statement and the contrapositive are also equivalent.

$$p \rightarrow q \equiv \sim q \rightarrow \sim p$$

EXERCISES 16.1–16.8

ESSAYS AND DISCUSSION QUESTIONS:

1. Who is the Father of Symbolic Logic, and what did he do?

2. What does symbolic enable us to do with logic?

3. What is a valid argument, and how did symbolic logic impact analysis of arguments?

4. What is the difference between a conditional and a bi–conditional statement.

5. Describe the difference between inclusive and exclusive OR statements.

PROBLEMS:

Translate each of the following verbal statements into symbolic form. Identify all statements.

1. I went to the store and I brought my friend.

2. If I go to the store then I will take a taxi.

3. If you build it they will come.

4. I will either go to college or start working.

5. She did not buy a dress and shoes.

6. It is either going to rain or snow.

7. John will go to college and study mathematics.

8. The car has a sunroof and a spoiler.

9. If you buy a pack of cigarettes you will smoke them.

10. Mary will either stay home or go to the store and buy a bathing suit.

11. Fred will either take a bus of take a plane home for Christmas and then drive back to school.

12. If Aaron files the case then he will sue them in court.

Write the negation of each statement.

13. All cats have whiskers.

14. All the planets in our solar system orbit the sun.

15. Some cultures created mathematics.

16. Some women study mathematics.

17. No dogs have wings.

18. No airplanes have legs.

19. All your damages are covered by your insurance.

20. All the money must come from you.

21. Some people believe that mathematics is created.

22. Nobody believes that mathematics is discovered.

23. Some mathematics is discovered.

24. All business' pay taxes.

25. No soldier will be injured.

26. At least one planet other than earth sustains life.

27. Some cell phones are defective.

28. All text messaging is included in the plan.

Find the truth values of the compound logical expressions with the given logical conditions.

29. *Deirdre will either text you or she will call you*. Assume that Deirdre called, but did not text you.

30. *Michael will pay his bill or his phone will be shut off*. Assume that Michael did not pay his bill, and his phone was not shut off.

31. *The Christmas tree has colored and white lights*. Assume that the tree has only colored lights.

32. *The car is equipped with a spoiler and mag wheels*. Assume that the car has both.

33. *The cell phone plan comes with unlimited texting and calling*. Assume that only calling is unlimited.

34. *The cell phone plan comes with either unlimited texting or unlimited calling*. Assume that the plan has unlimited calling only.

35. *Doris will either marry Michael or Bobby*. Assume that Doris marries Bobby.

36. *If Cynthia goes to Cancun she will take her friend Paula*. Assume that Cynthis went to Cancun, but that she did not take Paula.

37. *If your study then you will get an A on the test*. Assume that you studied, but that you did not get an A.

38. *I will buy a motorcycle or a car*. Assume that you bought both.

39. *If I go to the automotive dealership, then I will buy a car*. I did not go to the dealership and I did not buy a car.

Identify which statements are inclusive OR's and which are exclusive, and which could be interpreted either way.

40. *Amy will wear dress pants or a dress to the party*.

41. *Harry will play or sit on the sidelines Saturday*.

42. *Buy a CD or a portable MPG Player*.

43. *Take a job as a reporter or an engineer.*

44. *Eat a burger or a hot dog.*

45. *Ride home from the New Year's Eve celebration in either a limo or a cab.*

46. *Take a plane or a bus to Florida for vacation.*

47. *Sam will buy a sweater or a scarf from the store.*

48. *Jessica will either play WEII or dance at the party.*

49. *Tiger will either win or lose the tournament.*

Find the truth table for the logical expressions.

50. $p \wedge \sim q$

51. $\sim p \vee q$

52. $p \wedge (q \wedge p)$

53. $p \rightarrow \sim p$

54. $(p \wedge q) \rightarrow q$

55. $(p \vee q) \rightarrow \sim q$

56. $p \vee (q \vee \sim p)$

57. $\sim p \rightarrow (p \vee q)$

58. $(p \rightarrow q) \wedge \sim p$

59. $\sim(\sim q \vee p)$

60. $\sim q \rightarrow (q \vee \sim q)$

61. $(p \vee q) \wedge r$

62. $(p \rightarrow q) \wedge (q \rightarrow r)$

63. $\sim(p \vee r) \wedge (p \rightarrow \sim q)$

Determine if the arguments are valid or invalid using truth tables.

64. If math is discovered then it is not created.
    Math is discovered.
    Therefore, it is not created.

65. All men are pigs.
    Joe is a man.
    Therefore, Joe is a pig.

66. All men are mortal.

Socrates is a man.
Therefore, Socrates is mortal.

67. Princess Diana was assassinated or was killed in an accident.
Princess Diana was not assassinated.
Therefore, Princess Diana was killed in an accident .

68. All insects have wings.
Wood lice are insects.
Therefore, wood lice have wings.

69. If your are my women then I am your man.
I am your man.
Therefore, you are my women.

70. If John loves Linda then Milton does not love Linda.
Milton loves Linda
Therefore, John does not love Linda.

71. If Reba does not sing country then Garth does not sing country.
Garth sings country.
Therefore, Reba sings country.

72. If god exists then evil does not exist.
God does not exist.
Therefore, evil exists.

73. If you study for more than three hours then you will pass.
You did not study for three hours or more.
Therefore, you will not pass .

74. If you buy the car you do not buy the boat.
You do not buy the car.
Therefore, you buy the boat

75. If I owned a yacht I would be rich.
I do not own a yacht
Therefore, I am not rich.

Determine which statements are Materially Equivalent and which are Logically Equivalent.

76. If today is Sunday, then tomorrow is Monday.

77. The candidate becomes president if and only if he wins the election.

78. The player wins a trophy if and only if he wins the game.

79. Water falls from the sky if and only if it rains.

80. I'm breathing if and only if I'm alive.

81. You are permitted to vote if and only if you are registered.

82. I'll buy you lunch if and only if you pass the exam.

83. If it is raining then it is cloudy.

84. You can be a passenger on this plane if and only if you buy a ticket.

85. We'll meet outside tomorrow if and only if it is not raining.

86. Alice is tall and Jeremy is short if and only if Jeremy is short and Alice is tall.

87. If Danise is in London, then she is in Europe.

Given the direct statements, write the (a) converse, (b) inverse, and (c) contrapositive.

88. If Aaron buys a sports car, then it will be a Ford.

89. If it snows, then they will cancel school.

90. If two angles are congruent, then they have the same measure.

91. If today is Friday, then it is a weekday.

92. If you do not work, then you will not get paid.

93. If it rains, then I will stay home.

94. If you drink Pepsi, then you do not like Coke.

95. If this polygon is a triangle, then the sum of its interior angles is 180°.

96. If you can purchase a house, then your credit is good.

References

1. Beth, E.W., (1962). Formal Methods. An Introduction to Symbolic Logic and to the Study of Effective Operations in Arithmetic and Logic, Dordrecht-Boston.

2. Fitch, F. B., (1952). Symbolic Logic, An Introduction, New York.

3. Gabbay, D, and Woods J., (2009). Logic from Russell to Church, Amsterdam, Elsevier.

4. Gabbay, D, and Woods J., (2004). The rise of modern logic: from Leibniz to Frege, Amsterdam, Elsevier.

5. Van Heijenoort, J., Ed., (1967). From Frege to Gödel, A source book in mathematical logic, 1879-1931, Harvard University Press.

6. Lewis, C.I., (1918). A Survey of Symbolic Logic, Berkeley.

7. Lewis, C.I. and Cooper, H. L. (1959). Symbolic Logic, New York 1932, 2. ed. New York.

8. Whitehead, A.N., and Russell, B., (1910). Principia Mathematica, Cambridge University Press: Cambridge, England.

CHAPTER 17

The Search for the "Holy Grail" of Mathematics

*Mathematics is not a careful march down a well–cleared highway, but a journey
into a strange wilderness where the explorers often get lost.*

W.S. Anglin

17.1 Introduction

Mathematics, at is heart, is essentially a system based upon logic. This was the belief adopted by many mathematicians throughout history. Motivated by the contributions of Cantor in set theory and Boole in symbolic logic, mathematicians around the turn of the 20th century began to search for ways to show that all of mathematics could be explained by logic alone. One mathematician, Gottlob Frégé (1848–1925), a German mathematician, inspired by the earlier work of another German mathematician, Gottfried Leibniz (1646–1716), was the first to attempt to show a direct connection of mathematics to an exclusively logical foundation. For nearly half a century it was the quest of many prominent mathematicians, including Bertrand Russell (1872–1970) and Alfred North Whitehead (1861–1947), two British mathematicians, to prove the logical foundation of mathematics. It became the Holy Grail of mathematics, to prove that mathematics had a solid logical basis and foundation.

It was a quest that would eventually lead to one of the most important discoveries in all of mathematics and would change the face of mathematics forever.

17.2 The Origin of the Quest

Towards the end of the Renaissance, rationalism, the view that reason was the source of knowledge, had become a common system of belief. It was derived from the ideas of Descartes, Spinoza, and Leibniz. It was further believed that much of human reasoning could be reduced to some sort of "calculation" procedure, and that these calculations could resolve differences of opinion. Leibniz was one that believed that:

"The only way to rectify our reasonings is to make them as tangible as those of the Mathematicians, so that we can find our error at a glance, and when there are disputes among persons, we can simply say: Let us calculate, without further ado, to see who is right." Leibniz, *The Art of Discovery*, 1685

Leibniz (see picture to the left) believed that reason was supreme, and that mathematical reasoning was the highest form of reasoning. Thus, he was searching for a way to reduce all language and consequently all arguments to mathematical equations, so the answers could be found as easy as we solve equations. To further the whole process, Leibniz also designed and had built some of the first mechanical calculators for performing arithmetic. Many of Leibniz's findings were not published until well after his death. It is not certain whether or not Leibniz developed a rudimentary form of symbolic logic similar to what Boole created 140 years after his death, but he at least anticipated its birth long before others.

Leibniz is considered to be the most important logician between Aristotle (300 BCE) and George Boole (1847). Although Leibniz did not formally publish his work on logic when he was alive, he did leave behind many working drafts of papers. His logic was based upon the fundamental belief that:

1. All ideas are constructed from a small number of simple sentiments.
2. Complex ideas are created from these simple ideas using rules similar to arithmetic.

It was a system that was developing similar to what Boole created over a hundred years later. Although Leibniz did not create a formal symbolic logic his work did have a profound influence on those that followed him, especially in Germany.

In 1879 another German mathematician, Gottlob Frégé (see picture to the left) published *Begriffsschrift (Concept-Script)*, a book on formal logic. It was probably the most important complete publication on formal logic since Aristotle. Like Leibniz, Frégé was attempting to show that rational thought and analysis could be systematized.

To establish this even more concretely, Frégé focused his attention on mathematics, and was trying to show that mathematics could be built up from a few simple ideas which by logic alone could create all of mathematics. The belief that if you could show this for mathematics, and that mathematics was the fundamental language used in science, that you could ultimately show that the world could be founded on logic alone.

In 1893 and 1903 Frégé published his two-volume work *Grundgesetze der Arithmetik (Basic Laws of Arithmetic)* in which he attempted to show that arithmetic, the most fundamental form of mathematics, could be described by logic alone. He combined set theory and logic into a unified theory for the foundation of mathematics. It was a defining moment in mathematics. It had established what we have called the Holy Grail of Mathematics – An attempt to show that mathematics can be based upon logic alone. Since if it could be shown that mathematics was logical, and mathematics was fundamental to describing how the rest of the world worked, then the world in which we live should also be founded upon logical principles. It would establish a new world order, based exclusively on logic.

A darker side of genius.

Although Frégé was without a doubt a brilliant man, he also had a darker side to his personality. He was a follower of Kaiser Wilhelm II during World War I, and during that period developed a hatred towards parliamentary systems, democracy, socialism, and Marxism, as well as towards liberals, Catholics, French, and above all Jews, who he thought should be deprived of all rights and thrown out of Germany. Frégé even used his analytic skills to devise plans for expelling the Jews from Germany. Later in his life Frégé confided that "he had once thought of himself as a liberal and was an admirer of Bismarck, but now his heroes were General Ludendorff and Adolf Hitler."

Frégé was a highly bitter person who was filled with a lot of hatred, and convinced of his own genius, he became increasingly bitter at his lack of recognition. It is without doubt a dark and poisonous cloud consuming a highly intelligent person.

Frégé was well on his way towards his goal of obtaining the Holy Grail, when in the middle of preparing his second volume of the *Basic Laws of Arithmetic* for printing, he received a letter from Bertrand Russell pointing out that the system of axioms (rules for playing his logical game) he had chosen could lead to a contradiction. This crushed Frégé and after 1903 he took no further part in the development of mathematical logic.

17.3   A Bump in the Road – Russell's Paradox

In 1903 Bertrand Russell (see pictures below) had discovered a flaw in the logical foundations of Frégé's work. It was a fundamental flaw that had challenged Frégé's life's work (perhaps a little just deserve for his hatred and bigotry). What Russell had discovered was the other primary cause of

paradox, that of self–reference, in Frégé's work. It became know as **Russell's Paradox**. The mathematical explanation of the problem is a bit too complicated for this book, but layman's illustrations are available which describe it quite well. One is called the **Barber's Paradox** and another is the age–old **Liar Paradox**. They are essentially variations of the same idea.

The basic principle is that whenever what is being constructed can be used to try and make a claim about itself (self–reference) problems can arise.

The Barber's Paradox is an illustration of this, and reads as follows:

> In a certain village there lives a man who is a barber; this barber shaves
> all and only those men in the village who do not shave themselves.
> Does the barber shave himself?

We are in trouble if we say the barber shaves himself, and we are in trouble if we say he does not. We cannot answer the question without contradicting ourselves!

The simpler version of this form of a paradox is called the Liar's Paradox which was known well before Russell's paradox. It goes as follows:

> This sentence is False.

Again, we are in trouble if we say the sentence is false, and we are in trouble if we say it is true. We cannot come up with a solution that does not lead to a contradiction!

What Russell was able to show was that Frégé had actually built in the self–reference paradox into his formal logical system for arithmetic. This meant that Frégé's approach was fundamentally flawed.

Although Frégé gave up, Russell did not. He still believed that logic would remain supreme, and that the Holy Grail was still out there. Thus, together with his teacher and early mentor, Alfred North Whitehead (see picture to the right), Russell pressed onwards. Together with Whitehead, the two embarked on an ambitious retelling of mathematics, all through formal logic alone. It was an unbelievable task which resulted in the famous work *Principia Mathematica (Principles of Mathematics)* published in 1912. This was a three volume book on the principles of mathematics. They even chose the title of their work to be the same as the original ground breaking work of Newton's *Principia Mathematica*, believing it to be just as important and fundamental. Both Russell and Whitehead thought they were well on their way to achieving the Holy Grail.

The work was daunting, though. It nearly burned them out mentally. The level of concentration required to prove basic ideas that we take for granted was phenomenal. In fact, it took over 350 pages just to prove that $1 + 1 = 2$! We illustrate this with an excerpt from the book below. Notice the intense notation. Practically no one else, other than Whitehead and Russell, could verify the works accuracy.

*54·43 $\vdash :: \alpha, \beta \in 1 . \supset : \alpha \cap \beta = \Lambda . \equiv . \alpha \cup \beta \in 2$

   Dem

     $\vdash . *54 \cdot 26 . \supset \vdash :: \alpha = l' x . \beta = l' y . \supset : \alpha \cup \beta \in 2 . \equiv . x \neq y .$

     [*51·231] $\equiv . l' x \cap l' y = \Lambda .$

     [*13·12] $\equiv . \alpha \cap \beta = \Lambda$ (1)

     $\vdash . (1) . *11 \cdot 11 \cdot 35 . \supset$

          $\vdash :: (\exists x, y) . \alpha = l' x , \beta = l' y . \supset : \alpha \cup \beta \in 2 . \equiv . \alpha \cap \beta = \Lambda$ (2)

     $\vdash . (2) . *11 \cdot 54 \cdot 52 \cdot 1 . \supset \vdash . Prop$

From this propositon it will follow, when arithmetical addition has been defined, that $1 + 1 = 2$.

The lighter side of genius.

Bertrand Russell was definitely a man that loved life, and he lived a long and interesting one, too. He had major achievements in logic, mathematics, philosophy, epistemology, and literature (he won the Nobel Prize in 1950). He was imprisoned for his anti-war sentiments in 1918 and for anti-nuclear protest in 1961(even at the age of 88.) It is also not a secret that Russell loved the company of women. Women were attracted to him and he to them. He married four times and had numerous affairs including one with the wife of his student and friend, T.S. Eliot (Vivien Eliot). Russell was constantly in search of love. Russell died in 1970 and was nearly 98 years old, having lived a full and lively life.

It appeared that the Holy Grail would be found, and logic would be shown as the most fundamental aspect of reality. After all, we had two of the preeminent mathematicians working on it. However, a new twist was added that was to change mathematics forever.

17.4 Kurt Gödel – The New Dawn of Mathematics

Kurt Gödel (1906–1978) (see picture to the left), an Austrian mathematician was only six years old when the *Principia Mathematica* of Whitehead and Russell was first published. Little did they know the profound impact that this brilliant mind would have on the foundations of mathematics. In 1931, a year after finishing his doctorate degree at the University of Austria, Kurt Gödel published a paper titled: *Über formal unentscheidbare Sätze der Principia Mathematica und verwandter Systeme (On Formally Undecidable Propositions of Principia Mathematica and Related Systems)*. In this article and a following article later that year, Gödel essentially proved that what Russell and Whitehead were trying to accomplish was not possible – the Holy Grail did not exist!

Gödel proved what have become known as Gödel's Incompleteness Theorems. The theorems themselves are very complicated, and require a great deal of mathematical training to understand. We present versions of them below, however, we do not expect the reader to understand them. Instead we encourage you to look at the layperson's summaries beneath them, and the discussion that follows.

First Incompleteness Theorem:

Let T prove only true formulas. Then there exists a statement about natural numbers $\mathbb{N} = 1,2,3,...$ of the form $\forall x\, A x$ (for all x, Ax) of the system which is not provable in T, while for each individual number n, An is provable in T.

Second Incompleteness Theorem:

Let T prove only true formulas, and let T and Prov satisfy the so-called Hilbert-Bernays-Löb conditions:
  1. For each A, if T proves A, Then T proves prov(A),
  2. For each A, B, T proves $Prov(a \to b) \to (Prov(a) \to Prov(b))$,
  3. For each A, T proves $Prov(a) \to Prov(Prov(a))$.
Then T does not prove $\neg Prov(1=0)$.

> **Gödel's Incompleteness Theorems (Layperson's Summary 1, less accurate):** For any axiomatic system of sufficient complexity, such as arithmetic, there are things that are true, but cannot be proven true using the rules of the system. Also, there are things that are false that cannot be shown to be false, using these same rules.

> **Gödel's Incompleteness Theorems (Layperson's Summary 2, more accurate):** Given any system of axioms that produces no paradoxes, there exist statements about numbers which are true, but which cannot be proved using the given axioms.

> **Gödel's Incompleteness Theorems (Layperson's Summary 3, even more accurate):** Any effectively generated theory capable of expressing elementary arithmetic cannot be both consistent and complete. In particular, for any consistent, effectively generated formal theory that proves certain basic arithmetic truths, there is an arithmetical statement that is true, but not provable in the theory (Kleene 1967, p. 250).

Each of the summaries above attempt to describe what the original mathematical theorems state, but in a language more familiar to the average person. The problem here is that the more understandable the explanation becomes, the less accurate it becomes, and vice–versa.

The theorems tell us that arithmetic, and hence a lot of mathematics, is incomplete. For mathematicians this was a shocking result. This meant that you could devote your entire life to finding an answer about a mathematical question, but the answer could never be found. It might be undecidable, much like the Continuum Hypothesis that haunted Cantor into insanity. The theorems also put an end to both Logicism and Formalism as viable philosophies to explain what mathematics is. In essence they imply that there are possibly some things in mathematics that will always remain a mystery.

The fragile side of genius.

As a young child Kurt Gödel was known as "Mr. Why," due to his constant questioning of everything. He was naturally curious so it is not surprising that he gravitated towards mathematical logic which sought to provide answers. He would question, almost to an obsession and fault, though.

When he was applying for US citizenship in 1947 he studied the US Constitution in great depth. To his surprise he had found a flaw in it that could lead to a dictatorship. He confided this to his friends, Oskar Morgenstern (a famous economist), and Albert Einstein, who were both going to be his references at the citizenship hearing. They told Gödel not to voice this problem to the judge, as that might complicate matters. Everything was going fine until Judge Phillip Forman asked Gödel how he felt about living in a country that was not a dictatorship (Gödel had lived under the German occupation of Austria and Czechoslovakia). Gödel could not resist himself and proceeded to explain to the judge that this too could happen in the US, even under the rules of the Constitution. Eventually they calmed Gödel down and he was granted his US citizenship. The flaw he had found – The President could fill vacancies without Senate approval while the Senate was in recess, which Gödel reasoned, could lead to a dictatorship!

Later in his life Gödel became convinced that someone was trying to poison him, and would only trust eating food that his wife Adele had prepared for him. It so happened that Adele became ill and had to go into the hospital for six months. Gödel refused to eat anything, and eventually became ill as well. He weighed only 65 pounds when he was admitted to the hospital and where he eventually died in 1978. He had essentially starved himself to death due to his fragile mental state.

Studying mathematics can be hazardous to your mental, as well as physical, well-being!

Incompleteness Beyond Mathematics

Interestingly enough, Gödel's theorems have even been applied outside of mathematics, much to the dissatisfaction of many mathematicians. They have been applied to any and every possible rational or logic based system. The argument is that in any axiomatic system (a system of rules used to derive more complicated results) there are things that are true and false, but unprovable. The only way you can prove them is to "go outside" the system, by adding some additional axioms that you must assume to be true. This will then allow you to prove what you sought to prove in the first place. However, you have now created a new system which has new things that are both true and false, but unprovable.

In particular, Gödel's theorems have been applied to the field of artificial intelligence. Artificial intelligence by definition means you are creating a system of reasoning that depends upon rules to create the intelligence. The argument goes as follows: Artificially intelligent systems are created by people, and people can only manually create a system that is a subset of itself (Note: this does not apply to procreation), so the new system will have less than actual human intelligence.

Consequently, one will be able to distinguish between human intelligence and intelligence created artificially. Therefore, artificial intelligence cannot rise to the same level as human intelligence. Thus, it will never be possible to create artificially intelligent systems that equal human intelligence. Others, however, have argued very effectively that Gödel's theorems make no such assertions, which is true, but it doesn't mean that it is not possible. The jury is still out on this one. The theorems do, however, give more credence to the possibility, but in no way prove anything.

Another area we see the theorems used, is in the religious question of a whether or not there is a god. One argument is that since humans were created by some logical process or thing, then it stands to reason that whatever created humans was a more "complete" system, since it could only create a subset of itself, but not a copy. Therefore, there must be a supreme being, otherwise we could not have been created. However, some have pointed out an apparent contradiction with this approach, since we now have a new system (God) which would imply an even higher system above it, and so on. Again, many would deny the proof of anything even close to this, but it does add fuel to the philosophical fire.

The truth be told, Gödel's theorems only states that in arithmetic, an axiomatic system of sufficient complexity, you can construct a true statement that you cannot prove. Godel's theorem proves nothing outside of this. What is does do, however, is to hold open the possibility that something similar may or may not exist outside of the mathematical realm. Meaning if it can happen in mathematics, something that was once thought to be a logically self–contained and self–consistent subject, then what might be possible in other logically based systems, where our initial assuredness may not be as high? Outside of mathematics, however, the theorems do not prove anything, they only provide fuel for new and interesting speculations and questions.

Whatever your opinions are on any of this, it is still amazing that a mathematical theorem can cause such a discussion in the first place. It goes right to the heart, by seeking a deeper connection of mathematics to reality, and our place in it. Mathematics seems to raise as many questions as it answers. It is truly a magical mystery tour we have undertaken, so far.

We will end this part of our journey through logic by considering one last aspect. This has to deal with the gray areas between absolute truth and falsehood. Even Gödel himself looked into the prospect of a logic that had more than two truth values (true or false). One that could have varying degrees of truth. In a short paper in 1932 (*Zum intuitionistischen Aussagenkalkül*), Gödel used this new logic in a proof which would later become known as Gödel–Dummett intermediate logic (or Gödel fuzzy logic). It is the topic of Fuzzy Logic that we consider in the next chapter.

EXERCISES 17.1–17.3

ESSAYS AND DISCUSSION QUESTIONS:

1. What was the Holy Grail of Mathematics?

2. Who was Gottlob Frégé and how did he impact the search for a logical foundation for mathematics?

3. Who was Bertrand Russell, and what impact did he have on the foundations of mathematics?

4. Who was Kurt Gödel, and what impact did he have on the foundations of mathematics?

5. What is Russell's Paradox and what are some of its implications?

6. What is Kurt Gödel's Incompleteness Theorem and what does it say about mathematics as well as possibly to some areas outside of mathematics?

7. Where does the Holy Grail of Mathematics stand today?

369

References

1. Allen, (1957). Symbolic Logic: A Razor – Edged Tool for Drafting and Interpreting Legal Documents, Yale L. J. 833.

2. Allen, (1960). Law, Logic and Learning, Harvard L. Record, October 6, 1960.

3. Beth, E.W., (1962). Formal Methods. An Introduction to Symbolic Logic and to the Study of Effective Operations in Arithmetic and Logic, Dordrecht-Boston.

4. Fitch, F. B., (1952). Symbolic Logic, An Introduction, New York.

5. Gabbay, D, and Woods J., (2009). Logic from Russell to Church, Amsterdam, Elsevier.

6. Gabbay, D, and Woods J., (2004). The rise of modern logic: from Leibniz to Frege, Amsterdam, Elsevier.

7. Van Heijenoort, J., Ed., (1967). From Frege to Gödel, A source book in mathematical logic, 1879-1931, Harvard University Press.

8. Lewis, C.I., (1918). A Survey of Symbolic Logic, Berkeley.

9. Lewis, C.I. and Cooper, H. L. (1959). Symbolic Logic, New York 1932, 2. ed. New York.

10. Whitehead, A.N., and Russell, B., (1910). Principia Mathematica, Cambridge University Press: Cambridge, England.

CHAPTER 18

Fuzzy Logic

While, traditionally, logic has corrected or avoided it, fuzzy logic compromises with vagueness; it is not just a logic of vagueness, it is – from what Frege's point of view would have been a contradiction in terms – a vague logic.

Susan Haack

It is the mark of an instructed mind to rest satisfied with that degree of precision which the nature of the subject admits, and not to seek exactness where only approximation of the truth is possible.

Aristotle

18.1 Introduction

Amazingly, the story of logic does not end with Boole, nor does it end with the search for the Holy Grail of mathematics. It only expands to increase its applicability. All the logic we have considered previously, is based upon being able to assign a distinct true or false value to a particular statement. What about common statements such as – He is short. Is this true or false? How can we decide this question? This is what a new form of logic called fuzzy logic is all about. Sometimes things are not just true or false, but have degrees of truthfulness. Fuzzy logic was developed through the work of Lotfi Zadeh (1921–present) in the 1960's and 1970's to handle these sorts of situations.

The logic of Aristotle is based upon the Law of the Excluded Middle. This law basically states that something must be either true or false, and that these is no middle ground, or partial truth. Questions concerning degrees of truth are not easily studied using conventional logic. A logic based on the two truth values True and False is sometimes inadequate when describing human reasoning. To more effectively handle these wider reaching logical concepts, a new approach was needed. Fuzzy logic uses the "interval" between False and True to describe human reasoning. Fuzzy logic helps us handle the concept of partial truth. It allows us to consider values between "completely true" and "completely false." It is based upon the concept of something called a fuzzy set. (Note: Some logicians argue that there isn't a need for a new type of logic, and that partial truths can be accommodated using traditional logic.)

Fuzzy Set theory is based upon the idea that you can have a partial membership in a set. Membership in a set does not have to be 100%; it can instead be shared with other sets. This is a central principle of fuzzy set theory. Lotfi Zadeh developed fuzzy set theory to accommodate the partial membership question, and the logical principles that followed from redefining it in this way.

Since its creation in the 1960's and 1970's, fuzzy logic has seen a great deal of practical applications in the are of new "smarter" (decision based) consumer products. Everything from anti–lock breaks in cars, to auto–focusing lenses in cameras, to smart cook–tops and ovens (to name a few) have incorporated aspects of fuzzy logic into their design.

In this chapter we focus on a basic introduction to this new and exciting field. We begin with the fundamental concept; that of a fuzzy set.

18.2 Fuzzy Sets

Fuzzy set theory shares some of the same concepts as conventional set theory we introduced in Chapter 12. One of the concepts that distinguishes Fuzzy set theory from conventional set theory is the concept of membership in a set. A fuzzy set is defined in terms of the membership concept.

Membership

For example, if you consider the set of even numbers and the set of odd numbers, there is no ambiguity of whether a number is or is not in a particular set. However, if you consider the set of tall people and the set of short people, the boundaries on what is tall and what is short are not clear, they are in fact, "fuzzy". This is precisely what fuzzy sets are, sets without rigidly defined boundaries.

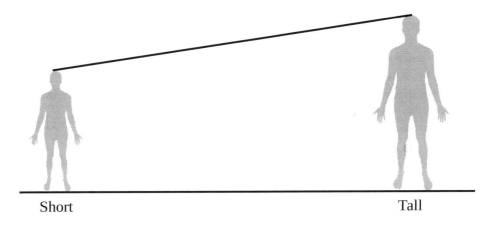

Short   Tall

When does a short person become tall or a tall person become short? It doesn't really make sense to consider anyone above a particular height tall and someone that is only a fraction of an inch shorter than this height to be considered short. We really have degrees of "tallness" or "shortness." There is not a distinct boundary between tallness and shortness. To capture this aspect of belonging or not belonging to a particular group, we use something called the membership concept.

In fuzzy set theory you can have degrees of membership in various set. Thus a person halfway between what has been defined as tall and short might have half a membership in the tall set and half in the short set. This is how the values in–between are handled. Consider the figure below. This graph illustrates the concept of membership.

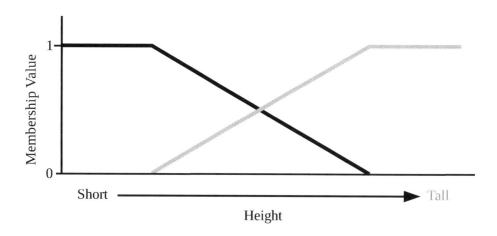

The bold black line indicates membership in the short set, which starts out as one (1) and then reduces down to zero (0) when you reach a certain height. The bold gray line indicates membership in the tall set, which starts out a zero (0) and then increases to one (1) when you reach the threshold height.

If you are below a predetermined height you are considered to be fully (100%) in the set of short people. This means you have a membership value of 1 in that set. However, as your height increases your membership in the short set increases, while your membership in the tall set increases. This happens until you reach a certain height, and then are fully in the tall set. Halfway between, you would have half a membership in the short set (value – 0.5) and half a membership in the tall set (value = 0.5).

We use the concept of membership to define a fuzzy set.

A **fuzzy set**, is a set whose elements have degrees of membership.

Examples of Fuzzy Sets.

1. The set of young people.
2. The set of overweight people.
3. The set of old people.
4. The set of rich people.

5. The set of fuel efficient cars.
6. The set of large dogs.
7. The set of tasty Indian food dishes.
8. The set of comfortable temperatures.

We need a way to numerically quantify membership in a set. We have done so above graphically, but we can also do so algebraically by creating a membership "calculator."

For example, let us say that any man 5 feet 6 inches and under is considered short and 6 feet and above, he is considered tall. Thus, their short membership below 5'6" is equal to 1, and their tall membership if they are above 6' is also 1. The question is how do we find the in–between values for membership? To do this we set up rules for calculating this value.

We'll use the letter "h" to identify the numerical height of an individual. We will use the symbolism of T(h) to represent the membership value in the Tall set, for a person of height h. T(h) is read as "t of h."

We will use the following rules or calculator for finding the tall set membership value:

1. If $h \geq 6$ then $T(h) = 1$

2. If $h \leq 5.5$ then $T(h) = 0$

3. If h is between 5.5 and 6 feet, then we use the following algebraic rule to find T(h):

$$T(h) = \frac{h - 5.5}{0.5}$$

We can use this to find T(h), the tall set membership value, for different values of h.

For h = 5.75 = 5'9"

$$T(5.75) = \frac{5.75 - 5.5}{0.5} = \frac{0.25}{0.5} = 0.5$$

For h = 5.9 ≈ 5'11"

$$T(5.9) = \frac{5.9 - 5.5}{0.5} = \frac{0.4}{0.5} = 0.8$$

For h = 5.6 ≈ 5'7"

$$T(5.6) = \frac{5.6 - 5.5}{0.5} = \frac{0.1}{0.5} = 0.2$$

The graph of the membership value for this set is shown below.

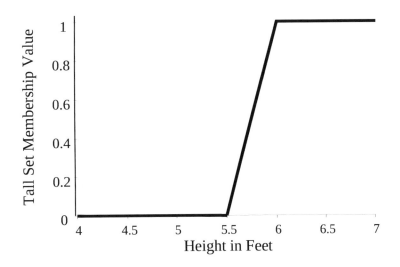

In a similar way we can define the Short Set membership value as:

1. If $h \leq 5.5$ then $S(h) = 1$

2. If $h \geq 6$ then $S(h) = 0$

3. If h is between 5.5 and 6 feet, then we use the following algebraic rule to find S(h):

$$S(h) = \frac{6-h}{0.5}$$

We can use this to find S(h), the short set membership value, for different values of h.

For h = 5.75 = 5'9"

$$S(5.75) = \frac{6-5.75}{0.5} = \frac{0.25}{0.5} = 0.5$$

For h = 5.9 ≈ 5'11"

$$S(5.9) = \frac{6-5.9}{0.5} = \frac{0.1}{0.5} = 0.2$$

For h = 5.6 ≈ 5'7"

$$S(5.6) = \frac{6-5.6}{0.5} - \frac{0.4}{0.5} = 0.8$$

The graph of the membership value for this set is shown below.

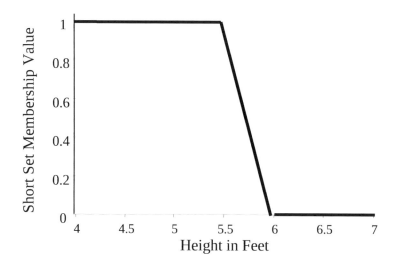

Comparison to Classical Set Theory

In classical (Boolean) set theory the graphs for tall and short membership would look like,

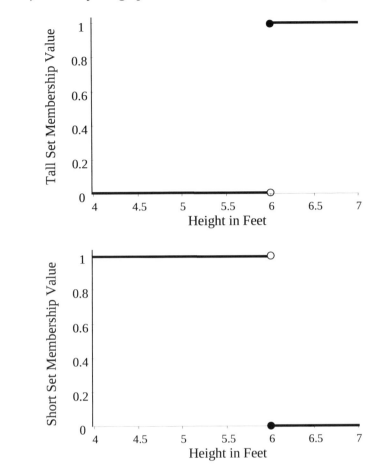

and

There is always a sharp boundary between sets in classical set theory. There is no ambiguity, you are either in the set or you are not in the set. Classical sets are also sometimes called "crisp" sets, because of the rigid boundaries. The problem is that classical set theory doesn't always accurately reflect reality. For example, consider a man that is 5'11.75" tall. In classical set theory he would be in the short set, while a man that is 6' tall would be in the tall set, although there is only 1/4 inch separating the two men in height.

Fuzzy set theory allows us to handle problems where rigid boundaries are not called for, by defining the membership calculator.

Membership Calculator, M(x)

Process for finding the membership calculator, $M(x)$, of a fuzzy set.

    A. For fuzzy sets whose members are all **greater than or equal** to a certain value, x.

1. Let "max" equal the upper threshold value for x, and "min" equal the lower threshold value.

2. Define upper test for the x–value.

$$x \geq max \text{ , membership value, } M(x) = 1$$

3. Define the lower test for the x–value.

$$x \leq min \text{ , membership value, } M(x) = 0$$

4. Define the formula for values of x between the max and min value, $min < x < max$.

$$M(x) = \frac{max - x}{max - min}$$

Graph:

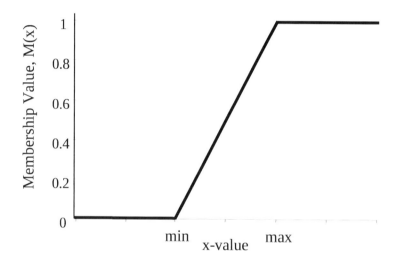

B. For fuzzy sets whose members are all **less than or equal** to a certain value, x.
1. Let "max" equal the upper threshold value for x, and "min" equal the lower threshold value.

2. Define lower test for the x–value.

$$x \leq min \text{ , membership value, } M(x) = 1$$

3. Define the upper test for the x–value.

$$x \geq max \text{ , membership value, } M(x) = 0$$

4. Define the formula for values of x between the max and min value, $min < x < max$.

$$M(x) = \frac{x - min}{max - min}$$

Graph:

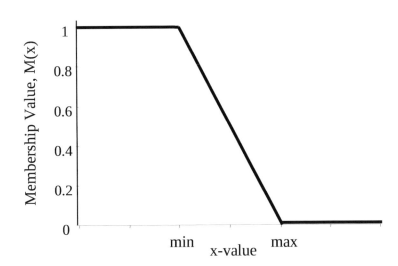

C. For fuzzy sets whose members are between certain value, x, $a < x < b$. (This is more complicated, and is beyond the scope of what we wish to present here in our introduction, so we treat this as an **OPTIONAL** topic, and include it for completeness.)

1. Let "max" equal the upper threshold value for x, and "min" equal the lower threshold value.

2. Define upper and lower test for the x–value.

   If $x \leq min$ or $x \geq max$, then membership value, $M(x) = 0$

3. Define the in–between test for the x–value.

   If $a < x < b$, then membership value, $M(x) = 1$

4. Define the formula for values of x between the min and a value, $min < x < a$.

$$M(x) = \frac{x - min}{a - min}$$

5. Define the formula for values of x between the b and max value, $b < x < max$.

$$M(x) = \frac{max - x}{max - b}$$

Graph:

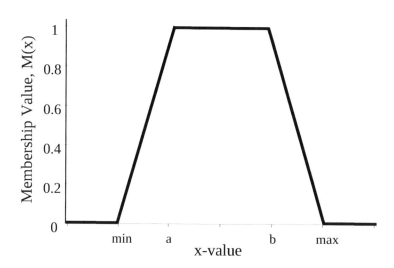

Example:

1. Find the membership calculator for the set of old people. A person is considered old if they are over 62 (the retirement age) and are considered not old (young) if they are 25 or under. Then find the membership value of a person that is 35, 45, and 55 years old.

   We are asked to find the membership calculator for a set that is greater than a certain age (x–value), so we use the A process above. We'll use A (age) for the x–value, and O (old) for M for the calculator.

   We can see from the problem that, max = 62, and min = 25. Using this we can find the membership calculator, O(A).

   Membership Calculator

   If $A \geq 62$ then $O(A) = 1$

   If $A \leq 25$ then $O(A) = 0$

   If A is between 25 and 62 $(25 < A < 62)$, then

   $$O(A) = \frac{A - min}{max - min} = \frac{A - 25}{62 - 25} = \frac{A - 25}{37}$$

   Now find the membership values O(35), O(45), and O(55)

   $$O(35) = \frac{35 - 25}{37} = \frac{10}{37} \approx 0.27$$

Meaning you are 27% in the old set.

$$O(45) = \frac{45-25}{37} = \frac{20}{37} \approx 0.54$$

Meaning you are 54% in the old set.

$$O(55) = \frac{55-25}{37} = \frac{30}{37} \approx 0.81$$

Meaning you are 81% in the old set.

2. Find the membership calculator for the set of poor people. A person is considered poor if they make $22,000 a year or less, and are rich if they earn $1,000,000 or more yearly. Also find the membership value for someone earning, $50,000, $100,000, and $350,000 a year.

We are asked to find the membership calculator for a set that is less than a certain income (x–value), so we use the B process above. We'll use s (salary) for the x–value, and P (poor) for M for the calculator.

We can see from the problem that, max = $1,000,000, and min = $22,000. Using this we can find the membership calculator, P(s).

Membership Calculator

If $s \leq \$22,000$ then P(s) =1

If $s \geq \$1,000,000$ then P(s) = 0

If s is between $22,000 and $1,000,000 $(22,000 < s < 1,000,000)$, then

$$P(s) = \frac{max - s}{max - min} = \frac{1,000,000 - s}{1,000,000 - 22,000} = \frac{1,000,000 - s}{978,000}$$

Now find the membership values P(50,000), P(100,000), and P(350,000)

$$P(50,000) = \frac{1,000,000 - 50,000}{978,000} = \frac{950,000}{978,000} \approx 0.97$$

Meaning you are 97% in the poor set.

$$P(100,000) = \frac{1,000,000 - 100,000}{978,000} = \frac{900,000}{978,000} \approx 0.92$$

Meaning you are 92% in the poor set.

$$P(350,000) = \frac{1,000,000 - 350,000}{978,000} = \frac{650,000}{978,000} \approx 0.66$$

Meaning you are still 66% in the poor set.

18.3 Operations on Fuzzy Sets

Now that we understand what a fuzzy set is and the concept of membership value, we can present the basic operations on sets we learned previously for classical sets: negation, intersection and union. We will also introduce a new operation called hedging, which we describe in detail below. The central idea in all cases is related to the membership value.

Negation (Complement)

To find the negation or the complement of a fuzzy set, you take the "opposite" membership value of the set you are negating. By opposite we mean, we subtract the membership value of the set we are negating from one (1). If $M(\overline{A})$ represents the membership values for being in the complement (negation) of set A, and $M(A)$ represents the membership values of being in set A, then the negation membership value can be expressed as,

$$M(\overline{A}) = 1 - M(A)$$

This is also shown graphically below.

In these graphs we show set A as the bold gray line, and the negation or complement of set A (~A or $\overline{A}$) in black.

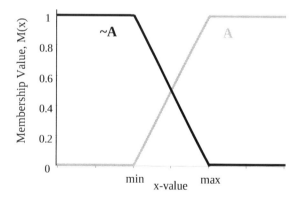 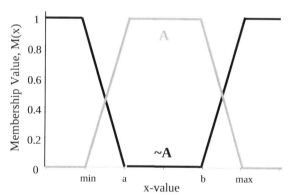

Intersection

To find the intersection of two fuzzy sets, you take the minimum membership value of the two sets together, that you are intersecting.

Here are two sets A and B. You trace the graph that is the lower of the two graphs A and B placed on top of each other. This is how the Intersection is found.

Sets A and B.

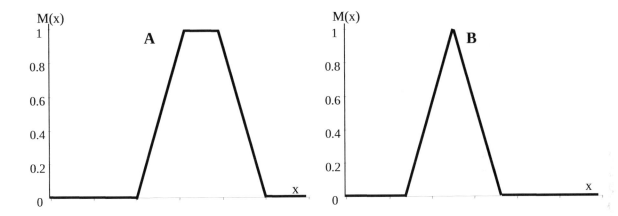

And their intersection, as shown in black below.

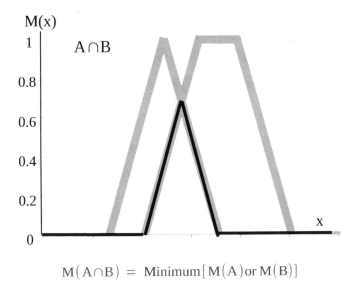

$$M(A \cap B) = \text{Minimum}[M(A) \text{ or } M(B)]$$

Union

To find the union of two sets, you take the maximum membership value of the two sets you are joining.

Here are two sets A and B. You trace the graph that is the higher of the two graphs A and B placed on top of each other.

Sets A and B.

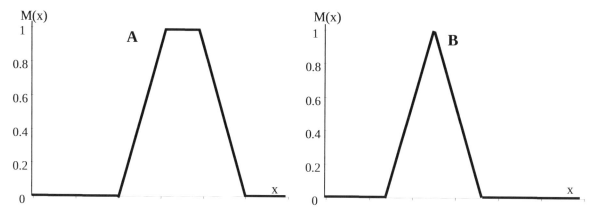

And their union, as shown in black below.

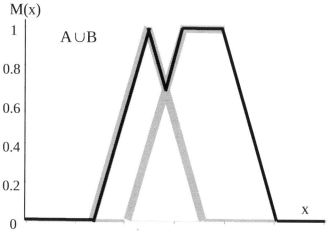

$$M(A \cup B) = Maximum[M(A) or M(B)]$$

383

Examples:

Given the following sets, A and B,

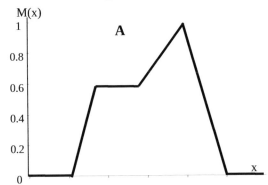 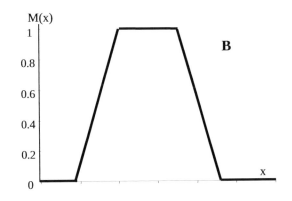

Find each of the following.

    1.   A∪B

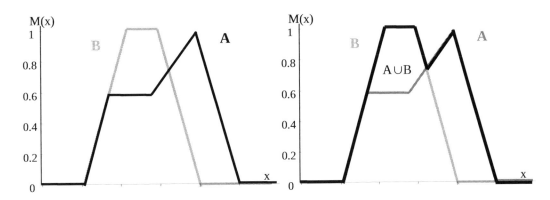

    2.   A∩B

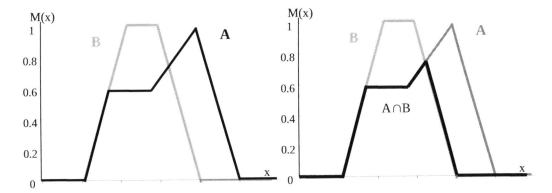

    3. $\overline{A}$

384

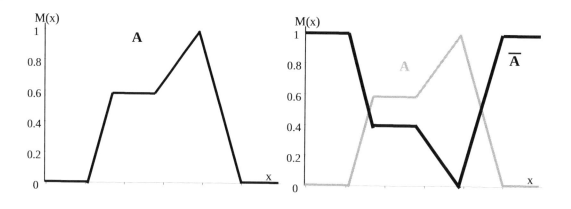

4. $\overline{B}$

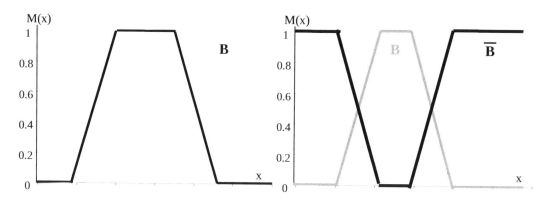

The Hedging Concept (Optional)

The other thing fuzzy sets enable, is a closer parallel to natural language. For example, adjectives like, very, mostly, somewhat, partly, fully, etc., fit naturally into this approach. Each of the previous words identifies a degree to which what they are describing belongs to a given condition or set. The real question is how can this be represented mathematically. The approach is to use common mathematical process such as squaring and the square root. Squaring an amount can enhance (concentrate) the value, while taking the square root can decrease (dilute) the trend.

Words like very and mostly can be represented by squaring the membership value, and words like somewhat and partly can be represented by taking the square root of the membership value. For example, compare the membership calculator for the two cases, somewhat and very, as shown below.

Diluting

In the first case we show the transition from being in a set to not being in a set, to illustrate the phrase, *somewhat in the set*. Notice, in the graph below, how the membership decreases more slowly from being in the set, to not being in the set, compared with the earlier graphs. Notice how the membership value decreases less rapidly from being in, to not being in the set.

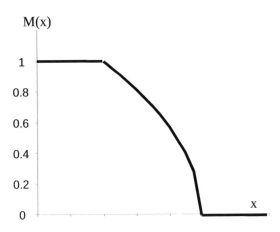

We represent this using the square root process. In this instance, we define the in–between values using:

$$M(x) = \sqrt{\left(\frac{max - x}{max - min}\right)}$$

Concentrating

In the next case, we show the transition from being in a set to not being in a set, to illustrate the phrase, *very much in the set*. Notice how the membership value decreases more rapidly from being in the set, to not being in the set from the earlier examples.

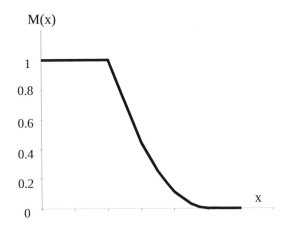

We represent this using the squaring process. In this instance, we define the in–between values using:

$$M(x) = \left(\frac{x-min}{max-min}\right)^2$$

Some specific words and their effect on a fuzzy set include:

Key Word	Effect on Set Membership
About, near, close to, approximately, almost	Approximate the set
not	Negate the set
Somewhat, rather, quite	Dilute the set
Very, extremely	Concentrate or intensify the set

18.4 Fuzzy Logic

Now that we have defined what a fuzzy set is, we can now discuss fuzzy logic.

> **Fuzzy logic** is logic that allows degrees of truth.

Something is no longer required to be either true or false. Instead it can be partly true, or more false than true. The law of the excluded middle is not enforced. It provides a method to formalize reasoning when vague terms or uncertainty is present.

Here is a simple example of fuzzy logic, showing its relation to Aristotle's deductive logic.

Aristotle's Deductive Logic	Fuzzy logic
I know all politicians are liars. I know that man is a politician. Therefore, I know that man is a liar.	I believe all politicians are liars. I believe that man is a politician. Therefore, I believe that man is a liar.

In fuzzy logic we do not have complete certainty in a state, idea or outcome. We instead have degrees of evidence to support our beliefs.

Fuzzy logic is a rule–based system of logic. This is related to the logic of the IF … THEN conditional.

IF premise (antecedent) THEN conclusion (consequent)

More than anything else, fuzzy logic provides a framework for developing the conditional, IF ... THEN, rules to obtain good decisions on vague or imprecise data.

Applications of Fuzzy Logic

Although fuzzy logic can be applied to analysis of propositional argument, as shown above, fuzzy logic has seen its largest implementation in applications in the industrial and business world. This is where it provides an excellent framework for developing consumer products that need to reliably accomplish complicated tasks.

General Applications areas:

> Quality Assurance
> Diagnostics
> Control
> Pattern Recognition

Some specific applications related to:
> Elevators
> Vacuum cleaners
> Hair dryers
> Cranes
> Robots
> Electric razors
> Camcorders
> TV's
> Showers

> Expert Systems:
>> Medical diagnosis
>> Legal
>> Stock Market Analysis
>> Mineral prospecting
>> Weather forecasting
>> Economics
>> Politics

Simple Examples of Fuzzy Logic In Industry:

Example 1: Controlling a Cooling Fan.

Classical Approach
> IF temperature > X, THEN run fan
> ELSE, stop fan

Fuzzy Approach
1. IF temperature = hot, THEN run fan at full speed
2. IF temperature = warm, THEN run fan at moderate speed

388

3. IF temperature = comfortable, THEN maintain fan speed
4. IF temperature = cool, THEN slow fan
5. IF temperature = cold, THEN stop fan

Determining if the temperature is either, hot, warm, comfortable, cool, or cold is accomplished through the membership value of the fuzzy sets associated with each of these conditions. Thus, the next step would be to construct the membership calculator for each condition being tested for.

Example 2: Controlling the speed of a motor.

We wish to control the speed of a motor. This is accomplished by changing the motor input voltage. The higher the applied voltage is, the faster the motor will run.

There are external conditions that can cause the motor to run at different speeds, however, we'd like to keep the speed of the motor within some predefined range. If the motor runs too fast, we need to slow it down by reducing the input voltage. If the motor runs too slow, the input voltage must be increased.

The monitored condition of the motor is speed, with three possible results:
Too Fast
Too slow
Just right

The corresponding (output) actions are:
Less voltage (Slow down)
No change
More voltage (Speed up)

Define the rule–base:
1. IF the motor is running *too slow*, THEN *more voltage*.
2. IF the motor speed is *about right*, THEN *no change*.
3. IF the motor speed is *too fast*, THEN *less voltage*.

The next step would be to define the membership calculator based upon a set of rules. For example, consider the following schematic that defines the rules graphically. The next step, which we will not show, would define the membership calculator for input and output variables using this schematic.

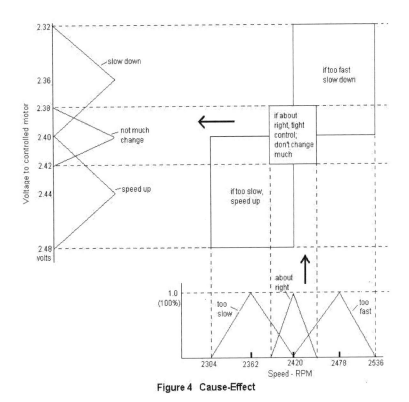

Figure 4   Cause-Effect

Membership Data to Construct the Calculator

Summary

What we intended to accomplish in this section is to provide a very brief introduction to fuzzy logic. This is not a comprehensive introduction, so interested readers should consult the references below for further study. The key feature to fuzzy logic is the membership function defined by fuzzy set theory. If we then apply the logical connectives and analyze arguments through fuzzy logic, we would see a much more complicated analysis than conventional logic, but an analysis which would be closer to human thought with all its ambiguities and partial truths.

EXERCISES 18.1–18.5

ESSAYS AND DISCUSSION QUESTIONS:

1. What is fuzzy set theory and how does it differ from conventional set theory?

2. Who is Lotfi Zadeh, and how is he related to fuzzy logic?

3. What is fuzzy logic?

4. What is a membership calculator?

5. What is hedging and how does it change the membership value of a set?

6. Where can fuzzy logic be applied?

PROBLEMS:

1. Create a membership calculator for each of the sets defined below.

   a. The set of household incomes, where $30,000 per year is considered poor and $1,000,000 per year is considered rich.

   b. The set of peoples ages, where 16 is considered young and 45 is considered old.

   c. The set of peoples ages, where 25 is considered young and 65 is considered old.

   d. The set of trips, where 20 miles is considered short and 100 miles is considered long.

   e. The set of the heights of men, where 5.5 feet is considered short and 6.5 feet is considered tall.

   f. The set of the heights of women, where 5 feet is considered short and 5.75 feet is considered tall.

2. Given the following fuzzy sets, A and B,

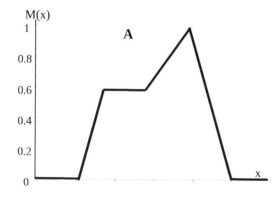 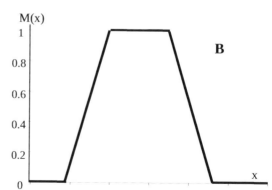

Find each of the following.
a. $A \cup B$ b. $A \cap B$ c. $\overline{A}$ d. $\overline{B}$

3. Given the following fuzzy sets, A and B,

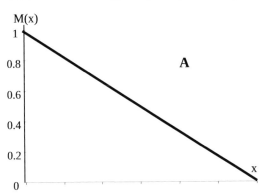
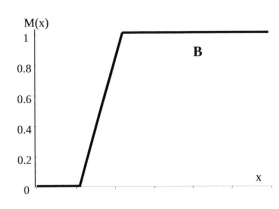

Find each of the following.

a. $A \cup B$ b. $A \cap B$ c. $\overline{A}$ d. $\overline{B}$

References

1. Von Altrock, C., (1995). Fuzzy logic and NeuroFuzzy applications explained. Upper Saddle River, NJ, Prentice Hall.

2. Biacino, L., Gerla, G., (2002). Fuzzy logic, continuity and effectiveness, Archive for Mathematical Logic 41 (7): 643–667.

3. Cox, E., (1994). The fuzzy systems handbook: a practitioner's guide to building, using, maintaining fuzzy systems. Boston.

4. Gerla, G., (2006). Effectiveness and Multivalued Logics, Journal of Symbolic Logic 71 (1): 137–162.

5. Goguen, J. A., (1968–69). The logic of inexact concepts, Synthese, 19: 325–373.

6. Hájek, P., (1998). Metamathematics of fuzzy logic, Dordrecht: Kluwer.

7. Hájek, P., (1995). Fuzzy logic and arithmetical hierarchy, Fuzzy Sets and Systems 3 (8): 359–363.

8. Hajek P., (2009a). On vagueness, truth values and fuzzy logics, Studia Logica, 91 (3): 367–382.

9. Hajek, P., Paris, J., and Shepherdson, J., (2000). The liar paradox and fuzzy logic, Journal of Symbolic Logic, 65: 339–346.

10. Hajek P., and Hanikova Z., (2003). A development of set theory in fuzzy logic

11. Halpern, J.Y., (2003). Reasoning about uncertainty. Cambridge, Mass., MIT Press.

12. Höppner, F., Klawonn, F., Kruse, R., Runkler, T., (1999). Fuzzy cluster analysis: methods for classification, data analysis and image recognition. New York: John Wiley.

13. Ibrahim, A.M., (1997). Introduction to Applied Fuzzy Electronics. Englewood Cliffs, N.J: Prentice Hall.

14. Klir, G.J., Folger, T.A., (1988). Fuzzy sets, uncertainty, and information. Englewood Cliffs, N.J: Prentice Hall.

15. Klir, George J., St Clair, Ute H., Yuan, Bo (1997). Fuzzy set theory: foundations and applications. Englewood Cliffs, NJ: Prentice Hall.

16. Klir, George J., Yuan, Bo (1995). Fuzzy sets and fuzzy logic: theory and applications. Upper Saddle River, NJ: Prentice Hall.

17. Kosko, Bart (1993). Fuzzy thinking: the new science of fuzzy logic. New York: Hyperion.

18. Kosko, Bart, Isaka, Satoru (July 1993). Fuzzy Logic, Scientific American 269 (1), 76–81.

19. Novák, V., (1989). Fuzzy Sets and Their Applications. Bristol, Adam Hilger.

20. Nguyen, H.T., and Walker, E., (1999). First course in fuzzy logic, Boca Raton, Chapman & Hall/CRC Press, second edition.

21. Passino, K.M., Yurkovich, S., (1998). Fuzzy control. Boston: Addison–Wesley.

22. Van Pelt, M., (2008). Fuzzy Logic Applied to Daily Life. Seattle, WA: No No No No Press.

23. Ross, T.J., (2004). Fuzzy Logic, With Engineering Applications, John Wiley & Sons

24. Shapiro, S., (2006). Vagueness in Context, Oxford: Oxford University Press.

25. Smith, N.J.J., (2008). Vagueness and truth degrees Oxford, Oxford University Press.

26. Turunen, E., (1999), Mathematics behind fuzzy logic (Advances in Soft Computing), Heidelberg: Physica Verlag.

27. Zadeh, L., (1965). Fuzzy sets, Information and Control, 8: 338–353.

28. Zadeh, L., (1994). Preface, in R. J. Marks II (ed.), Fuzzy logic technology and applications, IEEE Publications.

29. Zimmermann, H., (1991). Fuzzy set theory and its applications, Boston, Kluwer Academic Publishers.

Part III Summary – Putting it All Together

We have reached a plateau on our journey through the mystery of mathematics and its relation to our world. The goal of this exploration was to try and take a new look at mathematics, beyond the dry mechanical manipulation of numbers and formulas, and to see into a diverse world of wonder and beauty that so captures the imagination and hearts of mathematicians. Mathematics truly is a liberal art that touches just about everything in our world.

The first part of the book looked at the philosophical questions involved with trying to define and understand mathematics and the role it plays in the world. We saw how fundamental investigations about knowledge, consciousness, and life would often lead, or were connected to, questions about mathematics. In the more obvious area of number and quantity we saw how abstracting the concept of number led us down a path to the question of infinity, irrationality, and beyond, and to stretch the very boundaries of our imagination, via imaginary numbers. A stretch that caused some to break as they pushed the very limits of human understanding. We ended this part of the journey only to be amazed to see it all connected through a simple mathematical formula "discovered" by Euler.

The last part of this book focused on human reasoning and how it too is related to mathematics. It is rather amazing that something as fundamental as thought and logic would be so intimately connected to mathematics, and how only mathematics can lead us to certain provable suppositions. It is also fascinating that mathematics has shown us that certain things may forever be a mystery, with questions that cannot be answered with certainty.

Our journey has revealed to us that mathematics is so much more than just numbers and arithmetic, independent of humans. Mathematics truly is interconnected with just about everything we do. It is part of the fabric of our universe. Is it so intimately connected because we create it that way, or is it so relevant because it is the very framework on which our world is based, and we continue to discover it as we increase our understanding of it? We may never know the answer to this question, but it is an intriguing inquiry.

Finally, we saw how our study of mathematics brought to light questions regarding the very essence of reality. The mind–body question that has captured the imagination of great minds throughout our emergence from caves to enlightenment.

It truly is a mathematical world, whether we like it or not.

In the next part of the journey we explore the mathematics of patterns. The patterns that are all around us in the world. These patterns can arise from random processes, geometric causes, or simply be related to numbers. The interesting thing about patterns is that their structure is all related to the concept of symmetry. Mathematics may just be the study of symmetry.

ESSAYS AND DISCUSSION QUESTIONS:

Write an essay on each of the following, or be prepared to discuss these questions:

1. Summarize what you have learned from this book regarding mathematics and its relation to the world.

2. Do you think mathematics is discovered or created?

3. What are two of the most intriguing things that you took away from this tour through mathematics?

4. Has anything in this book changed your opinion towards mathematics? If yes explain, if no then discuss why.

5. Were you aware that mathematics has implications throughout the liberal arts? Explain either way.

396

Answers to Selected Exercises

5.2.1

<u>Translate Hindu-Arabic to Egyptian Hieroglyphs</u>

1)

3)

5)

7)

9)

11)

Translate Egyptian Hieroglyphs to Hindu-Arabic numerals

13) 313   15) 1,050,100   17) 3,021   19) 102   21) 3,000,010   23) 5.036

<u>Add the Egyptian Hieroglyphs</u>

25)

27)

29)

31)

33)

35)

<u>Subtract the Egyptian Hieroglyphs</u>

37)

39)

41)

43)

45)

47)

Multiply the Egyptian Hieroglyphs

49)

51)

53)

55)

57)

59)

Divide using the Egyptian Hieroglyph approach, but show the process using Hindu-Arabic numerals

61)

$$\frac{364}{4}$$

Yes	1	4
Yes	2	8
No	4	16
Yes	8	32
Yes	16	64
No	32	128
Yes	64	256
	$1+2+8+16+64=91$	$4+8+32+64+256 = 364$

398

63)
$$\frac{90}{15}$$

No	1	15
Yes	**2**	**30**
Yes	**4**	**60**
	2+4=6	**30+60** = 90

65)
$$\frac{28}{7}$$

No	1	7
No	2	14
Yes	**4**	**28**
	4=4	**28** = 28

67)
$$\frac{962}{74}$$

Yes	**1**	**74**
No	2	148
Yes	**4**	**296**
Yes	**8**	**592**
	1+4+8=13	**74+296+592** = 962

69)
$$\frac{2,675}{107}$$

Yes	**1**	**107**
No	2	214
No	4	428
Yes	**8**	**856**
Yes	**16**	**1712**
	1+8+16=25	**107+856+1712** = 2,675

71)

$\frac{1,204}{172}$

Yes	1	172	
Yes	2	344	
Yes	4	688	
	1+2+4=7	172+344+588 = 1,204	

5.2.2 Chinese Number System
<u>Translate Hind-Arabic to Chinese Numerals</u>

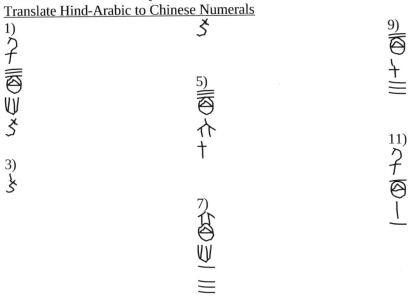

1)

3)

5)

7)

9)

11)

<u>Translate Chinese Numerals to Hind-Arabic</u>

13) 504   15) 466   17) 6,048   19) 26   21) 3,950   23) 85

5.2.3 Greek Numerals
<u>Translate Hind-Arabic to Greek Numerals</u>

1) ,ατμθ   3) ςθ'   5) τζξ'   7) χμγ'   9) σογ'   11) ,αρια

<u>Translate Greek Numerals to Hind-Arabic</u>

13) 83   15) 8,561   17) 222   19) 789   21) 355   23) 6,587

5.2.4 Roman Numerals
<u>Translate Hind-Arabic to Roman Numerals</u>

1) MCCCXLIX 3) XCIX   5) CCCLXVII 7) DCXLIII   9) CCLXXIII 11) MCXI

<u>Translate Roman Numerals to Hind-Arabic</u>

13) 1,909   15) 417   17) 2,011   19) 3,444   21) 938   23) 896

5.3 Positional Numbers
5.3.2 Babylonian Number System

<u>Translate Babylonian Numerals to Hindu–Arabic</u>
1) 633   3) 873   5) 18,082   7) 1,863   9) 1,277   11) 1,226

5.3.3 Mayan Numerals
<u>Translate Mayan Numerals to Hindu–Arabic</u>
1) 72   3) 1,819, or 2,019   5) 2,940, or 3,260   7) 6,485, or 7,205   9)172 11) 136

6.2 Egyptian Fractions
<u>Unit Fractions:</u> Hindu-Arabic to Egyptian

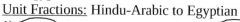

1)   7)   9)

3)

5)   11)

<u>Unit Fractions:</u> Egyptian to Hindu-Arabic

1) 1/100   3) 1/24   5) 1/10,000   7) 1/21   9) 1/2,100   11) 1/6

<u>Non – Unit Fractions:</u>

13)

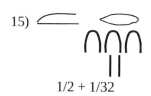

21)

2/3 + 1/6

23)

15)

1/2 + 1/20

1/2 + 1/32

17)

25)

1/2 + 1/22

19)

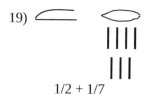

27)

1/2 + 1/7

2/3 + 1/15

6.3 Greek Fractions
<u>Unit Fractions</u>

1) ε" 3) ρκ" 5) η" 7) σζξ" 9) λβ" 11) ⅄κγ'

<u>Non-Unit Fractions</u>

13) β 'γ' 15) ιζ' λβ" 17) γ δ" 19) θ' ιδ" 21) ε' Ϝ" 13) Ϝ' τα"

6.4 Roman Fractions

1) • • 3) ƧƧ 5) • 7) Ƨ 9) Σ 11) Ɔ

6.5 Babylonian Fractions

1) $10/60 + 33/3{,}600 \approx 0.17583$

3) $14/60 + 43/3{,}600 \approx 0.245278$

5) $5/60 + 1/3{,}600 + 22/216{,}000 \approx 0.083712963$

7) $31/60 + 3/3{,}600 \approx 0.5175$

9) $21/60 + 17/3{,}600 \approx 0.354722$

11) $20/60 + 44/3{,}600 \approx 0.3455556$

8.2.1 Binary to Decimal

1) 19 3) 11 5) 64 7) 56 9) 58 11) 204

8.2.2 Decimal to Binary

1) 100100_2 3) 1000001_2 5) 10001001_2 7) 1001110_2 9) 111000001_2 11) 1000101_2

8.3.1 Quinary to Decimal

1) 38 3) 95 5) 269 7) 76 9) 130 11) 469

8.4.1 Octal to Decimal

1) 15 3) 127 5) 401 7) 2,072 9) 329 11) 2,609

8.4.2 Decimal to Octal

1) 111_8 3) 44_8 5) 570_8 7) 777_8 9) 6572_8 11) 1075_8

8.5.1 Duodecimal to Decimal

1) 81 3) 1,451 5) 971 7) 12,951 9) 4,834 11) 11

8.5.2 Decimal to Duodecimal

1) 13_{12} 3) 81_{12} 5) 323_{12} 7) 866_{12} 9) EE_{12} 11) $ET7_{12}$

8.6.1 Hexadecimal to Decimal

1) 70 3) 2,989 5) 57,005 7) 59,957 9) 48,813 11) 39,719

8.6.2 Decimal to Hexadecimal

402

1) $3A_{16}$ 3) FA_{16} 5) $CCC5_{16}$ 7) 2469_{16} 9) FFF_{16} 11) $F00D_{16}$

10.1-10.5 Exponential Numbers

1) (a) Arithmetic, (b) Geometric, (c) Arithmetic, (d) Arithmetic, (e) Geometric

3) 4.67 yrs (4 yrs 8 mths)   5) 28 yrs   7) 4.67%   9) 0.737%   11) $300, $2,300

13) $202.50, $3202.50   15) $17,500   17) $21,120.65

19) $1,582,213.87, $237, 332.10 using yearly compounding or $1,684,579.23, $252684.88 using continuous compounding

21) 109.5 Billion   23) 598 people/sq mi,   42,549,022,   17930 people/ sq mi

12.3.2 Basics of Set Theory

1) The set of odd numbers, {n| n is an odd number}

3) {red, yellow, blue}, {c| c is a primary color}

5) {Texas, Tennessee}, {s| s is a state whose name begins with a t}

7) The set of all vowels., {v| v is a vowel}

9) {General Motors, Ford, Chrysler}, {c| c is a US car company}

11) {Everest, K-2, Kangchenjunga}

13) infinite   19) infinite   25) infinite

15) finite   21) 26   27) 5

17) finite   23) 3   29) infinite

31) A Universal Set is a set containing all the elements in question. For example, the set of all the letters in the alphabet, if you are only considering letters in the English alphabet.

33) If we are only considering planets in the Milky Way Galaxy.

35) If we are only considering major league baseball players.

37) If we are only considering high cholesterol food.

39) not empty   49) not empty   59) False

41) empty   51) True   61) False

43) not empty   53) True   63) {1, 2, 3, 4, 5, 6, 7, 8, 9}

45) not empty   55) False   65) {1, 2, 3, 4, 5, 6, 7, 8, 9}

47) empty   57) True   67) {John, Joe}

69) {John, Joe Jake, James, Jeremy, Sue, Sally, Sarah, Samantha}

71) { }   79) No   87) {g, I}

73) Yes   81) {a, b, c, d, e, f, h}   89) { }

75) No   83) {b, d, f, g, h, I}   91) {a, b, c, d, e, f, g, h, I}

77) Yes   85) {a, c, e}   93) A and B, A and C

95) A, B, C are proper subsets of U

97) {(2,1), (2,2), (2,3), (4,1), (4,2), (4,3), (6,1), (6,2), (6,3)}

99) {(Sue, John), (Sue, Joe), (Sue, James), (Sue, Jeremy), (Sally, John), (Sally, Joe), (Sally, James), (Sally, Jeremy)}

101) {(John, Sue), (John, Sally), (Joe, Sue), (Joe, Sally), (James, Sue), (James, Sally), (Jeremy, Sue), (Jeremy, Sally)}

12.3.3 Visualizing Sets and Their Operations and Relationships

1)

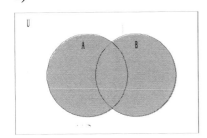

7)

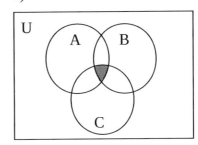

3)

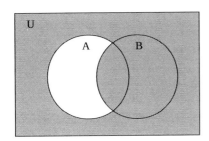

9)

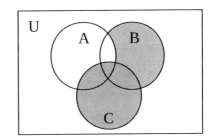

5)

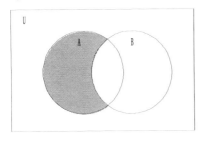

11)

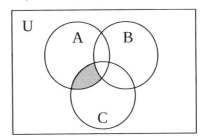

13) $\bar{B}$ 15) $(A\cup B)\cap C$ 17) $A\cup(B\cap C)$ 19) See textbook

12.3.4 Applications of Venn Diagrams

21) (a) 18, (b) 17, (c) 18, (d) 5   23) (a) 48, (b) 10, (c) 9, (d) 14   25) (a) 33, (b) 40, (c) 4, (d) 15

404
13.1-13.4 Imaginary Numbers

1) 2i   3) 3i   5) 7i   7) −6i

9)   11)

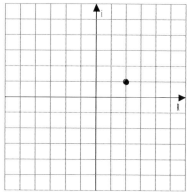   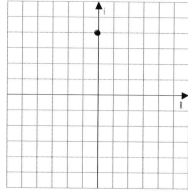

13)   15)

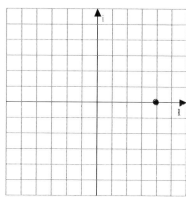   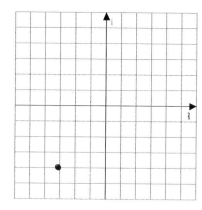

17)

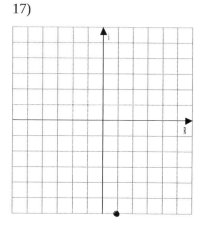

19) 3+6i   25) −1+4i   31) 26

21) 4   27) 2+9i   33) 12+8i

23) i   29) 3−i   35) −22+14i

37) $-2-6i$

41) $22+10i$

45) $-16+4i$

39) $36+17i$

43) $-3+12i$

47) $6i$

49) 13

57) i

63) $\dfrac{48}{29}+\dfrac{33}{29}i$

51) $64-8i$

59) $-\dfrac{1}{2}+\dfrac{1}{2}i$

53) $17-21i$

65) $\dfrac{21}{17}-\dfrac{29}{17}i$

55) $\dfrac{1}{2}+\dfrac{3}{2}i$

61) $\dfrac{49}{130}-\dfrac{63}{130}i$

14.1-14.2 Introduction to Types of Reasoning. What is Proof?

1) Inductive

8) Deductive

2) Deductive

9) Deductive

3) Inductive

10) Inductive

4) Inductive

11) Deductive

5) Deductive

12) Deductive

6) Inductive

13) Inductive

7) Inductive

14) car – can – van

15) sun – son – ton – top

16) bun – pun – pup – cup

17) red – rid – aid – add – odd – old – ole – one

18) run – rut – rot – pot – put – out

19) gun – gut – put – pup – pop – mop

20) moon – moot – boot – boat – boar – soar – star

21) mind – mend – meld – mold – mole – mode – bode – body

22) wine – line – lint – lift – loft – soft – sofa – soda

23) life – lift – loft – loot – moot – moor – poor

24) debt – dent – bent – best – lest – last – lash – cash

25) Challenge

26) Challenge

27) Challenge

29)

Let 2n represent any even number, where n is a natural number.

Let 2m+1 represent any odd numbers, where m is a whole number.

The product of the two numbers by the distributive property is:

$$2n(2m+1) = 4nm + 2n$$

Using the distributive property in reverse we can factor out a two (2) from both terms to obtain:

$$2(2nm+n)$$

We now note that the number 2nm+n is a natural number, which we'll call p:

2(2nm+n)=2p, which is an even number.

31)

We can represent any odd number as 2n+1.

Adding one (1) to this we obtain:

$$2n+1+1 = 2n+2$$

Using the distributive law in reverse we have:

$$2n+2 = 2(n+1)$$

Since n+1 is a natural number, call it m, we have:

$$2(n+1) = 2p, \text{ an even number.}$$

15.1-15.5 Aristotle – The Father of Propositional Logic

1) We have not been contacted by any aliens.

3) Therefore, we cannot know about life after death.

5) Acts of civil disobedience break the law.

7) If God existed, there would not be any evil in the world.

9) Premises: All dogs have four legs., All four–legged creatures have wings.
   Conclusion: All dogs have wings.

11) Premises: Government–funding of medical care is a form of socialism. Socialism is not good
   Conclusion: Government government funding of medicare should stop.

13) Premise: Humans are imperfect.
   Conclusion: You should not judge others unfairly.

15) Premise: Many of the ingredients in fast-foods have been shown to cause cancer in animals.
   Conclusion: It is reasonable to expect that eating at fast-food restaurants can lead to cancer.

17) Premises: Laura claims that he saw a flying saucer land in her back yard. But Laura is still in elementary school. She is completely unaware of what others have written about flying saucers and aliens
    Conclusion: Laura couldn't possibly be telling the truth

19) Premises: Killing is immoral, and to make the food available you must kill the animal.
    Conclusion: It is immoral to kill animals for food.

21) Premise: President Obama used too many executive actions.
    Conclusion: President Obama acted as a king.

23) Argument from Ignorance

25) Begging the Question

27) Ad–Hominum Attack

29) Hasty Generalization

33) Appeal to Inappropriate Authority

35) Gamblers

37) Straw Man

39) Argument from Ignorance

41) Hasty Generalization

43) Ad–Hominum Attack

45) Hasty Generalization

47) Gamblers

49) Hasty Generalization

51) Appeal to Inappropriate Authority

53) Begging the Question

55) Gamblers

57) Ad–Hominum Attack

59) Ad–Hominum Attack

61) Begging the Question

63) Appeal to Ignorance

65) Straw Man

67) Appeal to Ignorance

69) Appeal to Force

71) Gamblers

16.1-16.8 Beyond Aristotle – Symbolic Logic and the Insight of Boole

1)
p = I went to the store
q = I brought my friend
$p \wedge q$

3)
p = you build it
q = they will come
$p \rightarrow q$

5)
p = she bought a dress
q = she bought shoes
$\sim(p \wedge q)$

7)
p = John will go to college
q = John will study mathematics
$p \wedge q$

9)
p = you buy a pack of cigarettes
q = you will smoke them
$p \rightarrow q$

11)
p = Fred will take a bus home for Christmas
q = Fred will take a plane home for Christmas
r = Fred will drive back to school
$(p \vee q) \wedge r$

13) At least one cat does not have whiskers.

15) No cultures created mathematics.   17) At least one dog has wings.

19) Not all of your damages are covered by insurance.

21) All people do not believe that mathematics was created.

23) None of mathematics is discovered.   27. No cell phones are defective.

25) Some soldiers will be injured.

29) T	35) T	41) Both	47) Both
31) F	37) F	43) Both	49) Exclusive
33) F	39) T	45) Exclusive	

51)

p	q	~p	~p∨q
T	T	F	T
T	F	F	F
F	T	T	T
F	F	T	T

53)

p	~p	p→~p
T	F	F
F	T	T

55)

p	q	~q	p∨q	(p∨q)→~q
T	T	F	T	F
T	F	T	F	T
F	T	F	F	T
F	F	T	F	T

57)

p	q	~p	p∨q	~p→(p∨q)
T	T	F	F	T
T	F	F	F	T
F	T	T	T	T
F	F	T	F	F

59)

p	q	~q	~q∨p	~(~q∨p)
T	T	F	T	F
T	F	T	T	F
F	T	F	F	T
F	F	T	T	F

61)

p	q	r	p∨q	(p∨q)∧r
T	T	T	T	T
T	T	F	T	F
T	F	T	T	T
T	F	F	T	F
F	T	T	T	T
F	T	F	T	F
F	F	T	F	F
F	F	F	F	F

63)

p	q	r	p∨q	~q	~(p∨q)	p→~q	~(p∨r)∧(p→~q)
T	T	T	T	F	F	F	F
T	T	F	T	F	F	F	F
T	F	T	T	T	F	T	F
T	F	F	T	T	F	T	F
F	T	T	T	F	F	T	F
F	T	F	T	F	F	T	F
F	F	T	F	T	T	T	T
F	F	F	F	T	T	T	T

65)
p = is a man
q = is a pig

$p \rightarrow q$
p ___
q

$[(p \rightarrow q) \land p] \rightarrow q$

p	q	p→q	(p→q)∧p	[(p→q)∧p]→q
T	T	T	T	T
T	F	F	F	T
F	T	T	F	T
F	F	T	F	T

Tautology. A valid argument.

410

67)

p = Princess Diana was assassinated
q = Princess Diana was killed in an accident

p∨q
~p ___   [(p∨q)∧~p]→q
q

p	q	p∨q	~p	(p∨q)∧~p	[(p∨q)∧~p]→q
T	T	T	F	F	T
T	F	T	F	F	T
F	T	T	T	T	T
F	F	F	T	F	T

Tautology. A valid argument.

69)

p = you are my woman
q = I am your man

p→q
q ___   [(p→q)∧q]→p
p

p	q	p→q	(p→q)∧q	[(p→q)∧q]→p
T	T	T	T	T
T	F	F	F	T
F	T	T	T	F
F	F	T	F	T

Not a tautology. An invalid argument.

71)

p = Reba sings country
q = Garth sings country

~p→ ~q
q ___   [(~p→ ~q)∧q]→p
p

p	q	~p	~q	~p→ ~q	(~p→ ~q)∧q	[(~p→ ~q)∧q]→p
T	T	F	F	T	T	T
T	F	F	T	T	F	T
F	T	T	F	F	F	T
F	F	T	T	T	F	T

Tautology. A valid argument.

73)
p = you study more than 3 hours
q = you pass the exam

p→q
~p ___   [(p→q)∧~p]→~q
~q

p	q	p→q	~p	(p→q)∧~p	~q	[(p→q)∧~p]→~q
T	T	T	F	F	F	T
T	F	F	F	F	T	T
F	T	T	T	T	F	F
F	F	T	T	T	T	T

Not a tautology. An invalid argument.

75)
p = I own a yacht
q = I am rich

p→q
~p ___   [(p→q)∧~p]→~q
~q

p	q	p→q	~p	(p→q)∧~p	~q	[(p→q)∧~p]→~q
T	T	T	F	F	F	T
T	F	F	F	F	T	T
F	T	T	T	T	F	F
F	F	T	T	T	T	T

Not a tautology. An invalid argument.

77) Materially Equivalent (Provided there is another way to become president, e.g. assassination)

79) Logically Equivalent   83) Not bi-conditional   87) Logically Equivalent

81) Logically Equivalent   85) Materially Equivalent

89)   (a) If they cancel school, then it snowed.

(b) If it does not snow, then they did not cancel school.

(c) If they did not cancel school, then it did not snow.

412

91)   (a) If it is a week day, then it is Friday.

(b) If it is not Friday, then it is not a weekday.

(c) If it is not a weekday, then it is not Friday.

93)   (a) If I stay home, then it will rain.

(b) If it does not rain, then I will not stay home.

(c) If I do not stay home, then it will not rain.

95)   (a) If the sum of the interior angles is 180°, then the polygon is a triangle.

(b) If this polygon is not a triangle, then the sum of the interior angles is not 180°.

(c) If the sum of the interior angles is not 180°, then the polygon is not a triangle.

18.1-18.4 Fuzzy Logic

1)   (a) $M(x)=\dfrac{1,000,000-x}{970,000}$ (b) $M(x)=\dfrac{45-x}{29}$ (c) $M(x)=\dfrac{65-x}{40}$

(d) $M(x)=\dfrac{100-x}{80}$ (e) $M(x)=6.5-x$ (f) $M(x)=\dfrac{5.75-x}{0.75}$

2)

(a)   (b)

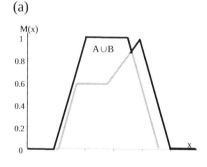   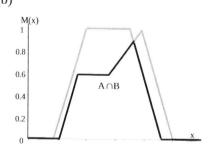

(c)   (d)

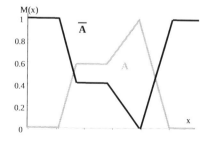   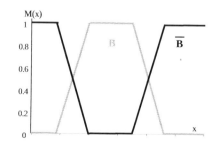

3)

(a)

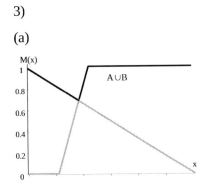

(b)

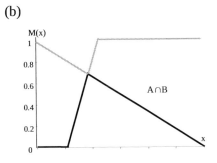

(c)

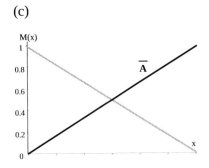

(d)

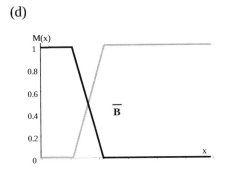

Alphabetical Index

Abacus.....92, 103, 104, 105, 119, 134, 137, 170

Achilles and the Tortoise Paradox..........257

Actual infinity.................................225

Albert Bartlett...............................191

Algebraic.....................................220

Alice's Adventures in Wonderland..314, 321

Analytic Proposition............................**54**

Ancient Greek Numbering System...........**87**

AND Connective................................**335**

Applications of Fuzzy Logic..................387

Argument......................................**307**

Aristotle 5, 7, 9, 10, 17, 20, 32, 53, 135, 136, 224, 225, 226, 227, 255, 256, 257, 258, 261, 302, 303, 304, 305, 308, 309, 322, 348, 361, 370, 386

Aristotle's (also known as Galileo's) Paradox....................................261

Arithmetic growth................................185

Axiom of choice................................262

Babylonian Number System......................**95**

Banach-Tarski Paradox.........................262

Barber's Paradox..............................362

Behaviorism...................................**58**

Belief..**52**

Benezet.......................................63

Bertrand Russell....9, 16, 50, 259, 322, 360, 362, 364, 368

Bi–conditional.............................326, **339**

Brouwer.........................17, 18, 19, 26

Bruno...226

Buddha.......................................4, 12

Buddhism..............................12, 29, 30

Butterworth..............................48, 63, 67

Cantor 15, 17, 220, 229, 230, 249, 251, 252, 253, 254, 255, 275, 276, 360, 365

Cardinality.........................232, 249, 254

Carrying capacity.............................193

Cartesian product.............................236

Chalmers.......................35, 36, 37, 48

Charles Dodgson..............................314

Complement of a set..........................234

Complex Conjugate............................**282**

Complex Numbers....................278, **279**

Compound interest............................199

Compound statement.........................**306**

Conclusion....................................**306**

Conditional..............................325, **337**

Confucius....................................4, 11

Conjunction...................................**323**

Consciousness............35, 36, 37, 39, 48, 49

Constructable.................................221

Constructivism...............14, 17, 26, **60**, 61

Continuous Compounding....................206

Continuum Hypothesis........................254

Contrapositive................................352

Converse.....................................352

Critical thinking..............**2,** iii, ix, 303, 304

De Morgan's Laws............................240

Decimal Fractions............................**124**

Deductive reasoning...........................**292**

Dehaene.....................48, 62, 67, 69

Devlin.....................48, 63, 67, 321

Dialectic.....................................**304**

Dichotomy Paradox...........................255

Disjoint sets..................................235

Disjunction...................................**324**

Doubling time................................193

Dualism......................................36

Duodecimal Number.........................**156**

Dyscalculia...................................**65**

Early Chinese Number System...............**86**

Egyptian Hieroglyphs........................**78**

Empiricism...................................53

Empty set....................................233

Epistemology.................................51

Euler's Formula...............................**287**

Euler's number...............................217

Exclusive OR.................................331

Existential quantifiers........................**327**

Exponential number..........................182

Fallacy.......................................**310**

Formalism....................14, 19, 276, 365

Fuzzy logic..............................370, 386

Fuzzy set.....................................**372**

Fuzzy set theory..............................371

Galileo.......xii, 226, 227, 228, 229, 250, 261

Geometric growth.............................185

George Boole.................................322

Gödel's Incompleteness Theorems..........365

Gottlob Frégé.................................1

Hersh.....................14, 20, 21, 22

Hexadecimal Number.................160
Hilbert's Hotel Paradox.................260
Hindu–Arabic Number System.................**92**
Hinduism.................11, 29, 137
Holy Grail of mathematics.................360
Humanism.................14, 20
Imaginary numbers.................278, **279**
Inclusive OR.................331
Inductive reasoning.................**291**
Inflationary Growth Formula.................207
Intuitionism.................**14, 17,** 18, **19,** 20, 276, **321**
Inverse.................352
Irrational number.................215
Justify.................**52**
Kant.................16, 53, 54, 55, 60, 275
Knowledge..xi, 13, 27, 29, 50, 51, 53, 54, 60
Kronecker.................17, 18, 72
Kurt Gödel.................**364**
Lao–Tzu.................11
Leibniz.................361
Lewis Carroll.................288, 293, 302, 314
Logicism.................14, 15, 16, 28, 276, 365
Materialism.................37
Mayan Number System.................**99**
Membership.................233
Membership Calculator.................375
Metaphysics.................3, 4, 13, 28
Mind–Body.................37
Negation.................**324**
Negation Operator.................**335**
Non–positional number...74, 76, 87, 92, 94, 120, 132, 137
Nonconstructable.................221
Octal Number.................**153**
OR Connective:.................**336**
Organon.................302, **304**
Paradox.................255
Parmenides.................5, 9, 255
Paul Ehrlich.................193
Perception.................40
Piaget.................58, 60, 61, 62
Plato. 4, 5, 7, 8, 9, 10, 14, 24, 28, 32, 40, 56, 220, 255, 321
Platonism.................14, 15, 27

Population Growth Formula.................207
Positional number 74, **92, 95, 100, 124,** 132, 133, **134,** 137, **144**
Potential infinity.................225
Premise.................**306**
Proof.................**289**
Proper subset.................235
Pythagoras 4, 5, 6, 7, 10, 14, 19, 24, 27, 135, 172, 213, 214, 215, 223, 227, 277
Pythagoreans 7, 8, 10, 14, 24, 213, 214, 215, 219
Quantum.................32, 41, 42
Quinary Number.................148
Rationalism.................37, 53
Rhetoric.................**304**
Roman Number System.................**89**
Russell's Paradox.................362
Simple interest.................197
Simple statement.................**306**
Skepticism.................6, 52, 291, 297
Socrates.................5, 8, 10, xii, 53, 60, 304
Statement.................**305**
Subset.................235
Synthetic Proposition.................**54**
Taoism.................11, 30
Tautology.................345, 348
Thales.................5, 6, 7, 10, 172, 213, 290, 291
The Arrow Paradox.................258
The Hedging Concept.................384
Thomas Robert Malthus.................190
Thompson's Lamp Paradox.................259
Transcendentals.................221
Truth Tables.................**333**
Uncomputable.................221
Undefinable.................221
Union.................233, 237, 240, 380, 382
Universal quantifiers.................**327**
Universal set.................232, 234
Valid Argument.................**308**
Venn diagram.................236
Word Ladder.................293
Zeno...5, 135, 143, 224, 255, 256, 258, 259, 276, 304, 313

36146915R00239

Made in the USA
San Bernardino, CA
14 July 2016